Physical Agents in the Environment and Workplace

Noise and Vibrations, Electromagnetic Fields and Ionizing Radiation

Physical Agents in the Environment and Workplace

Noise and Vibrations, Electromagnetic Fields and Ionizing Radiation

Edited by

Gaetano Licitra

Environmental Protection Agency of Tuscany Region, Italy

Giovanni d'Amore

Environmental Protection Agency of Piedmont Region, Italy

Mauro Magnoni

Environmental Protection Agency of Piedmont Region, Italy

CRC Press
Taylor & Francis Group
Boca Raton London New York

CRC Press is an imprint of the
Taylor & Francis Group, an **informa** business

CRC Press
Taylor & Francis Group
6000 Broken Sound Parkway NW, Suite 300
Boca Raton, FL 33487-2742

First issued in paperback 2020

© 2018 by Taylor & Francis Group, LLC
CRC Press is an imprint of Taylor & Francis Group, an Informa business

No claim to original U.S. Government works

ISBN-13: 978-0-367-57181-8 (pbk)
ISBN-13: 978-1-138-06517-8 (hbk)

Visit the Taylor & Francis Web site at
http://www.taylorandfrancis.com

and the CRC Press Web site at
http://www.crcpress.com

Contents

SECTION I   Noise and Vibrations

Gaetano Licitra

SECTION II   Electromagnetic Fields

Giovanni d'Amore

SECTION III Ionizing Radiation

Mauro Magnoni

Preface

The exposure to chemical, biological and physical pollutants in the living environment and in the workplace plays a major role in reducing well-being and in the development of many diseases.

The term *physical agents* usually refers to sources of energy the exposure to which may cause illness or injury to human health. They include noise and vibrations, electromagnetic fields and ionizing radiation.

This book intends to provide technicians, researchers and students with a comprehensive tool for the evaluation of human exposure to physical agents in order to assess compliance with regulation standards and laws by discussing all the relevant topics, from the physics to the comparison of measurement results with exposure limit values.

The chapters of the book can be broadly grouped into three sections according to the energetic contents of the considered physical pollutants. The first section, from Chapter 1 to 4, covers the complex phenomena of noise and vibrations in the living and work environments. In Chapter 1, after a brief introduction of the physics of sound and the main descriptors used in technical standards and legislations, the procedures and methods for the evaluation of noise levels are discussed with particular attention to the uncertainty estimation. On the other hand, it is well known how noise exposure affects human well-being, leading to detrimental health effects and increasing the risk of hypertension and cardiovascular disease. The World Health Organization (WHO) has classified the noise from road traffic as the second most harmful environmental stressor in Europe after air pollution. In this respect, the main effects of environmental noise on health and well-being are presented with reference to the mechanisms driving the human perception of noise, the subjective response and the metrics suitable for assessing the impact of noise exposure on daily life. After a short discussion about vibrations in the environment, their assessment and their impact on people, attention is focused on the environmental noise regulatory framework for industry and transportation infrastructures in Europe (Chapter 2). Finally, recent developments in characterizing noise sources in agglomerations and propagation phenomena in the European CNOSSOS-EU framework are presented, defining a very comprehensive picture of procedures and methods for the assessment of noise levels in the environment, the evaluation of their effects on health and the main techniques for exposure reduction. Noise mitigation techniques are consequently described. The first part of the book ends with a deep discussion of the very important topics of noise (Chapter 3) and vibrations (Chapter 4) in the workplace: their measurements, their effects on people and the strategies for their mitigation.

The chapters in the second section of the book (Chapters 5–9) are intended to provide the fundamental principles of electromagnetic field (EMF) protection, as well as the outcomes of recent research in this field.

Protection issues related to living and work environments are not treated in separate chapters but rather integrated with methods for measuring and calculating EMF exposure levels. Indeed, even if EMF sources used in workplaces have specific characteristics, occupational and general public exposure assessment are based on the same measurement instruments and prediction models.

Chapter 5 deals with the basic concepts of electromagnetic fields and the methods for exposure assessment. Furthermore, the main sources of EMF exposure are described with respect to their emission characteristics.

A review of the most important studies concerning the possible health effects of EMFs is presented in Chapter 6. A health risk analysis based on the results of these studies was performed in the context of the European Health Risk Assessment Network on Electromagnetic Fields Exposure (EFHRAN) project. The main results of the EFHRAN project are reported in terms of the strength of the evidence for health outcomes after exposure to electromagnetic fields.

Chapters 7 and 8 deal with extremely low frequency (ELF) and radiofrequency (RF) EMF sources, respectively.

In Chapter 7, after an analysis of the modelling of the most widespread ELF sources, an overview of the mitigation techniques to be adopted to lower magnetic fields is provided. Finally, methods for the assessment of exposure to sources generating non-sinusoidal or pulsed magnetic fields are discussed.

Radio frequency wireless communication systems are discussed in Chapter 8. After a summary of techniques for measuring human exposure to EMFs generated by second-, third- and fourth-generation (2G, 3G and 4G) mobile phone systems, the features of fifth-generation (5G) technology are discussed. We report on state-of-the-art literature on exposure to millimeter-wave radiation, which is emitted by 5G mobile communication systems, together with preliminary consideration of methods for assessing human exposure to RF radiation due to this new system network.

Chapter 9 describes measurement techniques for evaluating exposure levels to EMFs over the entire frequency range, from 1 Hz to 100 Ghz, and presents a comprehensive overview of the findings of monitoring surveys carried out by many authors. Typical levels of exposure to the main EMFs sources in the general public and in workers are presented and compared, and recommended limit values are offered.

The last three chapters of this book, Chapters 10 through 13, are devoted to ionizing radiation.

In Chapter 10, the fundamental concepts of the physics of radiation and radiation protection are introduced and discussed. The main features of the interaction of radiation with matter and the biological effects of radiation are briefly presented. The scope of radiation protection as a scientific and practical discipline aimed at protecting mankind and the environment, the harm caused by the use of radiation in various fields and the role of various international institutions, including UNSCEAR (a scientific committee of the United Nations), the International Commission on Radiological Protection (ICRP), the International Atomic Energy Agency (IAEA), the United Nations and the International Organization of Standardization (ISO), are also outlined.

Chapter 11 deals with the various sources of ionizing radiation and their application in different fields: nuclear, medical and industrial. The typical exposure of both the general public and workers arising from these sources are discussed.

In Chapter 12, the issue of environmental radioactivity is introduced and treated. After a brief historical preamble that recalls the main steps that gave birth to this scientific discipline, there is a detailed discussion of the two main environmental radioactivity components, the artificial one and the natural one. Some of the basic concepts of radioactivity (radioactive decay and secular equilibrium) are treated in some detail. In conclusion, theoretical and practical approaches to the study of environmental radioactivity are presented.

Finally, in Chapter 13, exposure to two particular natural sources is discussed: radon, which is the main source of ionizing radiation exposure for the world population, and naturally occurring radioactive materials (NORM), an emerging issue that is particularly related to the recycling of industrial wastes and by-products.

The authors are aware that the reader could probably find some redundancy in the treatment of the various issues presented in this third part of the book. However, we preferred to leave some repetition for greater clarity.

Writing a book about such a broad field as the evaluation of human exposure to physical agents in living and work environments has been a very challenging task and experience. It has been made possible thanks to the efforts and competence of all the experts involved and the willingness and support of the editors. We would like to thank all the authors for their invaluable contributions and time, and we hope this book can be a companion tool for the daily work of technicians, researchers and students.

Gaetano Licitra
Giovanni d'Amore
Mauro Magnoni

Editors

Gaetano Licitra is a physicist, Coordinator of the Coastal Area in Environmental Protection Agency of Tuscany Region (ARPA Toscana). Since 1988 his work has been focused on environmental physical agents. Most of his research is addressed to implement models and measurements to assess environmental pollution in urban areas according with 49/2002/EC directive and to find and experiment new solutions to mitigate noise exposure. He was involved in many EU projects and in working groups of WHO, EEA, DG Environment and ISO, setting guidelines that are a reference in the field. He has published 90 papers in international journals and is the author of a textbook on Noise Mapping published by Taylor and Francis, and is a member of the board of Italian Acoustical Association. Habilitated as Associated Professor of Applied Physics, he teaches Applied Acoustics in Pisa University since 1998.

Giovanni d'Amore is a physicist, Director of the Radiation Department of the Environmental Protection Agency of Piedmont Region (ARPA Piemonte). Since 1987 he has worked in control activity aimed to radiation protection of general public and workers. Main research interests are theoretical and experimental methods for electromagnetic field exposure assessment and non-ionizing radiation measurement techniques. On his research topics he has published various articles in journal papers and has been editor of special issues. He teaches "Electromagnetic fields" at Post Graduate School of Medical Physics at the University of Torino.

Mauro Magnoni was born in Milan, Italy, in 1963. He is a physicist and is currently leading the Ionizing Radiation Unit of ARPA Piemonte (Environmental Protection Agency of Piedmont). He has more than 25 years experience in environmental radioactivy and ionizing radiation protection. Since 1992 he is a Fellow of the Italian Radiation Protection Association in which he was elected President in 2015. He has published more than 20 peer reviewed papers and has regularly participated in a range of National Working Groups and Committees. Since 1997 he has also been a member of the CEVAD, the Italian Committee for the Management of Nuclear Emergencies.

Contributors

Sara Adda
Environmental Protection Agency of Piedmont
  Region
Italy

Laura Anglesio
Environmental Protection Agency of Piedmont
  Region
Italy

Giorgio Bertin
Engineering & Devices
Telecom Italia
Roma, Italy

Enrico Buracchini
Wireless Access Innovation
Telecom Italia
Roma, Italy

Aldo Canova
Politecnico di Torino
Dipartimento Energia 'G. Ferraris'
Turin, Italy

Emma Chiaramello
Istituto di Elettronica, Ingegneria
  dell'Informazione e delle Telecomunicazioni
  CNR IEIIT
Milano, Italy

Jose Luis Cueto
Acoustic Engineering Laboratory
University of Cadiz
Campus Rio San Pedro
Puerto Real, Spain

Giovanni d'Amore
Environmental Protection Agency of Piedmont
  Region
Italy

Guillaume Dutilleux
Acoustics Research Centre
Department of Electronic Systems
NTNU
Trondheim, Norway

Serena Fiocchi
Istituto di Elettronica, Ingegneria
  dell'Informazione e delle Telecomunicazioni
  CNR IEIIT
Milano, Italy

Luca Fredianelli
Consiglio Nazionale delle Ricerche
Istituto Nazionale per Studi ed Esperienze di
  Architettura Navale Vasca Navale
Roma, Italy

Peter Gajšek
Institute of Nonionizing Radiation (INIS)
Pohorskega bataljona
Ljubljana, Slovenia

Luca Giaccone
Politecnico di Torino
Dipartimento Energia 'G. Ferraris'
Turin, Italy

Paolo Gianola
Wireless Access Innovation
Telecom Italia
Roma, Italy

Daniele Giuffrida
Federal Authority for Nuclear Regulation
Abu Dhabi, United Arab Emirates

James Grellier
European Centre for Environment and
    Human Health
University of Exeter Medical School
Truro, United Kingdom

Gaetano Licitra
Environmental Protection Agency of Tuscany
    Region
Italy

Mauro Magnoni
Environmental Protection Agency of Piedmont
    Region
Italy

Cristina Nuccetelli
Center for Radiation Protection and
    Computational Physics
ISS (Italian National Institute of Health)
Rome, Italy

Tønnes Ognedal
Sinus
Stavanger, Norway

Diego Palazzuoli
Environmental Protection Agency of Tuscany
    Region
Italy

Lenzuni Paolo
Italian National Workers' Compensation
    Authority
Florence, Italy

Marta Parazzini
Istituto di Elettronica, Ingegneria
    dell'Informazione e delle Telecomunicazioni
    CNR IEIIT
Milano, Italy

Aslak Harbo Poulsen
Danish Cancer Society Research Center
Copenhagen, Denmark

Paolo Ravazzani
Istituto di Elettronica, Ingegneria
    dell'Informazione e delle Telecomunicazioni
    CNR IEIIT
Milano, Italy

Joachim Schüz
International Agency for Research on Cancer
    (IARC)
Lyon, France

Zenon Sienkiewicz
Public Health England
Centre for Radiation
Chemical and Environmental Hazards
Oxfordshire, United Kingdom

Rosabianca Trevisi
DiMEILA
INAIL (National Institute for Insurance against
    Accidents at Work)- Research Sector
Rome, Italy

I

Noise and Vibrations

Gaetano Licitra
Environmental Protection Agency of Tuscany Region, Italy

1

Noise and Vibrations in the Environment

Jose Luis Cueto
Universidad de Cádiz

Luca Fredianelli
Consiglio Nazionale delle Ricerche

Gaetano Licitra
Environmental Protection Agency of Tuscany Region

Diego Palazzuoli
Environmental Protection Agency of Tuscany Region

1.1 Introduction

Environmental noise pollution is a major concern of European citizens, especially for people living in agglomerations.

Noise exposure affects human well-being and can lead to detrimental health effects, increasing the risk of hypertension and cardiovascular disease. The World Health Organization (WHO) has classified the noise from road traffic as the second most harmful environmental stressor in Europe, behind only air pollution.

The main source of noise in agglomerations remains road traffic, with about 100 million people affected by harmful noise levels (above 55 dB L_{den}) in the EEA-33 member countries; of these, 32 million are exposed to very high noise levels (above 65 dB L_{den}). According to the recent European Environment Agency (EEA) Briefing Report (EEA Briefing 1/2017), railways are responsible for the second-highest level of noise, with 19 million people exposed above 55 dB L_{den}, followed by aircraft, which exposes more than 4.1 million people, and industrial noise within urban areas, with about 1 million people. The situation is dramatic if we consider that the EEA estimates that at least 10,000 cases of premature death in Europe each year are attributable to noise (Noise in Europe 2014, EEA Report 10/2014). As we can see in this chapter, WHO estimates the burden of disease due to environmental noise, expressed as disability-adjusted life years, the sum of the potential years of life lost due to premature death and the equivalent years of healthy life lost due to poor health or disability, to be '61,000 years for ischaemic heart disease; 45,000 years lost for cognitive impairment in children; 903,000 years for sleep disturbance;

22,000 years for tinnitus; 654,000 years for annoyance' (Science for Environment Policy, *FUTURE BRIEF: Noise abatement approaches*, April 2017, Issue 17, European Union 2017) each year in Western Europe.

On the other hand, it is also evident that vibrations induced by transportation infrastructures, especially railways, and building sites influence the perceived annoyance from noise and sleep quality. Compared with research and legislation on noise and its effects, vibrations in the environment are not as well investigated, and commonly accepted norms are still lacking in Europe. Many member states have their norms, but very often a harmonization seems to be needed.

Increasing the quality of life and reducing negative health effects for European citizens are priorities for policy makers. The efforts of the scientific community and technicians are fundamental for supporting policy in order to make better choices regarding a cost-benefit evaluation. With regard to noise and vibrations generated by transport infrastructures in the environment, a deep knowledge of physical principles and the possible effects of noise on health (with the definition of measurable indicators) is of primary importance.

This chapter is devoted to presenting to the technicians and researchers a comprehensive approach, even if obviously non-exhaustive, about the physics of noise, the main mechanisms through which it influences well-being, its effects on health and the useful descriptors for quantifying impacts. A review of vibrations in the environment is also provided with a deep insight into European ISO technical norms.

1.2 Environmental Noise

Acoustics is the science of sound. Acoustics studies revolve around the generation, propagation and effects of sound waves. In particular, when studying noise, we focus our concerns on protecting humans (and even other sensitive animal species) from pernicious noise pollution and preserving the positive components that sounds have in human lives.

The World Health Organization defines environmental noise as 'noise emitted from all sources except noise at industrial workplace' [1]. Transport, construction, leisure and industries are the main sources of environmental noise in urban areas, and an increasing number of people are exposed to noise levels that threaten their acoustical comfort.

Therefore, noise is an acoustic phenomenon and has to be studied as such in order to control it. Even though they are different, pleasant sounds and noises share the same physical laws. Any technician who begins to work in this field should have a minimum background that allows him or her to:

- Be aware of the impact of noise in human health and in the quality of life of people;
- Know the most usual acoustic magnitudes and noise descriptors;
- Be aware of the capabilities and limitations of the equipment that is used in noise measurements;
- Understand the principles of physics of noise propagation and its implications in environmental noise studies;
- Understand the details of the legislation and technical standards;
- Have the knowledge and skills to complete all the necessary steps to carry out outdoor noise measurements that meet the required quality standards;
- And be capable of processing the data obtained during the measurement campaigns.

This chapter is divided into several sections following the above items, trying to cover this basic knowledge in an introductory way.

1.2.1 Sound Principles

Basically, sound is produced by rapid pressure perturbations in the air propagating in the form of waves. In a more rigorous approach, sound waves in a fluid consist of time-dependent perturbations of the density of the medium associated with time-dependent changes of pressure, temperature, the position of fluid particles (very small elements of the medium containing a large number of fluid molecules) and their speed. In the simplest case, considering an ideal medium without attenuation and ignoring the effects of high amplitudes, the classic one-dimensional equation that describes the propagation of sound is given by:

$$\frac{\partial^2 \rho}{\partial x^2} = \frac{1}{c^2} \frac{\partial^2 \rho}{\partial t^2} \tag{1.1}$$

where $\rho(x, t)$ is the instantaneous variation in pressure at the point x at the time t and c is the speed of sound. This speed depends on the type of medium and its conditions. In a fluid, the direction of the acoustic wave is the same as that of the particle displacement (longitudinal waves). The sound pressure ρ is the difference between its equilibrium pressure value ρ_0 and the one caused by the acoustic wave itself. In the air ρ_0 is the atmospheric pressure, which value is normally considered around 100.000 Pa (Pascal). Considering the air and the temperature dependence of the pressure, the speed of sound is estimated using the equation $c = 331.5 + 0.61 \cdot T_C$ (m/s), where T_C is the temperature in Celsius degrees. The solution for the homogeneous problem of the one dimension perturbation is the sum of two waves traveling in opposite directions along the x-axis.

$$\rho(x, t) = g(ct \pm x) \tag{1.2}$$

in which g represents any arbitrary continuous functions. As the governing equations are linear, any steady harmonic perturbation travels without changes of frequency. In this kind of sinusoidal varying perturbation, ρ varies both in time, t, and space, x. If we take into account only the progressive component of a plane wave, we will obtain solutions of the form

$$\rho(x, t) = P_1 \cos(i\omega t - ikx) = |P_1| \cos(\omega t - kx + \varphi) \tag{1.3}$$

where P_1 is the complex amplitude of the plane perturbation traveling forward at the speed of sound c, $\omega = 2\pi f$ is the angular frequency (radians per second), f is the frequency (Hz – Hertz, or cycles per second) that accounts for the number of pressure fluctuations in 1 second in a given point, and $k = \omega/c$ is the wavenumber. φ is the phase (angle in radians) when $t = 0$. Also, the wavelength λ is the longitude of one cycle (the distance in meters between two consecutive points with the same phase) and T is the period in seconds (or the reciprocal of frequency f). All these magnitudes are related to the rest of the wave parameters by the relation

$$\lambda = c \cdot T = \frac{c}{f} = \frac{2\pi f}{\omega} = \frac{2\pi}{k} (m) \tag{1.4}$$

Any wave is a composition of sinusoids with different frequencies, amplitudes and phases.

$$\rho = \sum_{n=1}^{N} |P_n| \cos(\omega_n t + \varphi_n) \tag{1.5}$$

where φ_n is the phase of the complex amplitude P_n of the component 'n' that forms part of the perturbation.

1.2.2 Noise Magnitudes

Most environmental noise sources can be characterized by several simple descriptors. Throughout this section, a short discussion of the main quantities useful for characterizing noise sources, noise perception and the phenomena of sound generation and propagation will be presented in order to better understand the advanced topics.

1.2.2.1 Sound Pressure

Acoustic disturbance propagating in a fluid causes pressure to fluctuate. Pressure variations can be positive (compression) or negative (rarefaction) with respect to the equilibrium, and at a given point of space, its time average over a sufficiently large time interval is zero. Instead, the magnitude of interest characterizing the strength of the sound perturbation is the time average of the square average of the sound pressure, the mean square pressure. If the instantaneous value of the pressure at the time t is $p(t)$, then the root mean square (RMS) value of the time average of the square of the sound pressure over a time interval T is given by

$$p_{RMS} = \sqrt{\frac{1}{T} \int_0^T (|p| \cdot \cos 2\pi f t)^2 dt}$$

(1.6)

The RMS value of the pressure in Pa is related to the effective energy value of a sinusoidal signal. In the case of a monotone signal and integrating in a time period from 0 to 2π, $p_{rms} = |p|/\sqrt{2}$, which is a magnitude easy to measure directly.

1.2.2.2 Sound Intensity and Energy Density

The time averaged sound flow of energy that passes through the unit cross-sectional area (normal to the direction of propagation) in the unit time is called sound intensity I. For a plane wave or, in general, far from the sources and in the absence of reflections, the intensity of sound in the direction of propagation is:

$$I = p_{rms}^2/\varrho \cdot c$$

(1.7)

measured in watts per square meter (W/m^2), where ϱ is the density of the fluid and c the sound velocity. The product $\varrho \cdot c$ is known as the characteristic impedance of the fluid and for air at 20°C is about 415 kg m^{-2}s^{-1}, where the acoustic energy moves in the same direction of propagation. The energy density ϵ is equal to:

$$\epsilon = p_{rms}^2/\varrho \cdot c^2$$

(1.8)

When there are no losses in the propagation medium, any surface enclosing this sound source (for example, imagine a sphere) contains all the power passing through this enclosing surface, no matter how large the sphere is. However, in this case, we can verify that the power per unit of area (intensity) becomes smaller and smaller. W (the sound power in watts) emitted from a source results to be:

$$W = I \cdot S$$

(1.9)

S is the surface that surrounds the noise source. Sound power is a property of the source, whereas pressure levels depend on the distance and the propagating path. For a point source and spherical waves, the sound intensity is inversely proportional at a distance d according to:

$$I = \frac{W}{4\pi d^2}$$

(1.10)

whereas for a linear source of sound:

$$I = \frac{w}{2\pi d} \tag{1.11}$$

where w is the sound power per unit length.

1.2.2.3 The Decibel Scale

Since the mechanism of human hearing covers a very wide range of pressure variations, from about 20 µPa (human audition threshold) to sound pressures in the threshold of pain (more or less a range of 64–112 Pa), it is more appropriate to express such quantities on a logarithmic rather than a linear scale. The decibel scale expresses the ratio of two quantities

$$10 \log_{10}\left(\frac{A}{A_{ref}}\right)(dB) \tag{1.12}$$

where A_{ref} is the reference level and A is the measured one. Table 1.1 shows the expression in dB units of the main quantities used later in the book.

In terms of pressure levels, the decibel scale is not only more manageable but is better adjusted to the human ear's perception of changes in sound pressure level. We can use the definition of sound pressure level (SPL or L_p) of Table 1.1, simplifying it in the following way:

$$L_P = 20 \log_{10}(\rho_{RMS}) + 94(dB) \tag{1.13}$$

1.2.2.4 Frequency

Noise sources generally generate sounds composed of different frequencies. When observing the distribution of any sound in the time domain and in the frequency domain, both are different perspectives of the same acoustic phenomena, with the same energy. This is easy to see mathematically using the Fourier equations to transform a sound from time to frequency domain (analysis) and vice versa (synthesis, not shown).

$$X(\omega) = \int_{t_1}^{t_2} x(t) \cdot e^{-i2\pi ft} \, dt \tag{1.14}$$

The signal has been collected in a finite time window from t_1 to t_2. Rigorously speaking this signal has infinite bandwidth, but for practical reasons, we are going to focus on the audible frequency band that

TABLE 1.1 Definition of Main Quantities for Sound Level Description

Quantity	Expression	Reference Level
Sound power level - L_w	$10 \log_{10}\left(\frac{W}{W_{ref}}\right) dB$	$W_{ref} = 10^{-12}$ W
Sound intensity level - L_I	$10 \log_{10}\left(\frac{I}{I_{ref}}\right) dB$	$I_{ref} = 10^{-12}$ W/m²
Sound pressure level - L_p	$10 \log_{10}\left(\frac{\rho^2}{\rho_{ref}^2}\right) dB$	$\rho_{ref} = 20 \cdot 10^{-6}$ Pa

Note: Reference level According to the ISO 1683 standard.

spans from 20 Hz to 20,000 Hz. The range of sound frequencies up to the audible limit is known as 'ultrasound', and frequencies below the audio range are called 'infrasound'. All the frequency components have an amplitude and a phase. The amplitudes associated with each frequency represent the audio spectrum. The shape of this spectrum can reveal interesting things about the sound, its causes and its consequences (for example, a tone associated with the operation of a machine).

Another fact is that humans don't have the same frequency resolution in the low frequencies that they have in the high frequencies. That is why the description of noise frequency components is carried out using a bank of filters in which the respective center frequencies (f_c that identifies the band itself) are scaled logarithmically in 1/3 octave bands. Each band is characterized by its lower (f_1) and its upper (f_2) 'cutoff' frequencies, which define the bandwidth of every filter $\Delta f = f_2 - f_1$. In terms of filters, the upper and lower limits of each band occur where the frequency response of the band filter drops off by a factor of 3 dB. Different bands are labeled according to the center frequencies established in the standard ISO 266-1997 [2]. 1/1 and 1/3 octave frequency analysis is usually used in environmental and architectural acoustics.

The relationship that defines every filter is as follows: $f_2 = 2^n \cdot f_1$, where n = 1 for octave bands and n = 1/3 for third-octave bands. And the center frequency of the filter is obtained as follows: $f_c = (f_1 \cdot f_2)^{1/2}$. The definition of the filter bank is shown in the standard IEC-61260 [3]. In the *octave band analysis* (n = 1), $\Delta f/f_c = 0.707$, whereas in a *one-third-octave band analysis* (n = 1/3) $\Delta f/f_c = 0.232$. The bandwidth value of the octave band filter increases as f_c increases. So a tonal sound is defined as a sound characterized by a single-frequency component that emerges in an audible way from the total sound.

1.2.2.5 Frequency-Weighting Curves

Even in unimpaired young listeners, the audible frequency range of the human hearing system is not equally sensitive to all frequencies, as we don't perceive all third-octave frequency bands with the same strength. In order to consider such physiological behavior, various frequency weightings are used to define the relative strength of frequency components of environmental noise. This is called loudness, and it is an individual human characteristic that evolves during a lifetime. But when sampling a statistically significant number of individuals we can create a curve in the frequency domain that reproduces the weaknesses of the human ear. In technical standards, different weighting curves have been proposed for specific applications. The most widely employed weighting scale for environmental noise is the so-called 'scale A', which provides a very good correlation with the frequency sensitivity of human hearing [4], allowing evaluation of the noise disturbance.

Among others weighting curves, the 'C' curve stands out; it provides a flatter filtering at low frequencies. This weighting is often used in measuring noise peaks; moreover, it is used to compare A and C values, providing a quick quantification of low-frequency sound energy. Sound levels weighted by these curves [5] are reported in dB followed by the curve name, e.g. dBA or dBC.

1.2.2.6 Percentile Noise Levels

In order to better describe fluctuations in noise levels and the intermittent (discontinuous) character of some noises, *percentile sound pressure levels* L_n are used, where n is any integer number between 0 and 100. L_n is the value exceeded $n\%$ of the time. Low percentiles, such as L_1 or L_{10}, are used to represent the shortest and most intense noise events, which are an indication of the maximum levels reached during the measurement period. On the other hand, L_{90} or L_{99} represent the minimum noise levels (which sometimes match the residual levels in environmental evaluations). It is worthwhile to note that a comparison between percentile levels is useful for assessing the variation of noise in a specific time period. The distribution of noise levels during that period is enclosed between L_{AFmax} and L_{AFmin}, or maybe L_{ASmax} and L_{ASmin}. The 'F' (Fast) and the 'S' (Slow) indicate the lengths of two different time weightings used for that measurement. Fast corresponds to an exponential window of 125 milliseconds, and Slow corresponds to a larger window of 1 second that relates to smooth fluctuations of the noise data.

1.2.2.7 Environmental Noise Descriptors

Rarely, the laws of different countries match, adopting the same noise indicators. The indicators look to assess the harmful effects on well-being and health. But they match on the use of Sound Level Meters (SLMs) as the standard equipment to measure noise pollution. This is because sound pressure level may not be the best way to characterize noise, but the microphones are simple to use and cheaper than other sensors. SLMs are designed to mimic the human perception of sound. Microphones are sensitive to the fluctuating force of sound waves, transforming them into electrical signals. The same noise pressure levels expressed in Pa can be expressed in dB, covering the range of audible sound pressure levels. The most common sound descriptor is the equivalent continuous sound pressure level, which is normally assessed using the A-weighting $L_{Aeq,T}$ representing the total amount of acoustic energy over the representative time period (the dose):

$$L_{Aeq,T} = 10 \log_{10} \frac{1}{t_2 - t_1} \int_{t_1}^{t_2} \frac{\rho_A^2(t) dt}{\rho_{ref}^2} = 10 \log_{10} \left[\frac{1}{t_2 - t_1} \int_{t_1}^{t_2} 10^{\frac{L_{AP}(t)}{10}} dt \right] (dBA) \qquad (1.15)$$

where

$\rho_A(t)$ is the A-weighted instantaneous sound pressure (actually the RMS value) at running time t. Sound pressure in Pa;

$L_{AP}(t)$ is the set of instantaneous A-weighted sound pressure levels during the time interval $T = t_2 - t_1$ seconds. Sound pressure in dBA;

$L_{Aeq,T}$ is the time-averaged sound pressure level during the time interval $t_2 - t_1$ seconds. Sound pressure in dBA;

When $L_{Aeq,T}$ is obtained from a set of measurements taken in time intervals of length T_n, then we have

$$L_{Aeq,T} = 10 \log_{10} \left[\frac{1}{T} \sum_{n=1}^{N} T_n \cdot 10^{\frac{L_{Aeq,T_n}(t)}{10}} \right] (dBA) \qquad (1.16)$$

N is the total number of measurements

T is the total period of measurements $\sum T_n$

T_n is the time length of measurement number 'n'

L_{Aeq,T_n} is the A-weighting equivalent continuous sound pressure level measured during time interval 'n'. Sound pressure in dBA.

If L_{eq} represents an average noise level in a reference time interval, the *single level event* or *sound exposure level* (SEL) is a measure of the overall acoustic energy associated with an acoustic event compressed into a 1-second interval:

$$L_{A,E} = 10 \log_{10} \frac{\frac{1}{t_0} \int_{t_1}^{t_2} \rho_{Arms}^2(t) dt}{\rho_{ref}^2} (dBA) \qquad (1.17)$$

$\rho_{A,rms}(t)$ is the A-weighted instantaneous sound pressure at running time in Pa

$t_2 - t_1$ is certain time interval, which is long enough to consider all of the significant noise events (e.g. a gas release in a plant in which noise varies importantly during these short episodes in comparison with their absence)

t_0 is the reference time of 1 second.

In summary, SEL is the value that, if maintained for 1 second, would provide the same A-weighted acoustic energy as the actual event itself.

1.2.2.7.1 Rating Equivalent Continuous Sound Pressure Level

To establish levels of community annoyance more accurately, noise level measurements should be adjusted by certain characteristics of noise. These positive or negative corrections account for the certainty that noises with the same $L_{Aeq,T}$ do not always cause the same level of annoyance in the community. This is what we usually call noise indices or noise indicators. Trying to systematize the types of adjustments (penalties), we have found

- The time of day
- Some noise characteristics associated with specific noise sources

The truth is that noise is more disturbing at night than during the day, aircraft noise produces more discomfort than road noise, and tonality, impulsiveness and strong low-frequency content generate more complaints. The presence of peaks is also more negatively received by communities when comparing two residential areas that receive the same noise dose in the long term. For example, high values of L_{AFmax} could be considered a factor to penalize the railway's noise emissions. These facts, compiled over years of experience, have been introduced in numerous noise protection laws and standards. The A-weighted equivalent continuous sound pressure level is then substituted by $L_{AReq,T}$ to assess the impact of noise. Therefore, once these noise limits have been established, depending on the sensitivity of the receivers, the next step is to establish, through repeatable methods based on measurements and calculations, how to penalize the measurements in order to better describe the annoyance caused by noise.

Laws in different countries define the noise problem taking into account the difference in human beings' sensitivity at different periods of the day. So, T is the time period representing the day, night or evening. EU Directive 2002/49/EC 'relating to the assessment and management of environmental noise' [6] defines and prescribes to all Member States the indicators L_{den} (day-evening-night) and L_{night} for the assessment of exposure to environmental noise and the evaluation of its adverse effects. L_{den} is calculated from:

$$L_{den} = 10 \log_{10} \frac{1}{24h} [12h \cdot 10^{\frac{L_{day}}{10}} + 4h \cdot 10^{\frac{L_{evening}+5}{10}} + 8h \cdot 10^{\frac{L_{night}+10}{10}}](dBA) \qquad (1.18)$$

where L_{day}, $L_{evening}$ and L_{night} are the A-weighted long-term average sound levels as defined in ISO 1996-1 [7], determined, respectively, over all the day, the evening and the night period of a year (Appendix I, Directive 2002/49/EC). The 'day' period runs from 07:00 to 19:00, the 'evening' from 19:00 and 23:00 and the 'night' from 23:00 to 7:00. The weighting terms 5 and 10 dBA account for the different levels of annoyance and disturbance that the same noise level causes during the evening and the night. It is worth noting that both L_{den} and L_{night} are calculated over a period of a year, thus they are not suitable for assessing noise levels in the short term.

1.2.3 Environmental Noise Measurements

Regardless of their purpose, outdoor acoustic measurements have one thing in common: their difficulty. This is due to the large number of variables (some of them difficult to control) that influence the test and, therefore, the results. Unfortunately, there is no general method for accounting for and resolving all possible problems when designing a noise measurement campaign. This means that even when the objectives of a noise study have been well established, planning and performing the noise measurements is not an easy task. However, there are remarkable reviews that provide reference material to illustrate the acoustical measurement process, for example [8].

In order to facilitate the understanding of the complexity of the task, we proposed to break it down into simpler steps:

- Obtaining or generating the information relevant for designing the measurement campaign. This task includes the location and description of the temporal and spatial behavior of the specific and residual noise sources. We must also focus our attention on the sensitive receivers, describing their location (i.e. residential buildings, school playgrounds, etc.). We must also study the propagation conditions, both the static ones (due to the shape and impedance of the terrain and buildings) and those that vary over time (meteorology). With this information, we can calculate the probability that all possible combinations of source-propagation binomial will occur.
- Designing the temporal and spatial sampling. With the previous information, and following the requirements of the applicable standard (or test procedure), we can design the test. The design should indicate when to measure, where the measurement points are, and how long to measure at each point. The other decision concerns how to program the sound level meter and what magnitudes will be recorded. The design is highly conditioned by the number of measuring instruments and personnel that we can move to the measurement area.
- Performing measurements on site. On site, the operator should ensure the repeatability and reproducibility of the measures carried out. At the same time, he should guarantee that residual noise can be suppressed after data processing. This is why the operator should be able to redefine the predefined design of the noise measurement campaign, making decisions 'in situ'.
- Analyzing the data, providing answers to the questions posed and compiling a report. If everything is correct so far, we will be able to process and analyze the recorded data to gather evidence about specific noise characteristics and their impact on communities. Also, we can collect all the information necessary to calculate the uncertainty of the results.

The following sections seek to introduce the reader to some key knowledge regarding environmental noise measurements. In this document, we focus on short-term noise measurements in industry and any outdoor noise activity, with the exception of transportation. Although this scope might seem narrow, the range of studies that it encompasses can be quite wide. We can find examples where noise measurement campaigns support the development of strategic noise maps (e.g. when measuring large industrial areas to define their noise power emissions). Another common example is the inspections that determine whether the level of exposure of some residents to a certain outdoor noise source complies with the limits established by national legislation. To complete this information, the following sections are accompanied by international standards that propose a systematic approach to the noise measurement, the outdoor propagation calculation and the uncertainty estimation.

1.2.3.1 Noise Source and Its Influence on Time Sampling

Usually, the first step when studying noise sources consists of cataloging the noise emissions into 'noise classes' [9]. To carry out this classification, we have taken into account not only noise emission levels, but also other relevant characteristics of noise, as already mentioned: tonality, impulsiveness and low frequency. Each class is characterized by stable operating conditions which result in a small dispersion of noise emission levels. To define the set of noise classes, the most reasonable option is to use measurements taken near the noise source, in order to avoid the influence of propagation. However, in the case of noises composed of multiple noise sources, the microphone should be positioned far enough away to include the contributions of at least the most relevant ones.

The second step in the noise study is to find out the probability of occurrence of each noise class. Steps one and two allow the estimation of the noise emission in the long term.

In cases in which the purpose of the measurement campaign is to assess the noise problem in a residential area, then is necessary to characterize the worst noise scenario. This occurs when the highest noise emission levels coincide with favorable propagation conditions and impact the most exposed buildings in the residential area. When a national regulation uses a noise rating index, the worst case may not match the highest emission level because of the presence of certain noise characteristics that must be penalized.

The final L_{Aeq} (and the rest of the parameters) will be determined from the statistically significant amount of L_{Aeq} collected for every class. A set of 30 records is recommended, but when this is not possible, a number of not less than 10 will be considered acceptable if the records are homogeneous and their distribution is Gaussian.

1.2.3.2 Outdoor Sound Propagation and Its Influence on Time and Space Sampling

In the propagation path, sound waves encounter anomalies to an ideal homogeneous propagation medium. This can happen due to progressive changes in the acoustic impedance of the atmosphere or when an acoustic wave meets walls, the ground or any other object that causes abrupt changes in sound impedance. If there is a noise barrier, such as a wall, the noise energy received is partially absorbed by the barrier and partly reflected off its surface, whilst the remaining energy crosses the barrier and is transmitted again. In the case of very reflective surfaces, noise reflections can cause an increase in the noise level near the barrier. Also, as a consequence of the refraction in the atmosphere, the wave train can overcome obstacles; that is why it is usual to have higher levels than expected behind barriers.

The international standard ISO-9613-2 [10] has been commonly used to predict the attenuation of sound due to outdoor propagation and calculate the octave band sound pressure level at any receiver location. The standard ISO-9613-2 can complete the measurements carried out following another international standard, the ISO-1996-2 [9]. ISO-1996-2 includes a set of guidelines for the description, measurement and assessment of environmental noise, particularly applicable in community environments. In the case of excessive residual noise or complexity of the measurement scenario, it is recommended to complete and/or validate the on-site noise measurements with sound propagation calculations. Also, it is possible to use the ISO-9613-2 to extrapolate/interpolate the measurements in space, for example, when it is not possible to carry measurements in certain inaccessible places.

In the ideal free-field propagation condition, the sound level generated at a distance d by an omnidirectional point source (spherical wave propagation) of power W (in Watts) inserted in an isotropic and homogeneous media is defined as:

$$L_p = L_W - 10 \log_{10}(4\pi d^2) = L_W - 20 \log_{10} d - 11 \; (dB) \tag{1.19}$$

In which the term $20 \log_{10}(d) + 11$ is the geometrical divergence for spherical spreading (Figure 1.1a) caused by an omnidirectional source radiating in a free field. Doubling the distance from a point noise source, the sound pressure level is reduced by 6 dB. This means that the flow of energy through a sphere of radius 'd' that contains the noise source is proportional to $+4\pi$. This is why knowing the sound power

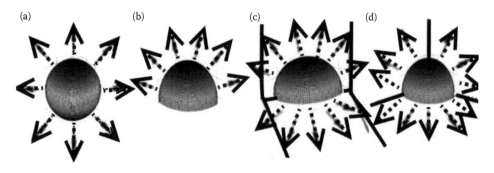

FIGURE 1.1 Directivity due to a noise point source combined with reflecting planes. From left to right, (a) spherical spreading, (b) hemispherical spreading, (c) half hemispherical spreading and (d) quarter hemispherical spreading.

emitted by the sound source is a fundamental aspect when predicting the outdoor noise levels. For determining the sound power levels emitted by outdoor machinery or equipment we can use the standards and method shown in the next Section (1.2.3.6).

If the source is constrained into a hemisphere with a perfectly reflecting floor (from $4\pi d^2$ to $2\pi d^2$), the sound pressure level can be estimated,

$$L_p = L_W - 20 \log_{10} d - 8 \ (dB) \tag{1.20}$$

For a point source the term $20 \log_{10}(d) + 8$, is the hemispherical spreading (Figure 1.1b).

When the point noise source is situated in the vertex of three perpendicular reflective planes (Figure 1.1d), the spreading term is $20 \log_{10}(d) + 2$, and if it is situated confined in the edge of two perpendicular planes (Figure 1.1c), then the spreading term is $20 \log_{10}(d) + 5$. This gives us an idea of directivity as a key factor when predicting noise levels. In fact, the same source with the same energy when radiated in the half of space doubles the intensity and increases the sound pressure level in 3 dB.

However, in real conditions, the propagation is attenuated. Considering this we can write that the sound pressure level is given by:

$$L_p = L_W + DI - GD - A \tag{1.21}$$

where

L_P is the sound pressure level at a certain point location,

L_W is the octave-band sound power of the noise source studied,

DI is the directivity index, which explains the directional dependence of a sound source emission. An omnidirectional point source creates a perfect sphere of equal sound pressure level because the sound radiates uniformly in all directions. The deviation from this pattern in dB is the DI. In the four cases (Figure 1.1) $DI = 0$ dB, $DI = 3$ dB, $DI = 6$ dB and $DI = 9$ dB.

$GD = 20 \log_{10}(d) + 11$ is the reference geometrical divergence that accounts for spherical spreading, being d the distance from the source to the receiver,

A is the total attenuation considering the real environmental conditions when the sound propagation deviates from ideal spreading. Normally attenuation is calculated in octave-band. In this document we use the followings terms:

- A_{Atm} is the attenuation term that counts for the atmospheric sound absorption,
- A_{Met} is the attenuation term that counts for the meteorological effects,
- A_{Gro} is the attenuation term that counts for the ground effects,
- A_{Bar} is the attenuation term that counts for the barriers and buildings.

A_{Atm}. This is the term that considers the air absorption that causes the degradation of sound energy into heat during its propagation in air. This term depends on frequency, humidity, pressure and temperature. The atmospheric attenuation coefficient α for octave band of sound propagation in an atmosphere temperature of 20°C and a relative humidity of 70% is reported in Table 1.2.

A_{Met}. Variations in weather conditions are another important effect that takes place in the atmosphere. This has an impact on the curvature of the sound path between noise source and receiver. Thus, meteorology is a key factor in sound propagation over long distances. However, the Equation 1.21 should be

TABLE 1.2   Frequency Dependence of Atmospheric Sound Absorption

Frequency in octave (Hz)	63	125	250	500	1 K	2 K	4 K	8 K
α Atmospheric absorption coefficient (dB/Km)	0.1	0.4	1.1	2.8	6.0	9.0	22.9	76.6

Note: The main mechanism that explains the air absorption is proportional to the square of the frequency.

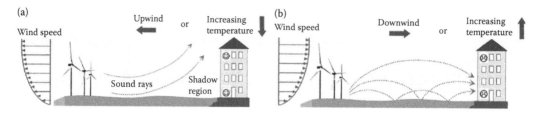

FIGURE 1.2 Propagation effects of sound waves in different weather conditions. From left to right, (a) unfavorable propagation due to refraction generated by upwind conditions or negative temperature gradient of the air column (during the day) and (b) favorable condition for propagation propitiated by a positive temperature gradient (during the night) and downwind conditions.

applied under certain conditions: it is only applicable to point noise sources and under favorable propagation conditions. Favorable conditions of propagation are met when:

$$d_p \leq 10(h_s + h_r) \tag{1.22}$$

d_p is the distance between noise source and receiver projected on an ideal flat ground,
h_s is the height of noise source over the ground,
h_r is the height of receiver over the ground.

Over long distances, the favorable and unfavorable propagation condition depends on the wind gradient conditions and the vertical temperature gradient. When a wave train passes from medium 1 to medium 2, the speed of sound can increase or decrease, bending the wave fronts and curving the sound rays.

- Favorable conditions for sound propagation (Figure 1.2b). If the speed of sound waves increases while ascending, the angle of a vector normal to the wave front (the sound ray) moves away from the vector describing the gradient between two media, so the sound path is refracted downward, producing an enhancement in sound levels. This situation takes place during downwind conditions (when the speed of wind increases with altitude) and at night, when the temperature profile of the atmosphere draws a positive gradient (temperature increases with altitude).
- Unfavorable conditions for sound propagation (Figure 1.2a). By day, with solar radiation at its maximum, the negative temperature gradient makes the speed of sound decrease with altitude. In this situation, the sound rays bend upward, reducing the sound level on the ground and forming a shadow area. Exactly the same phenomenon is produced in upwind situations.

The following data included in Table 1.3 is used to describe and classify the meteorological windows.

To program short-term noise measurements, the time intervals for measurements should cover all noise classes at favorable conditions. When talking about long distance, the recommended criteria is to select $T = 30$ minutes in order to get a minimum averaging time. If Equation 1.22 is fulfilled, 10 minutes are enough to describe any noise class.

A_{Gro}. The height of noise source and receiver with respect to the ground determines the interference between the direct sound (direct line of sight free of obstacles) from the source and the reflected sound from the ground, modifying the overall noise level at the receiver. The ground effect depends on the geometry of source–receiver profile and the sound impedance of ground surface and obstacles. The ground effects are particularly important when the source of noise and the receivers are very close to the terrain. ISO-9613-2 explains how to handle the height of the source and the receiver and the distance between them. The standard considers three zones: near the source, in the middle, and near the receiver. The acoustic characteristic of each zone is defined by the parameter G, varying from 0 (fully reflective surface) to 1

TABLE 1.3   Sound Propagation Condition in Classes

Meteorological Window	Verbal Description	Downwind Speed at 10 meters Height	Insulation (Clear Sky)
M1	Unfavorable	<1 m/s and upwind	Day
M1	Unfavorable	upwind	Night
M2	Neutral	1–3 m/s	Day
M3	Favorable	3–6 m/s	Day
M4	Very favorable	>6 m/s	Day
M4	Very favorable	Calm and downwind	Night

(fully absorbing ground). The total ground attenuation is then the sum of the attenuation due to the three zones (source, middle, and receiver).

A_{Bar}. Screening effects of a noise barrier are important when the horizontal dimension of the object normal to the source-receiver line is larger than the acoustic wavelength of the frequency band of interest. This can reduce the noise level at the receiver in its shadow zone where acoustic rays reach the receiver only because of the diffraction phenomena. The maximum attenuation [dB] performed by a barrier is about 20 dB (if ignoring the lateral diffraction). The attenuation can be evaluated by using the empirical Maekawa relation [11]:

$$A_{Bar} = 10 \log_{10}(3 + 20N)(dB) \tag{1.23}$$

where N represents the Fresnel number,

$$N = \frac{2\delta}{N} \tag{1.24}$$

and where δ is the difference between the diffracted path and the direct one, $\delta = a + b - c$ (Figure 1.3).

1.2.3.3 Residual Noise Sources and Their Influence on Time and Space Sampling

One of the most compromised aspects of the measurement process that the reader should be aware of is the handling of residual noise, since most of the occasions, we are only able to measure total noise present in the area, which is a combination of the residual noise and the specific noise we are trying to measure. It is well known that we can use the total noise measurement without any correction of residual noise when the residual noise is 10 dB lower than the total noise. On the other hand, this correction is possible when the residual noise is between 3 and 10 dB lower than the total noise. When the difference is less than that 3 dB,

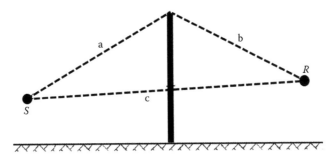

FIGURE 1.3   Propagation path between point noise source (S) and receiver (R) due to a noise barrier.

the uncertainty is too large and it is best to ignore the measurements. The correction of which we speak is a simple energetic subtraction:

$$L_{specific} = 10 \log_{10}[10^{L_{total}/10} - 10^{L_{residual}/10}] \tag{1.25}$$

where $L_{specific}$ is the estimation of specific sound, L_{total} the measured sound pressure level, and $L_{residual}$ the residual noise pressure level. When the residual noise is not negligible, it is necessary to study the (time and space) behavior of the relevant sources of residual noise with the same level of care as that dedicated to specific noise sources.

1.2.3.4 Noise Data

Some noise data are directly recovered from the sound level meter, and others must be calculated. Some of the most important are listed in Table 1.4.

There is no generally accepted method for detecting and rating impulsive, tonal and low-frequency sounds. It is very important to make sure that such negative noise characteristics are truly attributable to specific noise. Audible tones are generally detected using 1/3 octave band analysis without using any frequency-weighted network. The method consists in comparing the level of each frequency with its adjacent frequencies, for example using simple arithmetic operations: $L_{ton} = L(f) - ((L(f-1) + L(f+1))/2)$. Possible choices to penalize the noise source emissions could be selected separately for low-frequency band (25–125 Hz; $L_{ton} > 15$), middle-frequency band (160–400 Hz; $L_{ton} > 8$) and high-frequency band (500–10,000 Hz; $L_{ton} > 5$). The same thing happens with impulsiveness and low-frequency detection. A common indicator for evidence of impulsivity may be of the type: $L_{imp} = L_{AIm,T} - L_{Aeq,T}$ (for example, a value of $L_{imp} > 15$ can be used as a threshold to penalize the impulsive character of the noise source). In the case of low frequency: $L_{low} = L_{Ceq,T} - L_{Aeq,T}$ (again, a value of $L_{low} > 15$ can be used as a threshold to penalize the low-frequency character of the noise source).

However, some restrictions must be applied to the detection of low frequency and tonality in the low-frequency band, since the measurements should never be carried out inside rooms. Outdoors, the measurements should preferably be carried out under free field conditions or by mounting the microphone directly on a reflecting plane attached to the facade.

1.2.3.5 Uncertainty Estimation

Usually, regulations for noise measurement and assessment for the protection of communities are based in the standard ISO-1996-2. In many cases, these regulations reduce the work dedicated to temporal and

TABLE 1.4 Noise Magnitudes and Ratings of Interest for Noise Assessments

Magnitudes to Measure	Symbols Used in the Sound Level Meter
Maximum time-weighted sound pressure level	L_{AFmax}
N percentile exceedance level during the time interval T	$L_{AFN,T}$
Sound frequency-weighted exposure level during the time interval T	L_{AE}/L_{CE}
Equivalent frequency-weighted continuous sound pressure level during the time interval T	$L_{Aeq,T}/L_{Ceq,T}$
Equivalent continuous impulse level for the duration of the measurement as defined by IEC 60804	$L_{Im,T}$
Frequency content in one-third-octave bands (f_i) of equivalent continuous sound pressure level during the time interval T	$L_{Leq,fi,T}$
Rating sound exposure level	L_{RE}
Rating equivalent-continuous sound pressure level	$L_{Req,T}$

spatial sampling by limiting the number of noise classes and locations in which to determine it. Often, the noise inspection procedure includes only the identification and measurement of the worst-case scenarios, which usually occur in the most exposed locations with the highest levels of noise emissions and the best propagation conditions. This reduction very often simplifies the task, but the selection of the microphone location should:

- Maximize the relationship between specific sound and residual sound,
- Minimize the effect of the weather,
- Guarantee the representativeness of the noise measurements,
- And minimize (or at least estimate) the effect of the reflecting planes in the surroundings, taking into account that a free field location is not always available.

But even without the difficulty of meeting these criteria, the short-term noise measurements will be affected by uncertainty because any measurement process provides only an estimate of the actual magnitude that we want to measure. But that magnitude, the measurand, must be defined correctly and without any ambiguities. Uncertainty estimation provides a probability interval within which to find the measurand. This interval is based on the measurement and the factors that influence the measurement process. Inter alia, the sources of uncertainty are behind [9,12,13]:

- Changes in the operation of the noise source during the measurement time interval
- The distance between noise source and microphone
- Weather conditions between the source and the measurement position during the measurement time interval
- The measurement procedure
- The measurement effort (number of measurements, measurement time interval)
- The sound level equipment chain
- The influence of multipath reflections near microphones
- The behavior of the residual noise

Every noise measurement carried out L_m, is a function of the mentioned variables x_i,

$$L_m = f(x_1, x_2, \ldots, x_i, \ldots, x_N) \tag{1.26}$$

The model function identifies the sources of uncertainty affecting the measurement estimation and represents the influence of those variables on the final quantity shown in the display of the instrument L_m. If each variable has the standard uncertainty $u(x_i)$, the combined standard uncertainty is shown in the following equation:

$$u(L_m) = \sqrt{\sum_{i=1}^{N} (c_i \cdot u(x_i))^2} \tag{1.27}$$

where c_i is the sensitivity coefficient and it is given by

$$c_i = \frac{\partial f}{\partial x_i} \tag{1.28}$$

Basically, there are two types of uncertainty: type A and B. Type B uncertainties are caused by effects that can be predicted and modeled. The standard deviation of the repeatability (same source, equipment, method and operator within a short period of time) and the reproducibility (same source and measurement procedure, but different operators and equipment and a time interval long enough to guarantee that the measurements are independent) can be used as an A-type procedure. Both exercises are derived from the requirements that many laboratories follow to comply with good practices [14], especially the comparison between laboratories. In any case, the total uncertainty can be expressed through the combination of both types of standard uncertainties. It is a common practice to report the measurement with a coverage in which the measurand is within a confidence level of 95% and corresponds to a coverage factor $k = 2$. The output of measurement, in general, is expressed by the interval $L_{real} = L_m \pm k \cdot u_c(L_m)$, around the estimator L_m. The combined uncertainty can be expressed in a simplified way once the major sources of uncertainty according to the bibliography have been identified [9,13,15]:

$$L_{specific} = L'_{specific} + \delta_{slm} + \delta_{sou} + \delta_{met} + \delta_{pos} + \delta_{res} \tag{1.29}$$

where

- $L_{specific}$ is the measurand. Here it is preferable to use $L_{specific}$ instead of L_{real} to denote the value sought through the measurement process.
- $L'_{specific}$ is the estimation of the measurand, which coincides with L_m when the residual noise barely has any influence. Otherwise, it is necessary to use Equation 1.25 after measuring total noise L'_{total} and residual sound $L'_{residual}$.
- δ_{slm}. The uncertainty due to the measurement chain that affects the measurements of the total and residual noise.
- δ_{res}. The uncertainty that considers the influence of residual noise in the total noise measured.
- δ_{sou}. The uncertainty due to any deviation from the expected operating condition of the source.
- δ_{met}. The uncertainty due to any deviation from the ideal condition in meteorological conditions, bearing in mind the distance between receiver and source and their relative height from the ground.
- δ_{pos}. The uncertainty that accounts for any influence of the microphone's position.

δ_{slm}. Some guides and standards recommend directly applying a standard uncertainty for a sound level meter type 1, of 0.5 dB [12]. In the case of ISO-1996, the recommendation is 1 dB to quantify the reproducibility of an acoustic measurement [9]. Every time the sound level meter is going to be used, it is important to verify its sensitivity using a calibrator. A high deviation is a symptom of malfunction and the accuracy of the measurements could be questioned.

δ_{sou}. Short-time measurements of any industrial activities need to determine the standard uncertainty per noise class. A round of measurements can be taken close to the noise source to minimize the possible interference from meteorological conditions and residual noise. This method is a typical estimation of Type-A uncertainty. The statistical treatment of the standard uncertainty for the noise class can be calculated as follows:

$$u_{sou} = \sqrt{\sum_{i=1}^{N} \frac{(L_{m,i} - L'_m)^2}{N - 1}} \tag{1.30}$$

where

$L_{m,I}$ is the set of N measurements that represent the 'noise class'.
L'_m is the mean, time-averaged A-weighting sound pressure level, using N measurements. It is the arithmetic average of the set of measurements per 'noise class'.
N is the total number of independent measurements.

δ_{met}. Under favorable and very favorable propagation conditions (see Equation 1.22 and Table 1.3 to determine which conditions are considered favorable propagation conditions), the standard uncertainty of short-term measurements due to meteorological conditions [9,12,15] are set to $u_{met} = 2$ dB. This

TABLE 1.5 Microphones in Different Situations and the Standard Uncertainty Figures for Correction of Reflection

Microphone Assembly	Angle of Incidence of the Sound from Line Noise Source	Correction	Standard Uncertainty
Flux-mounted over a vertical reflective surface	Less than 60 degrees or Point Source	+6 dB	2 dB
0.5 to 2 meters from walls. Excluded low-frequency noise	Less than 60 degrees or Point Source	+3 dB	1 dB

estimation is increased when the distance between receiver and source d is more than 400 meters, employing the following formula:

$$u_{met} = 1 + \frac{d}{400} \tag{1.31}$$

δ_{pos}. The uncertainty introduced in the measurements carried out in an ideal free field is almost negligible. But in many practical situations, the presence of buildings makes it difficult to get an ideal situation like this. The location of the microphone could be a problem in residential and sensitive areas. In these cases it is better to flush-mount the microphone on the reflecting surface close to the building's facade. This involves nominally 6 dB of correction. If the microphone is positioned 4 meters from the ground and between 0.5 and 2 meters from a reflecting surface (a building facade), the correction to be applied should be 3 dB. When talking about industrial plants, point noise source or near point source could be considered. So the angle of sound incidence over the microphone and reflecting surface is very narrow and its influence on uncertainty is shown in Table 1.5 [9,12,15].

Other variables, like the impedance of walls, the geometry of the receiver area, etc., can also generate influence on uncertainty.

δ_{res}. Residual noise is assumed to be exactly the same during the total noise measurement and during the measurement of residual noise used for the correction. When L_{res} is between 3 and 10 dB below the total noise level L_{total}, the uncertainty of the application of Equation 1.25 needs to be determined. For that reason the sensitivity coefficient should be calculated as follows [15]:

$$c_{res} = \frac{10^{L_{res}/10}}{10^{L_{total}/10} - 10^{L_{res}/10}} = \frac{-10^{-(L_{total}-L_{res})/10}}{1 - 10^{-(L_{total}-L_{res})/10}} \tag{1.32}$$

that residual noise follows the same steps, and takes the same standard uncertainty, except for the source. Once again, to avoid confusion we have not used the nomenclature L_m since both values (total and residual) are measured. Throughout Tables 1.6 and 1.7 we provide an example of estimation of the uncertainty of a measurement campaign at a residential building, divided into two steps. Table 1.6 accounts for the residual noise measurement when the industrial activity has been turned off. Table 1.7 accounts for the total noise measurement when the industrial activity has been turned on. This second round of measurements is carried out under the premise that the residual noise behavior does not change in the two tests.

Being the estimation of specific noise (Equation 1.25) $L'_{specific} = 59.2$ dB, we can claim that the true value of the specific sound becomes $L_{specific} = 59.2 \pm 5.8$ dB. This means that we have a 95% level of confidence that the actual value emitted by the noise source is between 53.4 and 65 dB.

1.2.3.6 Noise Measurement for Sound Power Definition in Industrial Areas

The sound power level is the acoustical magnitude that informs about the total acoustic energy that a sound source (or a combination) radiates per time unit. For practical reasons is better to express the sound

TABLE 1.6   Example of Uncertainty Budget for a Short-Term Single Noise Measurement under Favorable Propagation Conditions

Variable	Estimator	Standard Uncertainty, in dB	Sensitivity Coefficient	Uncertainty Contribution in dB
Residual Noise $L'_{residual} = 52$ dB	$L'_{residual} = 52$ dB			
Sound level meter		$U_{SLM} = 0.5$	1	0.5
Noise source	Mixed traffic (C = 10) of n = 1600 vehicles	$C/\sqrt{n} = 0.25$	1	0.25
Meteorological contribution	Favorable	2	1	2
Microphone flush mounted on facade	Correction 6 dB	0.4 Wide angle of view on the road	1	0.4
Residual noise Uncertainty				2.11 Equation 1.27

Note: The first step involves the calculation of residual noise uncertainty due to traffic noise [12,15]. It is ensured that the residual sound behavior is the same and that the measurement is carried out identically during both measurement steps.

TABLE 1.7   Example of Uncertainty Budget for a Short-Term Single Noise Measurement under Favorable Propagation Conditions

Variable	Estimator	Standard Uncertainty, in dB	Sensitivity Coefficient	Uncertainty Contribution in dB
Total Noise $L'_{total} = 60$ dB	$L'_{total} = 60$ dB			
Sound level meter		$U_{SLM} = 0.5$	1	0.5
Noise source	synchronized measurements close to the noise source	0.17 Equation 1.30	1	0.17
Meteorological contribution	Favorable	2	1	2
Microphone flush mounted on facade	Correction 6 dB	2 Point source	1	2
Residual noise	$L_{res} = 52$ dB	2.11	0.23 Equation 1.32	0.48
Global uncertainty				2.91 Equation 1.27
Expanded uncertainty				5.82

Note: The second step is the calculation of total noise uncertainty due to industry noise [12,15]. Free field measurements with a second sound level meter located near the source have been performed under favorable propagation conditions. These measurements have been carried out synchronously with the main noise measurement in order to estimate the variability of noise source emissions.

power relative to the threshold of hearing (10^{-12} watts) in a logarithmic (dB) scale rather than to express it in watts.

$$L_W = 10 \log_{10}\left[\frac{W}{W_{ref}}\right] = 10 \log_{10} W + 120 \qquad (1.33)$$

where

L_W = sound power level (dB),
W = sound power in watts,
$W_{ref} = 10^{-12}$ watts (or 1 pW) is the reference sound power.

Sound power cannot be directly measured but can be estimated by conducting sound pressure measurements in specific conditions. Once the operation condition of the sound source is defined, the sound power level is a constant value. If the sound source is close to a flat (reflecting) surface, able to radiate acoustical energy in 'free field' conditions (actually a hemisphere), the general formula for determination of the sound power level from N sound pressure levels measurements is as follows [8]:

$$L_{A,W} = 10 \log_{10} \left(\frac{1}{N} \sum_{i=1}^{N} 10^{L_{Api}/10} \right) + K + 10 \log_{10} \left(\frac{S_m}{S_0} \right) \qquad (1.34)$$

$L_{A,W}$ = The A-weighting sound power in dB (total or per octave band),
N = The measurement positions (number of microphones) that describe a surface S,
$L_{A,pi}$ = The A-weighting sound pressure level measured at the ith microphone,
L'_{Ap} = The first term of the sum that appears between parentheses is the mean, time-averaged A-weighting sound pressure level, using N measurements that define the evaluation surface,
K = The correction factor that accounts for all the influences that have to be considered in the evaluation,
S_0 = The reference area of 1 m^2,
S = The measurement surface that corresponds to a hemisphere $S = 2\pi d^2$, and then $10 \log_{10}(S/S_0) = 20 \log_{10}(d) + 8$.

The dimension and position of the noise source with respect to the ground define the procedure of measurements (number and location of the measurement locations). When talking about industrial areas with multiple sound sources, we can choose between two different approaches. The first one consists of estimating the noise power of every relevant noise source, one by one; the second consists of considering the industrial area as only one source, taking measurements at its perimeter. Following the first approach and using ISO 3744 [16,17], K is the sum of two components $K_1 + K_2 \cdot K_1$ accounts for the correction due to background noise. One of the constraints of the use of this method is that the level of residual noise should be at least 6 dB below the noise source level. If $\Delta Lp > 15$ dB, then K_1 is equal to 0 and no correction is applied. If $\Delta Lp < 6$ dB, then $K_1 = 1.3$ decibel (dB) and the accuracy of the results may be reduced. K_2 accounts for the correction due to environmental conditions. This term is determined using a reference sound source which is compliant with ISO 6926 [18], and which is used to verify the influence of the environment on the measured sound power.

Following the second measurement approach, [19] all sources are treated as one single source located in the center of the area (Figure 1.4).

The microphone positions have to be located at a closed contour surrounding the industrial area and defining a regular polygon. Some standard requirements are difficult to achieve: for example, accessibility to all measurement points, and the height of the microphone determined by the relation: $h = H + 0.025 \cdot S_m \geq 5m$ (minimum height of measurements) where H is the average height of the plant's noise-emitting equipment and S_m is the measurement area. On the other hand, some problems may arise when residual noise sources are located adjacent to the measurement locations.

The uncertainty behind the practical application of this standard is composed by the variables considered for the general outdoor measurement processes and other than those specific to this method. These uncertainties arise from the irregular distribution of sound sources within the plant that influence in the spatial variations the sound pressure levels (averaged over time) at the different measurement positions along the contour. The second part is due to the number of microphones that can take measurements simultaneously.

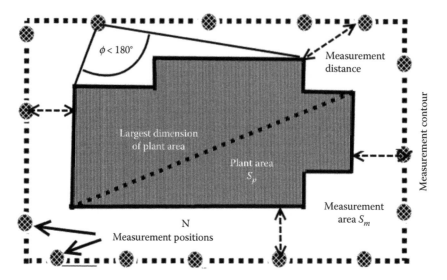

FIGURE 1.4   ISO 8297. Measurement contour definition. (From ISO 8297:1994: Acoustics – Determination of sound power levels of multisource industrial plants for evaluation of sound pressure levels in the environment – Engineering method).

References

1. 'Guidelines for community noise', WHO 1999.
2. ISO 266:1997, Acoustics – Preferred frequencies.
3. IEC 61260-1:2014. Electroacoustics - Octave-band and fractional-octave-band filters – Part 1: Specifications.
4. Acoustical Society of America. ANSI S1.42-2001 (R2011). *Design Response of Weighting Networks for Acoustical Measurements*. New York, NY: American Institute of Physics, 2011.
5. EC 61672-1:2013. Electroacoustic – Sound level meters – Part 1: Specifications.
6. Directive 2002/49/EC of the European Parliament and of the Council of 25 June 2002 relating to the assessment and management of environmental noise (http://eur-lex.europa.eu/legal-content/en/TXT/?uri=CELEX:32002L0049).
7. ISO 1996-1:2003. Acoustics – Description, measurement and assessment of environmental noise – Part 1: Basic quantities and assessment procedures.
8. Harris, C.M. *Handbook of Acoustical Measurements and Noise Control*. McGraw-Hill Inc., 1991.
9. ISO 1996-2:2007. Acoustics – Description, measurement and assessment of environmental noise – Part 2: Determination of environmental noise levels.
10. ISO 9613-2:1996. Acoustics – Attenuation of sound during propagation outdoors. General methods of calculation.
11. Palazzuoli, D., Licitra, G. Chapter 1. Fundamentals. In G. Licitra (ed.) *Noise Mapping in the EU: Models and Procedures (chapter 10)*. CRC Press. Taylor and Francis Group, 2013.
12. IMAGINE. 'Determination of Lden and Lnight using measurements.' IMAGINE report IMA32TR-040510-SP08, 11 January 2006, 2006, http://www.imagine-project.org/.
13. Cueto, J.L., Hernandez, R. Chapter 3. Measurements. In G. Licitra (ed.) *Noise Mapping in the EU: Models and Procedures (chapter 10)*. CRC Press. Taylor and Francis Group, 2013.
14. ISO/IEC 17025:2005. General requirements for the competence of testing and calibration laboratories.

15. Paviotti, Marco; Kephalopoulos, Stylianos; Jonasson, Hans. 'IMAGINE project: urban measurements of Lden and Lnight and calculation of the associated uncertainties.' ICA. 19th International Congress on Acoustics Madrid, 2007.

16. EN ISO 3744:1995. 'Acoustics – Determination of sound power levels of noise using sound pressure – Engineering method in an essentially free field over a reflecting plane'.

17. ISO 3744:2010. Acoustics – Determination of sound power levels and sound energy levels of noise sources using sound pressure – Engineering methods for an essentially free field over a reflecting plane.

18. ISO 6926:2016. Acoustics – Requirements for the performance and calibration of reference sound sources used for the determination of sound power levels.

19. ISO 8297:1994. Acoustics – Determination of sound power levels of multisource industrial plants for evaluation of sound pressure levels in the environment – Engineering method.

1.2.4 Human Noise Response

The human response to an audible stimulus can result in a physiological increase of well-being or in adverse health effects (the noise). They are two sides of the same physical phenomenon.

It is worth recalling that, according to WHO [1], an *adverse effect* is defined as a 'change in morphology, physiology, growth, development or life span of an organism, which results in impairment of the functional capacity to compensate for additional stress or increases in susceptibility to the harmful effect of other environmental influences'.

Usually, environmental noise levels are not so high as to cause hearing loss, but if the exposure is prolonged and the noise is above certain levels, some negative (non-auditory) health effects can be registered. Non-auditory health effects include annoyance, cardiovascular disease, sleep disturbance and learning impairment in children.

In the following section, a brief introduction to the main mechanism of human response to noise will be provided in order to achieve deeper comprehension of the adverse health effects discussed in the next chapter.

1.2.4.1 The Perception of Noise

The hearing apparatus represents the transduction system for the conversion of acoustic energy into the perception of sound in the brain. The comprehension of such a complex phenomenon involves physics, physiology and psychology, and the effects of any sound strongly depends on the listener's subjective interpretation.

The threshold of hearing depends on the frequency of the sound and is measured by determining the sound pressure level of a pure tone at a defined frequency that it is just detectable by human hearing apparatus. When such thresholds are measured in the free field, the levels are referred to as *MAF (Minimum Audible Field)*, and the psychophysical procedure is defined by international standards [2].

1.2.4.1.1 Loudness

Above the audibility threshold, when noise level increases, the apparent loudness (the subjective evaluation of the intensity of sound without any other psychological evaluation or influencing factor that could affect the annoying properties of the sound) also increases. In general, louder noise is considered to be more annoying, and the subjective judgment is driven by the interplay between physical characteristics of noise (frequency, level and its variability on time, duration, impulsive character) and some non-acoustical elements (time of day, economic factors, social evaluations, ability to control the sources).

The human ear is not equally sensitive to different frequencies. Sensitivity is generally higher at medium-high frequencies and decreases progressively at low and very high frequencies. The level of sound sensation is expressed in *phon*, numerically equivalent to the sound pressure value expressed in dB of 1 kHz sound that produced the same sound sensation. The *equal-loudness* contours according to the ISO 226 standard [2] are reported, the loudness level of a 1 kHz pure tone of 10 dB is defined as a 10 *phon*, and if the level of the same tone is raised by 10 dB it is perceived to be about twice as loud by the average listener. Also, equal loudness curves have been experimentally determined for noise by Stevens and Zwicker [3,4] and standardized in ISO 532 [5].

When noise level increases, physiological discomfort can be experienced and a further increase causes a tickling sensation (at about, 130 dB) that becomes pain at higher levels.

Exposure to high noise levels or long-time exposure to less intense levels may result in permanent or temporary effects on the threshold limit and can modify the subjective loudness and perception of sound. In the case of a non-permanent effect, noise exposure may elevate or shift the hearing threshold. The longer the exposure time, the greater the shift and the successive recovery time; when the sound is wideband, a temporary threshold shift usually occurs in a frequency range between 2 and 6 kHz.

1.2.4.2 Masking

When the presence of noise interferes with another sound, lowering its audibility, it is known as a *masking* phenomenon. As an example, speech becomes unintelligible with high levels of noise (cocktail party effect); on the other hand, broadband noise in open office workplaces can assure privacy during conversations. This effect is more pronounced when background noise has a frequency range between 200 Hz and 6 kHz.

In order to study the masking phenomenon, the hearing threshold is studied with a pure tone. In this case, when the test tone approaches the frequency of the masker one, the threshold shows a very steep rise, reaches a maximum at the masking frequency, and then drops again at the higher frequencies. As an example, a 100 dB masker tone of 1200 Hz raises the threshold of hearing of the pure tone one up to 70 or 80 dB in the frequency range above 1 kHz [6]. As a general consideration, the masking effect becomes stronger as the loudness of the masker increases and is perceived to be stronger at a frequency higher than that of the masker. On the other hand, the effect is stronger at frequencies very close to the masker frequency.

It is worth noting that the masking effect may be usefully used as a complementary measure to control and mitigate traditional noise by modifying the urban, annoying soundscape: the emission of a pleasant sound in the environment masks the unwanted noise, reducing annoyance. The *soundscape approach* to reducing the annoyance experienced by people exposed to unwanted noise could be effective in all those urban situations where noise level reduction is not viable. The main principles used for a masking effect are usually spectral/spatial masking and information masking [7], and in this respect recent experiences in 'Sonic Garden' in Florence (Italy) have shown the great potential of the soundscape approach for restoring acoustically degraded places in urban areas, increasing the perceived comfort without reducing noise levels [7].

The modification of the soundscape ('the acoustic environment at a place, like a residential area or a city park, as perceived and understood by people in context', an operative definition from the conference 'Designing Soundscape for Sustainable Urban Development' [8]) may be a useful tool for reducing annoyance and the negative effects of noise exposure in urban places. Soundscape design is not based on reducing the overall noise levels but on the modification of people's perception by providing sounds appropriate to a particular place [8].

In the soundscape approach, the key issue is the emission of a *preferred* sound in the environment, depending on the context, reducing negative perception of noise. Some artificial soundscapes, superimposed on a 'noise polluted' area, increase the noise level but transform the perception of people. Knowledge of masking phenomena and human noise perception are fundamental research goals for implementing 'soundscape-based' action plans.

1.2.4.3 The Human Response to Noise

According to the European Commission, in the EU 'more than 100 million citizens are affected by noise levels above 55 dBA L_{den} (a threshold at which negative effects on human health can be observed)' [9]. This is equivalent to about 14 million people annoyed by environmental noise and 6 million sleep-disturbed, [9] with very high social and hospitalization costs.

Noise exposure can cause temporary (acute) or long-term health effects which persist after the exposure; on the other hand, individual susceptibility to noise can have a key role in the severity of the outcome. Meanwhile, if it is simpler to recognize acute effects, in the long term many other variables (confounding factors) can be associated with the same health outcome.

The relationship between noise and stress is now well defined: noise exposure results in an activation of the sympathetic and endocrine systems. Noise activates the pituitary-adrenal cortical axis and the sympathetic adrenal medullary axis [10], reflecting a change in blood flow, blood pressure and heart rate. On the other hand, a release of stress hormones (catecholamines, adrenaline, noradrenaline and cortisol) is also observed [11].

According to Babish [10], the cause-effect of noise exposure consists in: noise – annoyance – physiological arousal (stress indicators) – (biological) risk factors – disease and mortality. Two paths are relevant in developing adverse health effects: *direct* and *indirect* arousal and stimulation of the human body.

The direct path involves the interaction of the acoustic nerve and the central nervous system, whereas the indirect pathway regards the interplay of emotional and cognitive perception [10] with the cortical and subcortical structures.

Physiological stress initiated by *direct* and *indirect* mechanisms causes negative health effects during long-term exposure, and low noise levels can induce health effects similar to those by high levels, particularly as regards sleep time and concentration. Disturbance via the *indirect* pathway is a stressor that can result in a more serious health outcome, enhancing the risk of cardiovascular disease and increasing mortality.

In Table 1.8 the effects and threshold levels of noise exposure are reported according to the Expert Panel on Noise (EPoN) of the European Environmental Agency [12].

TABLE 1.8 Effects of Noise on Health and Well-Being with Sufficient Evidence

Effect	Dimension	Acoustic Indicator[a]	Threshold[b]	Time Domain (Duration)
Annoyance disturbance	Psychosocial quality of life	L_{den}	42	chronic
Self-reported sleep disturbance	Quality of life, somatic health	L_{night}	42	chronic
Learning, memory	Performance	L_{eq}	50	Acute, chronic
Stress hormones	Stress indicator	L_{max}, L_{eq}	NA	Acute, chronic
Sleep (polysomnographic)	Arousal, motility, sleep quality	L_{max} (indoors)	32	Acute, chronic
Reported awakening	Sleep	SEL (indoors)	53	Acute
Reported health	Well-being clinical health	L_{den}	50	Chronic
Hypertension	Physiology somatic health	L_{den}	50	Chronic
Ischaemic heart disease	Clinical health	L_{den}	60	Chronic

Source: From 'Good practice guide on noise exposure and potential health effects', EEA Technical report No 11/2010, ISSN 17257–2237 [13].

[a] L_{den} and L_{night} are de ned as outside exposure levels. L_{max} may be either internal or external as indicated.

[b] Level above which effects start to occur or start to rise above background.

References

1. *Assessing Human Health Risks of Chemicals. Derivation of Guidance Values for Health based Exposure Limits.* Geneve: World Health Organization, 1994.
2. ISO 226:2003. Acoustics – Normal equal-loudness-level contours.
3. Stevens, S.S. 'Procedure for Calculating Loudness: Mark VI.' *The Journal of the Acoustical Society of America*, 1961; 33: 1577.
4. Zwicker, E. 'Subdivision of the Audible Frequency Range into Critical Bands (Frequensgruppen).' *The Journal of the Acoustical Society of America*, 1961; 33: 248.
5. ISO 532: 1975. Method for Calculating Loudness Level, International Organization for Standardization, Geneva 1975.
6. Maekawa, Z., Rindel, J.H., Lord, P. *Environmental and Architectural Acoustics.* Abingdon, Oxon: Spon Press, second edition, 2011.
7. Licitra, G., Cobianchi, M., Brusci, L. 'Artificial soundscape approach to noise pollution in urban areas.' in Proceeding of Internoise 2010, 13–16 June 2010, Lisbon, Portugal.
8. 'Designing soundscape for sustainable urban development', september 30–october 1, 2010, Sweden conference of the 'City of Stockholm' Environment and Health Administration.
9. Science for Environment Policy. Noise abatement approaches. Future Brief 17. *Produced for the European Commission DG Environment by the Science Communication Unit*, UWE, Bristol, 2017. Available at: http://ec.europa.eu/science-environment-policy.
10. Babisch, W. The noise/stress concept, risk assessment and research needs. *Noise Health*, 2002; 4: 1–11.
11. Babisch, W. Stress hormones in the research on cardiovascular effects of noise. *Noise Health*, 2003; 5(18): 1–11.
12. World Health Organisation. *Night Noise Guidelines for Europe.* World Health Organisation, Copenhage, 2009.
13. 'Good practice guide on noise exposure and potential health effects', EEA Technical report No 11/ 2010, ISSN 1725-2237.

1.2.5 Effects of Noise Exposure on Well-Being and Health

Exposure to high noise levels produces both physical and physiological outcomes, with negative effects on well-being and health. In particular, adverse effects on the nervous and cardiovascular systems are demonstrated nowadays.

Annoyance, sleep disturbance, cardiovascular diseases and impairment of cognitive performance in children are linked to noise exposure [1] and, in particular, according to WHO [1], 'traffic noise was ranked second among the selected environmental stressors evaluated in terms of their public health impact in six European countries'.

On the other hand, recent data on long-term exposure show that '65% of Europeans living in major urban areas are exposed to high noise levels ("High noise levels" are defined as noise levels above 55 dB L_{den} and 50 dB L_{night}), and more than 20% to night time noise levels at which adverse health effects occur frequently' [2], and one of the goals of the 7th EAP is ensuring a significant decrease, moving closer to WHO-recommended levels by 2020.

In the following sections, the non-auditory effects of noise exposure will be briefly presented, with a focus on the DALY metrics adopted by WHO in order to quantify the burden of disease from environmental noise [1].

1.2.5.1 Non-Auditory Effects of Noise Exposure

It is well known that prolonged exposure to high noise levels, exceeding 70 dBA [3], can lead to hearing loss and/or hearing impairment (internal hearing damage). Such high values are usually found in workplaces and, since environmental noise levels do not often exceed 70 dBA for long time periods, during recent years researchers have focused on developing dose-effect relationships between noise exposure and non-auditory effects. WHO has identified the main non-auditory health risks of noise in [4]:

1. Annoyance;
2. Sleep disturbance (and all its consequences on long- and short-term bases);
3. Cardiovascular disease;
4. Learning disability (cognitive impairment);
5. Tinnitus.

1.2.5.1.1 Annoyance

Annoyance is a very subjective feeling of discomfort, also linked to the meaning attributed to different sounds. Even if usually it has been defined as a 'feeling of resentment, displeasure, discomfort, dissatisfaction or offence which occurs when noise interferes with someone's thoughts, feelings or daily activities' [5], among acoustical researchers the term 'annoyance' assumes a wide range of meanings [6] and does not have a precise definition. However, according to the WHO definition of health [7], it has to be considered to have an adverse effect on health.

Some important factors increasing noise annoyance are:

- Fear perception from the source;
- A perceived lack of control of the source;
- Interference with cognitive tasks and activities;
- General noise sensitivity.

Annoyance increases when the sound pressure level rises, even if frequency contents and repetitive events play a major role in subjective perception. On the other hand, the perception that a noise source is important or that it could provide personal benefit decreases the associated annoyance [8,9].

After the first attempts to rate noise annoyance in terms of the percentage of 'highly annoyed' (%HA) people by different transportation sources in 1978 [10], dose-effect curves have been updated [11–13]. In the EU Position Paper on the dose-response relationship between transportation noise and annoyance [14], different curves for aircraft, road traffic and railway noise are summarized.

The functions are based on data from 54 studies from Europe, North America and Australia about noise annoyance from road traffic, aircraft and railways, analyzed by Schultz [10] and Fidell [11].

In terms of L_{den}, the percentages of 'highly annoyed' people for different transportation noise sources are:

Aircraft:

$$\%HA = -9.199 \cdot 10^{-5}(L_{den} - 42)^3 + 3.932 \cdot 10^{-2}(L_{den} - 42)^2 + 0.2939(L_{den} - 42)$$

Road traffic:

$$\%HA = -9.868 \cdot 10^{-4}(L_{den} - 42)^3 - 1.436 \cdot 10^{-2}(L_{den} - 42)^2 + 0.5118(L_{den} - 42)$$

(1.35)

Railways:

$$\%HA = -7.239 \cdot 10^{-4}(L_{den} - 42)^3 - 7.851 \cdot 10^{-3}(L_{den} - 42)^2 + 0.1695(L_{den} - 42)$$

excluding data below 45 dBA and above 75 dBA (L_{den}).

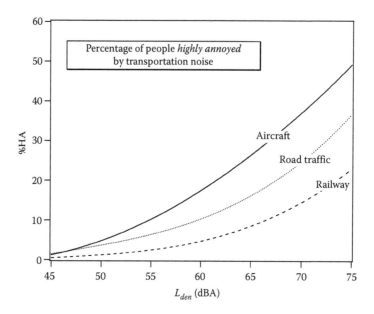

FIGURE 1.5   Percentage of highly annoyed people as a function of Lden according to Equation 1.35.

In an equivalent way, if we consider L_{dn} as a noise descriptor, according to Miedema and Oudshoorn [13] such curves are replaced by:

Aircraft:

$$\%HA = -1.395 \cdot 10^{-4}(L_{dn} - 42)^3 + 4.081 \cdot 10^{-2}(L_{dn} - 42)^2 + 0.342(L_{dn} - 42)$$

Road trafffic:

$$\%HA = 9.994 \cdot 10^{-4}(L_{dn} - 42)^3 - 1.523 \cdot 10^{-2}(L_{dn} - 42)^2 + 0.538(L_{dn} - 42)$$

(1.36)

Railways:

$$\%HA = 7.158 \cdot 10^{-4}(L_{dn} - 42)^3 - 7.774 \cdot 10^{-3}(L_{dn} - 42)^2 + 0.163(L_{dn} - 42)$$

In Figure 1.5 the relations (1.35) are reported. It can be seen that aircraft noise is perceived as the most annoying among the transportation sources and that railway noise is perceived as the least annoying.

1.2.5.1.2 Cardiovascular Disease

Noise exposure is associated with cardiovascular disease including hypertension and ischaemic heart disease [15,16].

Psychological stress following the discomfort reaction to noise (indirect pathway in the general stress model) and non-conscious physiological stress (direct pathway) can affect the endocrine system by stimulating the release of stress hormones (adrenaline, noradrenaline and cortisol), increasing blood pressure and heart rate, which may have, in the long term, effects on the cardiovascular system [17]. Exposure to noise increases the risk of arteriosclerosis and hypertension that may give rise to serious complications such as myocardial infarction and stroke.

In order to account for the different acoustic characteristics of various transportation noises, different dose-response curves are developed for the exposure to road traffic noise, railway noise and aircraft noise

[18,19]. The association between noise exposure and cardiovascular disease are based on meta-analyses, excluding many potential confounding variables.

Hypertension   Many recent studies have focussed on the possible link between exposure to transportation noise and hypertension [20–23]. In particular, the large multinational study HYENA [24] (Hypertension and Exposure to Noise near Airports) was carried out between 2005 and 2006 in order to assess the relationship between hypertension and noise [20]. In the study, both aircraft and road traffic noise sources were evaluated.

The results from HYENA showed links between hypertension and road traffic noise at all times of day and aircraft noise during the night time [25].

Such results reflect the adult population, and in this respect, it is worth mentioning a recent research study in Belgrade [22] addressing school children aged 7–11 years. Children in noisy schools had higher systolic blood pressure than those attending school in quite areas.

All the studies agree that there is a positive correlation between the exposure to transportation noise and hypertension, but the degree of association is still unclear. There is not enough data to provide a clear dose-response relationship, but an increased risk for hypertension is associated with road noise levels exceeding 65 dBA during the day and 55 dBA at night.

Ischaemic Disease   Many researchers have reported a clear correlation between environmental noise exposure and ischaemic heart disease [26,27]. In particular, there is evidence that the risk of ischaemic heart disease is increased by the exposure to road traffic noise [27–29] in areas with an average daytime sound pressure level above 60 dBA. On the other hand, in the document *Night Noise Guidelines for Europe* [30], WHO reports limited evidence of an increased risk of cardiovascular disease due to night noise levels above 55 dBA.

The risk associated with road noise exposure has been defined using the statistical approach of meta-analysis, and all the studies considered refer to myocardial infarction (MI) as a proxy of ischaemic heart disease.

A polynomial function for the dose-response curve has been defined, explaining 96% of the variance in the results. The exposure-response function for MI refers to the daytime noise indicator $L_{day,16h}$ and is valid for noise levels in the range from 55 dBA to about 80 dBA:

$$OR = 1.63 - 0.000613 \cdot (L_{day,16h})^2 + 0.00000736 \cdot (L_{day,16h})^3 \qquad (1.37)$$

considering the odd ratio (OR) value as an estimation of the relative risk. It is possible to use the conversion from the indicator L_{den} to $L_{day,16h}$ according to '*Good practice guide on noise exposure and potential health effects*' [31]:

$$L_{day,16h} \approx L_{den} - 2dBA \qquad (1.38)$$

where $L_{day,16h} \approx L_{den} \leq 60$ dB(A) is considered as a reference category (relative risk = 1) [31].

1.2.5.1.3 Cognitive Impairment in Children

As defined by WHO [1], noise-related cognitive impairment is 'the reduction in cognitive ability in school-age children' (7–19 years) 'that occurs while the noise exposure persists and will persist for some time after the cessation of the noise exposure'. Many studies, starting with the RANCH study [32], have shown that environmental noise exposure has a negative effect on children's learning outcomes, and a comparison between the reading ability, memory and cognitive performance of children exposed and those not exposed to transportation noise at school has shown poorer results in the first group [33–36]. In order to avoid negative effects on children, 'Community noise guidelines' [37] suggest that during lessons noise levels in school should be less than 35 dBA L_{Aeq}.

Hygge [38] developed a hypothetical dose-response curve for the percentage of children with cognitive performance deteriorated by noise. Such an approach assumes that 100% of children exposed to noise are cognitively affected at noise level of 95 dBA L_{dn} and that none are affected at 50 dBA L_{dn}. The exposure-response function is defined as a straight line connecting these two extremes. The proposed cut-off level is $L_{dn} = 50$ dBA. Because the difference between L_{dn} and L_{den} are usually small, in a study about the effects of road traffic noise L_{dn} values are approximated with the L_{den} ones.

1.2.5.1.4 Sleep disturbance

Sleep disturbance is, according to WHO [39], the most important negative effect on well-being and health. It reduces daytime readiness and work performance, deteriorating the quality of life and health [40]. During sleep, noise levels as low as Lamax 33 dBA can induce physiological reactions, altering the quality of the sleep itself [40].

In addition to short-term effects, such as an increase in daytime sleepiness, altered mood and cognitive activity, noise exposure during the night could be relevant in developing cardiovascular disease [20]. In the Night Noise Guidelines [30] WHO recommended nocturnal noise levels of less than L_{Aeq}, outside 40 dB a long-term goal for the prevention of noise-induced health effects.

In Directive 2002/49/EC [41] EU adopted L_{night} as the most appropriate metric for assessing sleep disturbance that can be assessed electrophysiologically or by self-reporting in epidemiological studies using questionnaires.

According to Miedema [42], the percentage of 'highly sleep-disturbed' persons (%HSD) as a function of noise exposure during the night is given by:

$$
\begin{aligned}
\text{Aircraft:} &\quad \%\text{HSD} = 18.147 - 0.956(L_{night}) + 0.01482(L_{night})^2 \\
\text{Road traffic:} &\quad \%\text{HSD} = 20.8 - 1.05(L_{night}) + 0.01486(L_{night})^2 \\
\text{Railways:} &\quad \%\text{HSD} = 11.3 - 0.55(L_{night}) + 0.00759(L_{night})^2
\end{aligned}
\tag{1.39}
$$

The relationships are based on data in the L_{night} (outside, maximally exposed facade, range 45–65 dBA) and can be extrapolated to lower exposure (40–45 dBA) and higher exposure (65–70 dBA).

1.2.5.1.5 Tinnitus

There is no consensus in the scientific community about the cause of tinnitus, and very often it is present concomitantly with hearing loss. WHO [1] considers tinnitus as an independent outcome of environmental noise that can be the cause of psychological distress, annoyance and a reduction in well-being. Even if there is not a single explanation of the occurrence of such disease, chronic exposure to noise can induce incapacitating tinnitus [43].

As regards the relationship between noise exposure and tinnitus, there is still not a valid dose-response curve, and survey-based studies are considered to estimate its prevalence [1]. This approach makes it possible to define the percentage of tinnitus that is attributable to environmental noise exposure.

According to WHO [1], a plausible value for the percentage of tinnitus attributable to the environmental noise is 3%.

1.2.5.2 The DALY Metric

It is necessary to adopt a suitable quantifiable indicator based on the estimation of the probability of health risk in order to assess the impact of noise on health. The disability adjusted life year (DALY) was developed by WHO and the World Bank in their Global Burden of Disease Study to estimate disease burden on people [44–46]. In this section, only a brief and operational introduction of this metric is provided; interested readers are referred to specialized books and articles [1].

DALY combines the time lived with disability and time lost due to premature mortality [44]; it is 'the sum of the potential years of life lost due to premature death and the equivalent years of "healthy" life lost by virtue of being in states of poor health or disability' [1]:

$$DALYs = YLD + YLL \qquad (1.40)$$

where YLL represents the years of life lost due to premature mortality and YLD the years lived with disability, combining with several social preferences and time discounting rate [44]. One DALY is equivalent to one year of healthy life lost.

1.2.5.2.1 YLL Calculation for Premature Death: Myocardial Infarction

In the simplified method [47] YLL and YLD can be calculated in a simple way as [1]:

$$YLL = \sum_i (N_i^m \cdot L_i^m + N_i^f \cdot L_i^f) \cdot AF \qquad (1.41)$$

where $N_i^m (N_i^f)$ is the number of deaths of males (females) after myocardial infarction in age group i multiplied by the standard life expectancy of males $L_i^m (L_i^f)$ (females) at the age at which death occurs [1] and AF is the 'attributive fraction' (WHO [1]) defined as:

$$AF = \left\{ \sum (P_i * RR_i) - 1 \right\} \Big/ \sum (P_i * RR_i) \qquad (1.42)$$

with P_i the proportion of the population in exposure category i and RR_i the relative risk at exposure category i compared to the reference level. The attributable fraction RR due to road traffic noise can be evaluated by applying the polynomial Equation 1.37 (odds ratios OR are estimates of the relative risk).

1.2.5.2.2 YLD Calculation

For each non-fatal health outcome WHO suggested including in DALY calculation, the value of YLD can be calculated by:

$$YLD = I \cdot DW \cdot D \qquad (1.43)$$

where I 'is the number of incident cases multiplied by a disability weight (DW)', D is an average duration in years and 'DW is associated with each health condition and lies on a scale between 0 (indicating the health condition is equivalent to full health) and 1 (indicating the health condition is equivalent to death)' [1,48]. The disability weight value expresses the importance of the degradation of the health (and well-being) status.

It is worth noting that in the calculation of YLD due to non-fatal myocardial infarction the same AF used for YLL has to be considered. The prevalence of tinnitus in Europe for people 15 years old and over is estimated by WHO (based on population data extracted from the European health for all database) [49]. Moreover, the fraction of tinnitus attributable to environmental noise, according to WHO, is 3% [1].

1.2.5.2.3 DALY Calculation: Some Experiences

The publication of the guidelines 'Burden of disease from environmental noise' in 2011 [1] and 'Methodological guidance for estimating the burden of disease from environmental noise' in 2012, [16] and the many data at the disposal of researchers from the implementation of the END Directive [41] in 2012, has allowed researchers to quantify the impact of noise in a standardized method by applying the DALY indicator. In this respect it is worth citing, among different experiences, the cases of Paris in 2011 [50], the city of Warsaw in 2015 [51], the agglomerations of Firenze, Prato, Livorno and the municipality of Pisa in Italy in 2016 [52] and the application of DALY calculations to the railway system in

TABLE 1.9   Transport Infrastructures Considered in the Studies [50–53] for the Evaluation of Noise Exposure

Noise Source	Paris	Italy (Prato, Livorno, Firenze and Pisa)	Warsaw	Sweden
Road	X	X	X	X
Railway	X	X		X
Aircraft	X	X		
Industrial noise				

Sweden in 2017 [53]. The noise sources considered in these different studies are reported in Table 1.9. All the studies estimated the DALYs due to noise according to the procedures recommended by WHO [1].

In the calculus of DALYs, different health endpoints and diseases have been considered. The number and severity of considered diseases deeply characterize DALY values, and a comparison between studies is not so simple. In Table 1.10 the health endpoints considered in the four studies are reported.

The DALYs per year in the Paris agglomeration [50] have been estimated to be 66,000 healthy lives lost (not considering tinnitus), annoyance is the second health disturbance, accounting for 38% of DALYs, sleep disturbance accounts for 60% and ischemic diseases accounts for 22%. Road traffic is the main source of burden of disease from environmental noise, representing 87% of DALYs. Focussing only on roads as a noise source, annoyance accounts for 38% of the total 58,000 DALYs, sleep disturbance represents 61% and ischaemic disease the remaining 1%.

In Warsaw [51] noise exposure from road infrastructure is responsible for around 26,000 DALYs per year. Annoyance, sleep disturbance, and ischaemic diseases account for 49%, 38% and 13%, respectively, of the total DALYs.

In agglomerations considered in Italy [52] road noise represents the cause of 85% of the total calculated DALYs (without tinnitus), with annoyance accounting for 60% of DALYs from road noise.

The Swedish study [53] presents an evaluation of DALY, considering exposure to noise throughout the country. The exposure distribution is based on data from the noise mapping of 13 agglomerations in the second round of the application of END requirements extrapolated to national figures. Also, in this case, roads represent the major source of noise, responsible for 90% of DALYs. Sleep disturbance has been found to be the main contributor to DALYs (54%), followed by annoyance (30%) and cardiovascular diseases (16%) (Table 1.10).

In comparing results from different studies, many potential confounding factors must be considered. Firstly, the data from noise mapping have to be improved, and the standard exposure-response could lead to inaccuracies, both over- and under-estimation [54]. On the other hand, the choice of the number of health endpoints and their disability weight (DW) has a very strong influence on the results and makes comparison very difficult.

TABLE 1.10   Different Health Outcomes and Diseases Considered in the Studies [50–53] for the Evaluation of Noise Exposure

Health Endpoint/ Disease	Paris		Italy (Prato, Livorno, Firenze, and Pisa)		Poland (Warsaw)		Sweden	
	YLL	YLD	YLL	YLD	YLL	YLD	YLL	YLD
Sleep disturbance		X		X		X		X
Annoyance		X		X		X		X
Cardiovascular diseases	X (road)	X			X (road)	X	X (road and railway)	X
Tinnitus		X		X				

The DALY indicator could, however, be very useful, both in the evaluation of the impact of noise on people and for the assessment of cost-benefit during the optimization of the action plans or remediation according to national regulations.

References

1. 'Burden of disease from environmental noise – Quantification of healthy life years lost in Europe', *WHO* 2011 (and references therein).
2. Decision No 1386/2013/EU of the European Parliament and of the Council of 20 November 2013 on a General Union Environment Action Programme to 2020 'Living well, within the limits of our planet' Text with EEA relevance. (http://eur-lex.europa.eu/legal-content/EN/TXT/?uri= CELEX:32013D1386).
3. Murphy, E., King, E.A. 'Environmental noise pollution', Elsevier, 2014.
4. Noise and health resources. WHO website. http://www.euro.who.int/noise/. Evans, G.W. Child development and the physical environment. *Annual Review of Psychology*, 2006; 57: 423–451.
5. Concha-Barrientos, M., Campbell-Lendrum, D., Steenland, K. *Occupationnal Noise: Assessing the Burden of Disease from Work-Related Hearing Impairment at National and Local Levels.* Geneva, Switzerland: WHO (http://www.who.int/quantifying_ehimpacts/publications/en/ebd9.pdf).
6. Guski, R., Felscher-Suhr, U., Schuemer, R. 'The concept of noise annoyance: how international experts see it.' *Journal of Sound and Vibration*, 1999; 223(4): 513–527.
7. *Guidelines for Community Noise.* Geneva: World Health Organization, 1999 (http://whqlibdoc. who.int/hq/1999/a68672.pdf).
8. Fields, J.M. 'Effect of personal and situational variables on noise annoyance in residential areas.' *Journal of the Acoustical Society of America*, 1993; 93(5): 2753–2763.
9. Flindell, I.H., Stallen, P.J.M. 'Non-acoustical factors in environmental noise.' *Noise & Health*, 1999; 3: 11–16.
10. Schultz, T.J. 'Synthesis of social surveys on noise annoyance.' *Journal of the Acoustical Society of America*, 1978; 64(2): 377–405.
11. Fidell, S., Barber, D.S., Schultz, T.J. 'Updating a dosage-effect relationship for the prevalence of annoyance due to general transportation noise.' *Journal of the Acoustical Society of America*, 1991; 89(1): 221–233.
12. Miedema, H.M.E., Vos, H. 'Exposure-response relationships for transportation noise.' *Journal of the Acoustical Society of America*, 1998; 104(6): 3432–3445.
13. Miedema, H.M.E., Oudshoorn, C.G.M. 'Annoyance from transportation noise: relationships with exposure metrics DNL and DENL and their confidence intervals.' *Environ Health Perspect*, 2001; 109: 409–416.
14. European Commission. *Position Paper on Dose Response Relationships between Transportation Noise and Annoyance.* Luxembourg: Office for Official Publications of the European Communities, 2002 (http://ec.europa.eu/environment/noise/pdf/noise_expert_network.pdf).
15. WHO Regional Office for Europe. *Burden of Disease from Environmental Noise-Quantification of Healthy Life Years Lost in Europe.* Copenhagen: WHO Regional Office for Europe, 2011a.
16. WHO Regional Office for Europe, Copenhagen. *Methodological Guidance for Estimating the Burden of Disease from Environmental Noise.* Copenhagen: WHO Regional Office for Europe, 2012.
17. Tomei, F. et al. 'Epidemiological and clinical study of subjects occupationally exposed to noise.' *International Journal of Angiology*, 1995; 4: 117–121.
18. van Kempen, E., Babisch, W. 'The quantitative relationship between road traffic noise and hypertension: a meta-analysis.' *Journal of Hypertension*, 2012; 30: 1075–1086.
19. Babisch, W. 'Road traffic noise and cardiovascular risk.' *Noise Health*, 2008; 10: 27–33.

20. Jarup, L. et al. 'Hypertension and exposure to noise near airports: the HYENA study.' *Environmental Health Perspectives*, 2008; 116(3): 329–333.
21. Dratva, J. et al. 'Transportation noise and blood pressure in a population-based sample of adults.' *Environmental Health Perspectives*, 2012; 120(1): 50–55.
22. Belojevic, G. et al. 'Urban road-traffic noise and blood pressure in school children.' Proceedings of ICBEN 2008, Foxwoods, CT, 2008.
23. Babisch, W., van Kamp, I. 'Exposure-response relationship of the association between aircraft noise and the risk of hypertension.' *Noise & Health*, 2009; 11(44): 161.
24. HYENA project http://www.hyena.eu.com/.
25. Babisch, W. et al. 'Exposure modifiers of the relationships of transportation noise with high blood pressure and noise annoyance.' *Journal of the Acoustical Society of America*, 2012; 132(6): 3788–3808.
26. Babisch, W., Beule, B., Schust, M., Kersten, N., Ising, H. 'Traffic noise and risk of myocardial infarction.' *Epidemiology*, 2005; 16: 33–40.
27. Babisch, W. 'Transportation noise and cardiovascular risk: Updated review and synthesis of epidemiological studies indicate that the evidence has increased.' *Noise & Health*, 2006; 8: 30.
28. Babisch, W. 'Traffic noise and cardiovascular disease: epidemiological review and synthesis.' *Noise & Health*, 2000; 2(8): 9–32.
29. van Kempen, E.E.M.M. et al. 'The association between noise exposure and blood pressure and ischaemic heart disease: a meta-analysis.' *Environmental Health Perspectives*, 2002; 110: 307–331.
30. *Night Noise Guide Lines for Europe.* Copenhagen, Denmark: World Health Organisation (WHO), 2009. http://www.euro.who.int/__data/assets/pdf_fi le/0017/43316/E92845.pdf.
31. 'Good practice guide on noise exposure and potential health effects', EEA Technical report No 11/2010, ISSN 1725-2237 (https://www.eea.europa.eu/publications/good-practice-guide-on-noise/download).
32. Stansfeld, S.A., Berglund, B., Clark, C., Lopez-Barrio, I., Fischer, P., Öhrström, E. et al. 'Aircraft and road traffic noise and children's cognition and health: a cross-national study.' *Lancet*, 2005; 365: 1942–1949.
33. Evans, G., Hygge, S. Noise and performance in adults and children. In L. Luxon, D. Prasher (eds) *Noise and Its Effects.* London: Whurr Publishers, 2007.
34. Hygge, S., Evans, G.W., Bullinger, M. 'A prospective study of some effects of aircraft noise on cognitive performance in schoolchildren.' *Psychological Science*, 2002; 13: 469–474.
35. Bronzaft, A.L. 'The effect of a noise abatement program on reading ability.' *Journal of Environmental Psychology*, 1981; 1: 215–222.
36. Lercher, P., Evans, G.W., Meis, M. 'Ambient noise and cognitive processes among primary schoolchildren.' *Environment and Behavior*, 2003; 35: 725–735.
37. Berglund, B., Lindvall, T., Schwela, D.H. *Guidelines for Community Noise.* Geneva: World Health Organization (WHO), 1999. http://whqlibdoc.who.int/hq/1999/a68672.pdf (accessed April 03, 2017).
38. Hygge, S. Environmental noise and cognitive impairment in children. Document prepared for the WHO working group on Environmental Noise Burden of Disease.
39. Fritschi, L., Brown, A.L., Kim, R., Schwela, D.H., Kephalopoulos, S. (eds) *Burden of Disease from Environmental Noise.* Bonn: World Health Organization, 2011.
40. Muzet, A. 'Environmental noise, sleep and health.' *Sleep Medicine Reviews*, 2007; 11: 135–142.
41. Directive 2002/49/EC of the European Parliament and of the Council of 25 June 2002 relating to the assessment and management of environmental noise. *Official Journal of the European Communities*, 2002, L 189:12–25.
42. Miedema, H.M.E., Passchier-Vermeer, W., Vos, H. *Elements for a position paper on night-time transportation noise and sleep disturbance.* Delft, TNO, 2003 (Inro Report 2002-59).
43. Plontke, S.K.R. et al. 'The incidence of acoustic trauma due to New Year's firecrackers.' *European Archives of Oto-Rhino-Laryngology*, 2002; 259: 247–252.

44. Murray, C.J.L. 'Quantifying the burden of disease: the technical basis for disability-adjusted life years.' *Bull World Health Organ*, 1994; 72: 429–445.

45. Murray, C.J.L., Lopez, A.D. *Global Health Statistics: A Compendium of Incidence, Prevalence and Mortality Estimates for Over 200 Conditions.* Cambridge, MA: Harvard University Press, 1996a.

46. Murray, C.J.L., Lopez, A.D. (eds) *The Global Burden of Disease: A Comprehensive of Mortality and Disability from Diseases, Injuries, and Risk Factors in 1990 and Projected to 2020.* The Global Burden of Disease and Injury Series Boston, MA: Harvard University Press, for Harvard school of Public Health on behalf of the World Health Organization and the World Bank, 1996b.

47. Murray, C.J.L., Ezzati, M., Flaxman, A.D., Lim, S., Lozano, R., Michaud, C. et al. 'The Global Burden of Disease Study 2010: design, definitions, and metrics.' *Lancet*, 2012; 380: 2063–2066.

48. Salgot, M., Huertas, E. *Integrated Concepts for Reuse of Upgraded Wastewater. WP2-Aquarec-Guideline for Quality Standards for Water Reuse in Europe.* University of Barcelona, 2006.

49. European health for all database (HFA-DB) [online database]. Copenhagen, WHO Regional Office for Europe, 2010 (http://data.euro.who.int/hfadb/, accessed 30 July 2010).

50. 'Health impact of noise in the Paris agglomeration: quantification of healthy life years lost – position paper on the application to the Paris agglomeration of the WHO method to determine the burden od disease from noise', *Bruitparif*, 2011.

51. Tainio, M. 'Burden of disease caused by local transport in Warsaw, Poland.' *Journal of Transport & Health*, 2015; 2: 423–433.

52. Palazzuoli, D., Licitra, G., Ascari, E. 'The impact of noise exposure on citizen health in some Tuscany cities: what DALY can suggests to local administrators.' in Proceedings of the '23rd International Congress on Sound & Vibration', Athens, Greece, 10–14 July, 2016.

53. Eriksson, C., Bodin, T., Selander, J. 'Burden of disease from road traffic and railway noise – a quantification of healthy life years lost in Sweden.' *Scandinavian Journal of Work, Environment & Health*, – online rst. doi:10.5271/sjweh.3653

54. Lercher, P. 'Road and railway traffic noise exposure and its effects on health and quality of life.' in Proceedings of '44th Convegno Nazionale dell'Associazione Italiana di Acustica', Merano, Italy, 2013.

1.2.6 Vibrations in the Environment

Sound and vibrations are different facets of the same mechanical phenomenon: a particle displacement producing a wave motion that is transmitted in a medium.

Considering vibrations should be of primary importance when the environmental impact of transportation infrastructures must be evaluated. Among sources of ground-borne vibration, trains, heavy trucks and buses on rough roads have to be considered with construction sites.

Vibrations from transportation infrastructures are rarely strong enough to cause damage to buildings or annoyance.

In the following, only a brief description of the main issues about vibrations from transportation infrastructures will be provided, and the interested reader is referred to specialized books.

1.2.6.1 Descriptors

Vibrations can affect the human body by different paths, through machines or surfaces. Depending on possible paths, mechanical vibrations can be grouped in *whole body* or *part of the body*, such an organ or limb.

The transmission of ground vibrations to the whole body occurs on a surface which is vibrating (e.g. a seat or a bed) or on a vibrating floor in a building.

Low-level vibrations of the whole body can cause discomfort, whereas high levels can interfere with cognitive performance, reducing visual and manual control.

In describing vibrations, the reference quantity is the motion of a particle *in* or *on* the ground or building. The response of a structure to vibrations depends on the particle displacement, its velocity and its acceleration. Accelerometers, the instruments used to measure the vibrations, directly measure velocity and acceleration and vibratory motion is commonly described by the *peak particle velocity* or the *peak particle acceleration*.

The peak particle velocity is usually used to evaluate possible building damage, whereas for the human response an average vibration amplitude (the root mean square value, r.m.s.) is more appropriate, since the body responds to an average vibration amplitude. The average is usually calculated over a 1-second period. The ratio of the peak amplitude to the r.m.s. amplitude is called the *crest factor*.

Displacements are typically measured in millimeters (mm), velocity in millimeters per seconds (mm/s) and acceleration in mm/s per second (mm/s^2).

There are many different (national) technical norms for the evaluation of vibrations regarding human response. In general, all the evaluation methods recommend expressing vibration exposure by a single figure value with weighting factors accounting for the frequency dependence of human perception of vibrations.

The international standards ISO 2631 and ISO 5349 series provide methods and procedures for the evaluation of whole-body and hand-transmitted vibration exposure. In particular, ISO 2631-1:2014 [1] and ISO 2613-2:2003 [2] are the relevant standards for the assessment of the exposure of people to vibrations, whereas ISO 5349-1:2001 [3] accounts for hand-transmitted vibrations.

Such technical norms state that the acceleration has to be measured in the frequency range of 0.5–80 Hz for the whole body and between 8 and 1000 Hz for the hand-harm system. On the other hand, as regards the evaluation of the effects of vibrations on buildings, the reference quantity is the particle velocity measured in the range 1–250 Hz [4–6].

Particle velocity and acceleration are both very often expressed as *decibel* (dB) values. Vibration velocity level (L_v) in dB is defined as:

$$L_v = 20 \log_{10}(v/v_{ref}) \, \text{dB} \tag{1.44}$$

where v is the r.m.s. velocity amplitude value and $v_{ref} = 10^{-9}$ m/s the reference velocity amplitude. The vibration acceleration level L_a, is:

$$L_a = 20 \log_{10}(a/a_{ref}) \, \text{dB} \tag{1.45}$$

with internationally accepted value $a_0 = 10^{-6}$ m/s^2. It is important to note that quite often, vibration levels are expressed in decibels with reference to different values, e.g. $v_0 = 10^{-8}$ m/s, $a_0 = 10^{-5}$ m/s^2, etc.

In addition to international standards (ISO 2631-1:2014, ISO 2631-2:2003), several national technical norms define relevant quantities and measurement procedures to assess vibrations in buildings.

A very complete analysis of different standards was carried out in 2012 in the framework of the RIVAS project [7] in order to highlight common features and peculiarities. The following is a very short summary of the deliverables of the project.

According to Deliverable D1.4 (Del. 1.4 – Review of existing standards, regulations and guidelines, as well as laboratory and field studies concerning human exposure to vibration (Jan 2012), all the international standards use the same vibration quantities defined in the ISO standards [7]:

- The r.m.s. values;
- The maximum running r.m.s. values;
- The fourth-power vibration dose values.

and the amplitude of vibrations is that measured at the location where it assumes the highest value.

In the ISO standards the primary quantity used to assess the intensity of vibrations is the *acceleration* (except for very low frequencies and amplitudes, for which velocity measurements are preferred), whereas it varies in national norms. A quite similar situation is valid for the frequency weighting factors and for the time constant used in running r.m.s. measurements.

1.2.6.1.1 The ISO 2631 Standard

The principal target of ISO 2631 is the assessment of human exposure to vibrations with reference to health and comfort, the threshold of perception and possible negative effects on well-being, even though it doesn't specify any exposure limits.

In this standard, three different exposure descriptors are defined and proposed with different frequency weightings, depending on the aims of the evaluations.

Comfort and perception are assessed by using frequency-weighted r.m.s. acceleration in the three orthogonal directions at the surface where the person is. The weighting factors W_k and W_d have to be used for vertical and horizontal directions, respectively.

The standard states that the relevant quantities are both the weighted r.m.s. acceleration values in x, y and z directions (a_{wx}, a_{wy}, a_{wz}) and the combined one a_v:

$$a_v = \sqrt{a_{wx^2} + a_{wy^2} + a_{wz^2}} \tag{1.46}$$

where a_{wx}, a_{wy}, a_{wz} are the r.m.s. values over a measurement period T calculated from the weighted instantaneous values $a_{wx}(t)$, $a_{wy}(t)$, $a_{wz}(t)$:

$$a_{wi} = \sqrt{\frac{1}{T} \int_0^T a_{wi}^2(t) dt} \tag{1.47}$$

with $i = x,y,z$.

When the crest factor of $a_w(t)$ is greater than 9, ISO 2631 suggests two other different descriptors that should be reported with the energy equivalent r.m.s. acceleration:

The maximum transient vibration value (MTVV), defined as the maximum of a running r.m.s. of $a_w(t)$, and

The vibration dose value (VDV) that better accounts for peaks in acceleration time history.

The MTTV is calculated as:

$$MTTV = \max[a_w(t_0)] = \sqrt{\frac{1}{\tau} \int_{t_0-\tau}^{t_0} a_w^2(t) dt} \tag{1.48}$$

with τ equal to 1 s corresponding to a *slow* integration time constant. The maximum value is calculated over the measurement period.

The VDV value is instead calculated as:

$$VDV = \sqrt[4]{\int_0^T a_w^4(t) dt} \tag{1.49}$$

with T the evaluation period.

The human exposure to vibration in buildings is considered in ISO 2631-2:2003 and, as in ISO 2631-1:1997, limit values for acceptable levels of vibration exposure are not provided. ISO 2631-2 states that vibrations must be evaluated in the directions of maximum magnitude by using a different frequency-weighting factor W_m.

1.2.6.1.2 Some Open Questions

Many national standards have been defined on the basis of the ISO 2631 series. Zhang et al. [8] (2013) collected and analysed several national standards assessing vibrations in order to provide comfort in buildings, with particular reference to timber floors.

Nowadays, the synergetic effects of noise and vibrations on the perception of vibrations and the consequent evaluation of annoyance and discomfort are becoming ever more relevant. As an example, in the Norwegian Standard NS 8176.E:2008 the combined effects of noise and vibrations are discussed in an informative annex stating that 'no method has been found anywhere in the world for the measurement or evaluation of the total annoyance when combined effects are included'. Also in this respect, it is useful to note that many research projects about the assessment of vibrations in the environment and their mitigation have been recently funded, especially regarding railway infrastructures. Among others it is worth mentioning:

- RIVAS (Railway Induced Vibration Abatement Solutions, www.rivas-project.eu), a project aimed at reducing the environmental impact of ground-borne vibrations from rail traffic in residential areas (2011–2013).
- CARGOVIBES (Attenuation of ground-borne vibration affecting residents near freight railway lines, www.cargovibes.eu), about developing measures to ensure acceptable levels of vibration for residents living in the vicinity of freight railway lines (2011–2014).

1.2.6.2 The Main Vibration Sources

Among different classification criteria, vibrations can be grouped by their *duration* and *temporal* characteristics:

1. Continuous vibrations (from machinery, road traffic, continuous construction activities)
2. Impulsive vibrations (e.g. dropping of heavy equipment, blasting)
3. Intermittent vibrations (railway traffic, passing heavy trucks, impact pile-driving, drilling).

Trains and traffic (especially heavy trucks) on rough roads are common outdoor *continuous* vibration sources, even if road vehicles, in normal conditions, are not efficient sources of vibration, thanks to their suspension systems and tires. On the other hand, when pronounced discontinuities are present in the road, heavy trucks can cause *intermittent* vibration peaks lasting for a fraction of a second or a few seconds.

Rail traffic, and in particular freight trains, can instead be a significant source of ground vibrations. In this case the main parameters influencing the generation of rail vibrations are the roughness of rails and wheels, the stiffness of the rail supporting system (sleepers and ballast) and the vibrations from the source that propagate through the ground and are transmitted into buildings.

The propagation path and attenuation of ground vibrations, such as those from transport infrastructures, propagating by means of longitudinal and surface waves through soil and rock strata to the foundations and structures of buildings, depend both on their physical parameters (frequency, direction and amplitude) and on the characteristics of the ground (soil layers, local discontinuities such as ground water). All these factors make the prediction of vibration levels at a receiver very difficult. On the other hand, as a general rule, at least for surface waves, the attenuation with distance is in the range of −3 to −6 dB per doubling distance [9].

1.2.6.3 Impact of Vibrations on People

Human response to vibrations is frequency-dependent [10] and consistently non-linear [11–13]. A large number of studies have been carried out in order to determine the absolute perception thresholds or the equal sensation contours, but a definitive answer is still lacking. For a seated person it seems that the highest sensitivity to vertical vibrations generally occurs at around 5–6 Hz, whereas for horizontal vibrations the human body is more sensitive in the 1–2 Hz range. The perception threshold in the vertical direction is almost flat, with acceleration values above around 10 Hz, [14] while equal comfort contours for different postures and vibration directions follow the reciprocal of the perception threshold.

Vibrations from road and railway traffic may cause annoyance and discomfort in people. In assessing the effects of vibrations, one has to consider that they can enter the human body along different orthogonal paths: x-axis (back to chest), y-axis (right side to left side) or z-axis (foot to head). For each relevant axis, a different frequency weighting has to be considered. As reported in ISO 2631-2, [2] 'Human response to vibration in buildings is very complex'. Many studies have shown that the perception of vibrations is strongly subject-dependent [15,16] and influenced by an individual's sensitivity.

The standard ISO 2631-1:2014 [1] reports curves for the threshold of vibration perception as a function of frequency, also defining the equal human response with reference to annoyance and negative effects on comfort.

ISO 2631 further indicates that the degree of annoyance cannot be explained simply by the magnitude of vibrations; very often, complaints are associated with vibration levels lower than the perception threshold. In this respect, the evaluation of human response has to consider not only measurable physical parameters but also other related effects, such as the time of day in which vibrations are perceived, visual effects, ground-borne noise, etc.

References

1. ISO 2631-1:1997/Amd.1:2010(en) – Mechanical vibration and shock – Evaluation of human exposure to whole-body vibration – Part 1: General requirements AMENDMENT 1.
2. IS/ISO 4866: Mechanical Vibration and Shock – Vibration of Fixed Structures – Guidelines for the Measurement of Vibrations and Evaluation of Their Effects on Structures.
3. ISO 5349-1:2001 Mechanical vibration – Measurement and evaluation of human exposure to hand-transmitted vibration – Part 1: General requirements.
4. UNI 9916:2014 Criteria for the measurement of vibrations and the assessment of their effects on building.
5. ISO 2631-2:2003 Mechanical vibration and shock – Evaluation of human exposure to whole-body vibration – Part 2: Vibration in buildings (1 Hz to 80 Hz).
6. BS 7385-2:1993 Evaluation and measurement for vibration in buildings. Guide to damage levels from groundborne vibration.
7. http://www.rivas-project.eu, Deliverable D1.4 'Del. 1.4 – Review of existing standards, regulations and guidelines, as well as laboratory and field studies concerning human exposure to vibration (Jan 2012)'.
8. Zhang, B., Rasmussen, B., Jorissen, A., Harte, A. 'Comparison of vibrational comfort assessment criteria for design of timber floors among the European countries.' *Engineering Structures*, 2013; 52: 592–607.
9. Shioda, M. *J. Low Freq. Noise Vib.* 1986; 5(2): 51–59.
10. Miwa, T. 'Evaluation methods for vibration effect. Part 1: Measurements of threshold and equal sensation contours of whole body for vertical and horizontal vibrations.' *Industrial Health*, 1967; 5: 183–205.
11. Nawayseh, N., Griffin, M.J. 'Non-linear dual-axis biodynamic response to vertical whole-body vibration.' *Journal of Sound and Vibration*, 2003; 268(3): 503–523.
12. Nawayseh, N., Griffin, M.J. 'Non-linear dual-axis biodynamic response to fore-and-aft whole-body vibration.' *Journal of Sound and Vibration*, 2005; 282: 831–862.
13. Matsumoto, Y., Griffin, M.J. 'Dynamic response of the standing human body exposed to vertical vibration: influence of posture and vibration magnitude.' *Journal of Sound and Vibration*, 1998; 212(1): 85–107.

14. Morioka, M., Griffin, M.J. 'Magnitude-dependence of equivalent comfort contours for fore-and-aft, lateral and vertical whole-body vibration.' *Journal of Sound and Vibration*, 2006; 298(3): 755–772.

15. Griffin, M.J. *Handbook of Human Vibration.* London: Elsevier Academic Press, 1990.

16. Jakubczyk-Galczynskaa, A., Jankowskib, R. 'Traffic-induced vibrations. The impact on buildings and people.' The 9th International Conference 'Environmental Engineering', 22–23 May 2014, Vilnius, Lithuania, Selected Papers (http://enviro.vgtu.lt).

2

Industrial and Transport Infrastructure Noise

Luca Fredianelli
Consiglio Nazionale delle
Ricerche

Gaetano Licitra
Environmental Protection
Agency of Tuscany Region

Guillaume Dutilleux
Acoustics Research Centre

Jose Luis Cueto
Universidad de Cádiz

2.1 The EU Environmental Noise Directive

In 2002 the European Union approved Environmental Noise Directive 2002/49/EC (END) [1] in order to assess and manage environmental noise. END represents the main EU instrument for identifying noise pollution levels and outlines the actions needed to solve the related issues at the EU level. The directive defines a common approach in order to avoid, prevent or reduce the harmful effects, including annoyance, of exposure to environmental noise. Three actions form the basis of the European policy on noise, which aims to develop measures to reduce noise emitted by the major sources:

a. Determination of exposure to environmental noise through noise mapping
b. Information on environmental noise and its effects on the public
c. Adoption of action plans based upon noise-mapping results, with a view to preventing and reducing environmental noise where necessary, particularly where exposure levels can induce harmful effects on human health, and preserving environmental noise quality where it is good

a. *Strategic Noise Mapping*: *Noise maps* are defined as 'the presentation of data on an existing or predicted noise situation in terms of a noise indicator, indicating breaches of any relevant limit value in force, the number of people affected in a certain area, or the number of dwellings exposed to certain values of a noise indicator'. Thus, noise maps not only show noise levels in the area but also contain information on limit exceedances and the number of people and dwellings exposed to environmental noise. *Strategic noise mapping* is differently defined as 'a map designed for the global assessment of noise exposure in a given area due to different noise sources or for overall predictions for such an area'. Thus, while noise mapping is primarily a presentation of noise data, strategic noise mapping is more focused on noise exposure assessment. Indeed, the mapping requirements of the directive are primarily concerned with strategic noise mapping, because Annex IV of the directive points towards a strategic noise map as highlighting both the presentation and assessment exposure–related information.

The way to assess exposure to environmental noise is to use strategic noise maps for major roads, railways, airports and agglomerations using the harmonised noise indicators L_{den} (day–evening–night equivalent sound pressure levels) and L_{night} (night-time equivalent sound pressure levels) defined in Annex I. Strategic noise maps were compiled before June 2007 for all agglomerations with more than 250,000 inhabitants and for all major roads (those with more than 6 million vehicle passages a year), railways (with more than 60,000 train passages a year) and major airports (with more than 50,000 movements a year) within the territories. The second phase of the directive (June 2012) requires the Member States to prepare and publish, every five years, noise maps and noise management action plans for:

- Agglomerations with more than 100,000 inhabitants
- Major roads (more than 3 million vehicles a year)
- Major railways (more than 30,000 trains a year)
- Major airports (more than 50,000 movements a year)

b. *Population Exposure*: The determination of exposure levels to environmental noise through the common indicators L_{den} and L_{night} is the second key element of the END. The directive requires competent authorities to estimate the number of people living in dwellings exposed to noise. Values of L_{den} and L_{night} are provided in various categories, at the most exposed building facade and separately for road, rail, air traffic and industrial noise. In addition, where the information is available, people living in dwellings having special insulation against environmental noise or a quiet facade should also be reported. Then, the strategic noise maps include relevant assessment data detailing the exposure level for each area under consideration and can take the form of graphical plots or numerical data in table or electronic form.

c. *Noise Action Planning*: One year after they have created noise maps, the directive requires competent authorities to provide *noise action plans* for the major roads, railways, airports and agglomerations specified in the previous chapter. According to the END, *action plans* refer to 'plans designed to manage noise issues and effects, including noise reduction if necessary' that should be reviewed every five years after the initial date of approval by the competent authority, or when there is a major development. The directive also introduces the notion of *acoustical planning*, which is defined as 'controlling future noise by planned measures, such as land-use planning, systems engineering for traffic, traffic planning, abatement by sound insulation measures and noise control of sources', which has direct relevance to the development of action plans for noise. By including the word 'planning', the END points directly towards the role that can be played by national planning systems as a means of mitigating environmental noise in the future and in structured programs.

Annex V of the END clearly establishes a set of minimum requirements for noise action planning, as well as pointing to a series of measures that may be used for the mitigation of environmental noise. In addition to this description of agglomeration and noise sources, limit values, summary noise mapping results, evaluation of human exposure, actions to be taken in next five years and estimates of the reduction of the people affected are laid out.

Finally, public participation represents a fundamental basic principle of noise action planning under the terms of the directive. Authorities are required to 'ensure that the public is consulted about proposals for action plans' and 'given early and effective opportunities to participate in the preparation and review of the action plans', and they must make sure 'the results of participation are taken into account and that the public is informed on the decisions taken'. Indeed, one of the central objectives of the directive is to ensure that 'information on environmental noise and its effects are made available to the public' [2].

Quiet Area: The directive also clearly states the need to preserve 'environmental noise quality where it is good', as well as to preserve quiet areas, as first introduced in the Green Paper on Future Noise Policy [3]. As reported by many Member States [4], the END's regulation of quiet areas is limited: Article 8 states that action plans for agglomerations with more than 250,000 inhabitants 'shall also aim to protect quiet areas against an increase in noise'. This is followed up by the requirement in Annex V to report on actions or measures that the competent authorities intend to take to preserve quiet areas. Even if it is specified that actions may include land-use planning, systems engineering for traffic, traffic planning and noise control of sources, the directive does not specify any requirements regarding the protection of quiet areas in open country. The END's weak focus on quiet areas has led to heightened activity in this field, especially in the field of soundscapes, the study of how people perceive the acoustic environment. Several Member States have initiated or intensified their policies with respect to quiet areas, and the resulting knowledge gained has led to the 'Good practice guide on quiet areas', EEA Technical Report No 4/2014 [5]. Among other aspects, this document offers an economic overview of the quiet areas' value and methods for identifying them. The document concludes that 'the issue of quiet areas remains under development' and 'there is a need for further in-depth research into the field'.

2.1.1 CNOSSOS-EU (Directive 2015/996/EC)

The presentation of consistent and comparable data on the number of people exposed to excessive noise levels across the European Member States [6,7] was a difficult task due to:

1. Incompleteness of the required strategic noise mapping in Europe [8]
2. Different quality and format of data reported at EU level
3. Different assessment methods used
4. Different strategies adopted concerning the selection of, e.g., roads to be mapped and distribution of the population within the buildings
5. Unavailability or outdated dose-response curves to use for health impact assessment

The END has not yet reached its objectives, but the review of its implementation in 2011 [9] allowed the identification of some achievements. Among the main implementation problems identified were: delays in implementation by the Member States, the non-enforcement of national noise limit values, strategic noise maps and action plans with poor quality, inconsistent use of approaches in noise mapping, different approaches to quiet areas and insufficient communication with and involvement of the public in the noise assessment and mitigation process. Inconsistent approaches to noise mapping led to a non-reliable comparison of the population's exposure to noise in the EU.

A common methodology for noise assessment in Europe would, therefore, represent a step forward in this respect. For this reason, in 2008, the EC initiated the development of harmonised methods for assessing noise exposure in Europe. The core of the common noise assessment methodological framework in Europe was developed in the project 'Common Noise Assessment Methods in the EU' (CNOSSOS-EU) [10], focusing on strategic noise mapping and balancing the need for harmonisation with the principle of proportionality and sectoral specificities in Member States. In 2015, based on a revised Annex II of the END, CNOSSOS-EU became a new Commission Directive, (EU) 2015/996 [11], mandatory for all EU Member States after 31 December 2018. The Annex to this Directive sets out the common assessment methods and specifies common prediction methods for the transportation and industrial noise sources covered by the END. Thus, CNOSSOS-EU replaces the so-called interim methods specified in the original END. Moreover, from this date on, the 'equivalent' methods permitted by the original END shall no longer be allowed.

Briefly, CNOSSOS-EU is defined in octave bands from 63 to 8,000 Hz and is composed by two harmonized methods and two groups of noise sources: aircraft and terrestrial sources. Aircraft sources are not treated differently in CNOSSOS-EU than they were in the ECAC DOC 29 method [12]. The general approach for outdoor sound propagation of CNOSSOS-EU is to split the noise sources into equivalent point ones, to identify the relevant propagation paths between one equivalent source and a receiver and to do a point-to-point calculation for each identified path. The contribution of an individual path to the sound pressure level L_p at the receiver is given by the sound power level L_W of each source minus the attenuation $L_p = L_W - A$.

The terrestrial part of CNOSSOS-EU concerns road and rail traffic and industrial noise. An emission model is provided for traffic, which relates input data like vehicle speed to L_W with a frequency spectrum dependent on vehicle speed. Due to the variability of industrial sources, only a general framework is defined for industrial noise in CNOSSOS-EU. Regarding road traffic noise, CNOSSOS-EU specifies several correction terms that can be added to the rolling and engine noise terms for acceleration/deceleration, road gradients, studded tyres and temperature. For industrial noise, only four different type of sources have tabulated values of L_W, so dedicated measurements remain the best way to properly model an industrial site.

In CNOSSOS-EU the propagation of terrestrial sources is accounted for in a simplified form of the NMPB2008 French method [13], thus the total attenuation is the sum of three components: a geometrical spreading one (A) due to the distance source-receiver, an absorption from atmosphere one (A_{atm}) as in ISO 9613-1 [14] and A_{bnd} relating to boundary characteristics and obstacles like buildings or noise barriers.

In conclusion, from a terrestrial noise source point of view the new unified prediction method in CNOSSOS-EU represents a major step toward consistency and comparability of the data delivered by the Member States to the EU. However, the representativeness provided by the data coming out is limited, and guidance would be needed when it comes to collecting or producing complementary input data, particularly for road pavements, rail rolling stock and occurrences of favourable conditions.

2.1.2 Actual Implementation of the END and Future Perspectives

In 2016, the 2nd Commission published its 'implementation report on noise directive' document [15], reviewing the current situation on application of the END and planning the following evaluation of the directive. The report concluded that noise pollution continues to represent a major environmental health problem in Europe, with all the scientific evidence related to prolonged exposure to high or low noise levels. According to the World Health Organisation (WHO), noise pollution leads to a disease burden that is second only to air pollution among the environment-related causes in Europe. In 2012, the number of people exposed to environmental noise in Europe was >55 dB L_{den}; inside and outside agglomerations were reported in Figure 2.1.

From a regulation point of view, the implementation of the END is in progress, but it varies according to Member States' levels of ambition and allocated resources. Transposition into national legislation has been done correctly in all Member States, either by implementing new regulations or adjusting the existing ones. No significant problems were identified in the designation of applicable major infrastructures and agglomerations, but in some Member States, producing noise maps presented differences between national bodies and local authorities. Left free to establish source-specific common limit values, only 21 Member States have set them. In strategic noise maps, Member States have used the required indicators and other national noise indicators for special cases. Data between reporting rounds and countries has not been comparable, as Member States may still use adapted national methods to assess noise. This is a known problem that the CNOSSOS-EU model will try to solve from 2018 onward. Table 2.1 reports on the completeness of reporting for the current round of noise mapping and action planning for Member States.

Lastly, the directive is increasingly drawing attention to the harmful effects of noise on health, but noise population exposure data has not yet been used to design legislation about noise at the source.

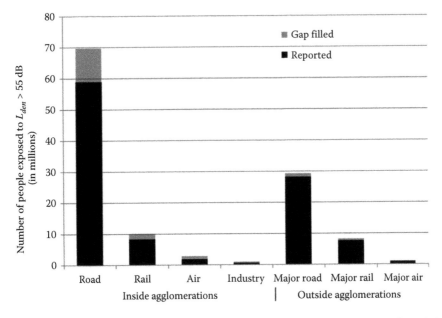

FIGURE 2.1 People exposed to environmental noise in Europe >55 dB L_{den}. (From European Commission. 2017. Report from the Commission to the European Parliament and the Council on the Implementation of the Environmental Noise Directive in accordance with Article 11 of Directive 2002/49/EC. COM/2017/0151 final.) [15]

TABLE 2.1 Completeness of Reporting for the Current Round of Noise Mapping and Action Planning for Member States

	Inside Agglomerations				Outside Agglomerations		
Entity	Road Noise	Railway Noise	Aircraft Noise	Industry Noise	Major Roads	Major Railways	Major Airports
Noise maps completed	78%	75%	52%	69%	79%	73%	75%
Action plans completed	49%				47% (average)	41 (average)	43%

Source: European Commission. 2017. Report from the Commission to the European Parliament and the Council on the Implementation of the Environmental Noise Directive in accordance with Article 11 of Directive 2002/49/EC. COM/2017/0151 final [15].

In the near future, the Commission will continue improving implementation via targeted enforcement actions and provide scientifically sound guidance to Member States, in particular with regard to the assessment of harmful effects with the help of dose-effect relationships. The Commission will also work with Member States to improve the reporting of monitoring mechanisms and reporting obligations in environmental policy and to reconsider their implementation arrangements, including the designation of quiet areas.

Recent scientific evidence suggests that harmful health effects can occur at lower levels than those so far addressed by the directive. Thus, the next challenge will be how to assess these issues without unnecessarily reducing the flexibility of Member States to define their own levels of ambition or choice of approaches.

Mitigation measures that directly address noise have high initial costs and it takes a long time to recover the financial investment, compared to solutions with a broader spectrum of action. However, they are highly efficient when comparing their cost to the societal benefit. For this reason, the Commission will continue to stimulate and encourage activities to mitigate excessive noise in urban areas by facilitating the exchange of good practices and supporting research and innovation in this field. Furthermore, the Commission will support the implementation of noise-mitigating measures that improve the urban environment or develop and improve environmental friendliness, such as low-noise transport systems.

References

1. European Parliament and of the Council. 2002. Directive 2002/49/EC of the European Parliament and of the Council of 25 June 2002 relating to the assessment and management of environmental noise.
2. Murphy, E. and King, E. A. 2010. Strategic environmental noise mapping: Methodological issues concerning the implementation of the EU Environmental Noise Directive and their policy implications. *Environment international* 36.3, 290–298.
3. EC, 1996, Future Noise Policy, European Commission Green Paper (COM(1996) 540 final of 4 November 1996).
4. Milieu, 2010, Final Report on Task 1, Review of the Implementation of Directive 2002/49/EC on Environmental Noise, Milieu Ltd, Brussels.
5. Licitra, G., Van den Berg, M., De Vos, P. (eds). Good practice guide on quiet areas. European Environmental Agency, EAA Technical Report n. 4/2014, June 2014.
6. Kephalopoulos, S. and Paviotti, M. Common noise assessment methods for Europe (CNOSSOS-EU): implementation challenges in the context of EU noise policy developments and future perspectives. Proceedings of the 23rd International Congress on Sound and Vibration (ICSV23).
7. Licitra, G. 2015. Differences between the Principles of the European National Noise Laws and those of the Environmental Noise Directive. Euronoise Conference, Maastricht.
8. Alberts, W. 2012. Anomalous Data of END Noise Mapping for Major Roads on the Website of the European Environment Agency. *Procedia – Social and Behavioral Sciences*, 48, 1284–1293. Transport Research Arena 2012.
9. Report from the Commission to the European Parliament and the Council on the implementation of the Environmental Noise Directive in accordance with Article 11 of the Directive 2002/49/EC, COM(2011) 321 final, Brussels, 1 June 2011.
10. Kephalopoulos, S., Paviotti, M., Anfosso-Lédée, F. 2012. *JRC Reference report on Common NOise SSessment MethOdS in EU (CNOSSOS-EU)*. ISBN (on-line). Publications Office of the European Union. Luxembourg.
11. COMMISSION DIRECTIVE (EU) 2015/996 of 19 May 2015 establishing common noise assessment methods according to Directive 2002/49/EC of the European Parliament and of the Council, OJ of the European Communities, July 2015.
12. ECAC. 2016. *Doc 29 – Report on Standard Method of Computing Noise Contours around Civil Airports*.
13. Besnard, F., Dutilleux, G., Defrance, J., Bérengier, M., Ecotière, D., Gauvreau, B., Junker, F. and Van Maercke, D. (2009) Road noise prediction 2-Noise propagation computation method including meteorological effects. ISRN: EQ-SETRA–09–ED32–FR+ENG, Sétra, France.
14. ISO 9613-1 Acoustics – Attenuation of sound during propagation outdoors – Part 1: Calculation of the absorption of sound by the atmosphere. 1993.
15. European Commission. 2017. Report From The Commission To The European Parliament And The Council On the Implementation of the Environmental Noise Directive in accordance with Article 11 of Directive 2002/49/EC. COM/2017/0151 final.

2.2 Sources and Propagation: The CNOSSOS-EU Framework

2.2.1 General Considerations

What follows is an introduction to the so-called CNOSSOS-EU* environmental noise prediction method as adopted by the European Commission in Directive 2015/996/EC [1]. Even if the reader is not directly involved in the implementation of CNOSSOS-EU, the following is worth reading for an overview of the main parameters and phenomena of noise emissions and propagation from terrestrial transportation and industrial activities.

Directive 2015/996/EC is a revision of Annex II in Directive 2002/49/EC – the so-called Environmental Noise Directive [2] – and specifies a common prediction method for transportation and industrial noise sources covered by the END. Strategic noise mapping (SNM) as directed by Directive 2015/996/EC must be implemented by all EU Member States (MS) no later than 1 January 2019. Thus, CNOSSOS-EU will replace the so-called interim methods specified in the original END. Moreover, from this date on, the so-called 'equivalent' methods permitted by the original END shall no longer be allowed. Although not mandatory for Action Plans (AP), CNOSSOS-EU is likely to be used by several MS for the sake of continuity with SNM, and may also be chosen by some MS as the reference method for the implementation of national regulations regarding noise impact studies.

2.2.2 Structure and Basic Principles of CNOSSOS-EU

Because it consists of more than 800 pages and has no table of contents, Directive 2015/996/EC may feel difficult to grasp. The overall structure can be summarized in Figure 2.2, from which it is readily seen that there are indeed two harmonized methods and two groups of noise sources: aircraft and 'terrestrial sources'.

Since the aircraft part is very close to the well-established ECAC DOC 29 method [3], the following will focus on terrestrial sources and acoustical aspects. Moreover, population assignment is not addressed here.

CNOSSOS-EU is defined in octave bands from 63 to 8000 Hz. As in most engineering models for outdoor sound propagation, the general approach of CNOSSOS-EU is to break down the physical noise sources into equivalent point sources, to identify the relevant propagation paths between one equivalent source and a receiver, and to do a point-to-point calculation for each identified path. The method for identifying paths is not specified in CNOSSOS-EU and is left to software developers. It will rely either on ray-

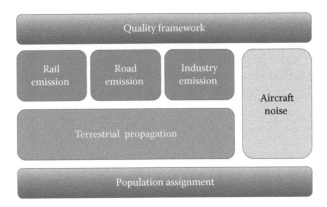

FIGURE 2.2   Overall structure of Directive 2015/996/EC, also known as CNOSSOS-EU.

---

* Common NOise aSseSsment methOdS in EUrope.

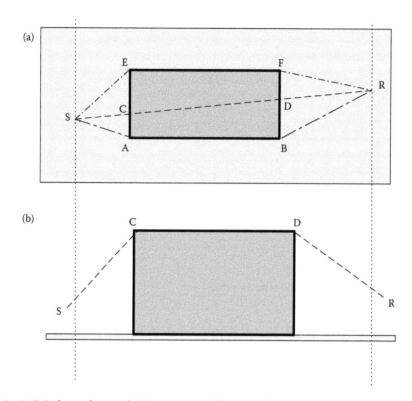

FIGURE 2.3 2.5D Path search example (a) 1st step in the horizontal plane three paths are found (SABR, SCDR, SEFR), (b) 2nd step in the vertical plane containing SCDR, which is the reason why S and R are not on the vertical dotted lines.

tracing [4] or on image source methods [5]. The most common practice in environmental noise prediction is that paths are located using the so-called 2.5D approach (i.e. intermediate between 2D and 3D). The 2.5D expression means that propagation paths existing between a point source and a receiver are first considered in a projection of the site on a horizontal plane. Once all the propagation paths are identified, each one is processed in two dimensions in a vertical plane. In the case of paths that have reflections or diffractions on vertical edges, the vertical planes are flattened, like an unfolding Chinese screen (Cf Figure 2.3). A practical consequence of this is that some source receiver configurations or propagation paths will not be addressed properly, such as openings through buildings or a receiver that is located below a viaduct, but at least for SNM purposes and large-scale calculations, this is a minor issue.

A sound power level L_W is attached to the sound source, the geometry of each path to an attenuation A. From this one can compute the contribution of an individual path to the sound pressure level L_p at the receiver:

$$L_p = L_W - A \tag{2.1}$$

The next sections provide more detail, first on the calculation of L_W for road, rail and industrial sources (see Emission) and second on the different components of A (see Propagation).

The total sound pressure level at the receiver is the energetic sum of all contributions from all sources through all possible paths to the receiver:

$$L_p = 10\log_{10}\left[\sum_{i\in I} 10^{L_{p,i}/10}\right] \tag{2.2}$$

2.2.3 Quality Framework

A major preoccupation in CNOSSOS-EU is to ensure consistency and comparability of the data delivered to the EU by the MS in the END framework. In order to achieve this, in addition to using a common prediction method, it is necessary to control the accuracy of the vast amount of input data required by the method for SNM. Examples of input data are digital terrain models, vehicle speeds, pavement types, rail roughness and distributions of populations. CNOSSOS-EU requires that the uncertainty associated with individual source input parameters be within $+/-2$ dB in terms of emission level, *ceteris paribus*. However, the confidence level attached to this uncertainty is not specified. In many cases, ensuring the ubiquitous availability of certain types of input data is challenging. Default input values are not allowed by CNOSSOS-EU in principle. They are tolerated, though, when data acquisition leads to disproportionately high costs.

Since CNOSSOS-EU will be implemented in software, quality insurance is a key aspect in order to guarantee that different software implementations will provide similar results, if not identical ones. At the time of writing the ISO/TC43/SC1/WG56 working group on quality insurance in noise prediction software is drafting an *ad hoc* technical report for CNOSSOS-EU, which will be part of the ISO 17534 series [6,7]. This technical report will be the opportunity to correct a few errors in CNOSSOS-EU and to disambiguate certain sections.

2.2.4 Emission

The terrestrial part of CNOSSOS-EU addresses road traffic, rail traffic and industrial noise sources at different levels of detail. For the first two, an emission model is provided which relates input data like vehicle speed to L_W. Arguably the emission model for road traffic is more complete and directly usable than the one for rail traffic. For industrial sources, it is not possible to provide such a model due to the variability of sources, so only a general framework is defined in CNOSSOS-EU. The emission part of CNOSSOS-EU was mainly influenced by Harmonoise and Imagine EU research projects [8,9].

2.2.4.1 Road Traffic

The emission model in CNOSSOS-EU is the energetic sum of two components: rolling noise and propulsion noise. The former is noise from tyre-road interaction, the latter is engine and transmission noise. The general expressions for the model are

$$L_{WR,i,m} = A_{R,i,m} + B_{R,i,m}\log_{10}\frac{v}{v_{ref}} + \Delta L_{W,R,i,m} \tag{2.3}$$

for rolling noise (subscript R) and

$$L_{WP,i,m} = A_{P,i,m} + B_{P,i,m}\frac{v - v_{ref}}{v_{ref}} + \Delta L_{W,P,i,m} \tag{2.4}$$

for propulsion noise (subscript P), where i is the index of octave band and m is the vehicle category (see Vehicles). A and B are tabulated coefficients, $v_{ref} = 70$ km/h and v is vehicle speed in km/h. The $\Delta L_{W,R,i,m}$ and $\Delta L_{W,P,i,m}$ are corrections to take into account deviations from the following reference conditions:

- Constant speed
- Horizontal road
- Air temperature 20°C
- *Virtual reference pavement*, which is a combination of DAC* 0/11 and SMA[†] 0/11 aged between 2 and 7 years in a representative maintenance condition
- Dry pavement
- No studded tyres

---

* Dense Asphalt Concrete.
[†] Stone Mastic Asphalt.

In other words, if these conditions are satisfied, which is a theoretical case since the so-called virtual reference pavement does not exist by definition, $\Delta L_{W,P,i,m}$ and $\Delta L_{W,R,i,m}$ vanish.

CNOSSOS-EU specifies several correction terms ΔL that can be added to the rolling noise and the engine noise terms to account for acceleration/deceleration, road gradients, studded tyres and temperature.

A notable feature of the emission model is that its frequency spectrum depends on vehicle speed.

In contrast to other existing models [10], the age of the pavement is not taken into account in CNOSSOS-EU. Deviations from constant speed are only considered in the vicinity of traffic lights and round-abouts, although there are other situations where unsteady flow-type is observed.

2.2.4.1.1 Vehicles

Five vehicle categories are identified in CNOSSOS-EU, and details on their definition are given in Table 2.2.

Although the composition of the traffic may vary from one country to another, the assumption is that the acoustic properties of vehicles in a given category are valid for all MS.

Whatever the vehicle category, a vehicle is represented by an omnidirectional point source at a height 0.05 m above the pavement. But CNOSSOS-EU is not designed to simulate individual vehicle pass-bys. In practice, road traffic on a lane is represented by a source-line at a height of 0.05 m. The computer representation of such a source-line is a series of aligned point sources.

2.2.4.1.2 Pavements

Pavement types are quite often country-specific. CNOSSOS-EU provides two correction terms to take real-world pavements into account:

$$\Delta L_{WR,road,i,m} = \alpha_{i,m} + \beta_m \log_{10} \frac{v}{v_{ref}}$$

$$\Delta L_{WP,road,i,m} = \min\left(\alpha_{i,m};\ 0\right)$$

(2.5)

Tabulated values for $\alpha_{i,m}$ and β_m are given for CNOSSOS-EU, but they are of limited use, since most of them correspond to pavements used in the Netherlands. Thus, many MS need to derive their own coefficients for the pavements in use in their countries. For each new pavement, a set of coefficients is needed for each vehicle category except powered two-wheelers, since in CNOSSOS-EU their emission is made of propulsion noise only.

TABLE 2.2 CNOSSOS-EU Road Vehicle Categories

Category	Name	Description		Vehicle Category in EC Whole Vehicle Type Approval[a]
1	Light motor vehicles	Passenger cars, delivery vans ≤3,5 tons, SUVs[b], MPVs[c] including trailers and caravans		M1 and N1
2	Medium heavy vehicles	Medium heavy vehicles, delivery vans >3,5 tons, buses, motorhomes, etc., with two axles and twin tyre mounting on rear axle		M2, M3 and N2, N3
3	Heavy vehicles	Heavy duty vehicles, touring cars, buses with three or more axles		M2 and N2 with trailer, M3 and N3
4	Powered two-wheelers	4a	Two-, three- and four-wheel mopeds	L1, L2, L6
		4b	Motorcycles with and without sidecars, tricycles and quadricycles	L3, L4, L5, L7
5	Open category	To be defined according to future needs		N/A

[a] Directive 2007/46/EC.
[b] Sport Utility Vehicles.
[c] Multi-Purpose Vehicles.

The number of degrees of freedom left for adapting CNOSSOS-EU to a particular pavement formulation is 9, *i.e.* 8 for $\alpha_{i,m}$ and 1 β_m. This means that, although adapting the rate of variation of the rolling noise component as a function of vehicle speed is permitted, this correction will be the same for all octave bands, and difficulties could arise because of this limitation. The sensitivity of sound power level to these degrees of freedom is illustrated in Figures 2.4 and 2.5. The companion guidance to CNOSSOS-EU recommends using statistical pass-by measurements [11] to obtain pavement-specific values, but other approaches could also be used [12–14].

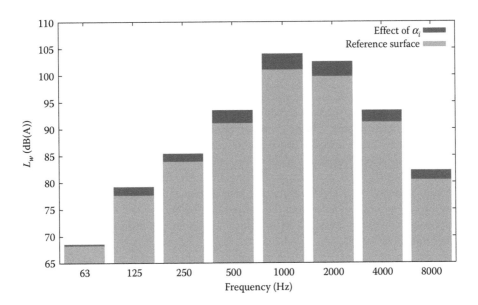

FIGURE 2.4   Pavement correction coefficients $\alpha_{i,m}$, effect on L_W.

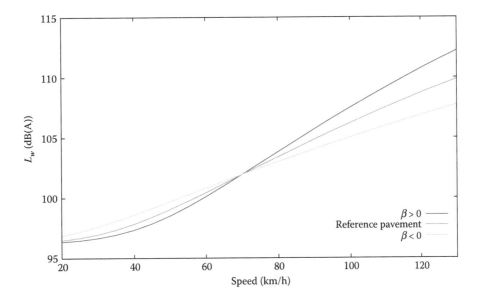

FIGURE 2.5   Pavement correction coefficient β_m, effect on L_W. The neutral point is the reference speed.

2.2.4.2 Rail Traffic

CNOSSOS-EU represents railbound vehicles by two equivalent sources at heights of 0.5 m (source A) and 4 m (source B) above the running plane (Figure 2.6). As for roads, source heights do not represent individual vehicles, but traffic and two parallel source-lines at a constant height above the ground shall be used to model a railway line.

Unlike road sources, railway sources are highly directional. Both source heights share the same figure-of-eight horizontal directivity, but they have different vertical directivities. Source A has a frequency-dependent directivity and B has a cardioid-like frequency-independent directivity. They add to L_W to obtain directional sound power.

CNOSSOS-EU takes six different train noise generation mechanisms into account. They contribute either to source A (rolling noise, bridge noise, impact noise, squeal noise) or to both A and B sources (traction noise and aerodynamic noise) (see Figure 2.6). When a mechanism contributes to both A and B, the distribution depends on the vehicle type.

- Rolling noise depends on rail and wheel roughness, on the so-called *contact filter* between wheel and rail and on vehicle speed. It is radiated by rail sleepers and ballast/slab (Figure 2.7), wheels/bogies and superstructure. The contribution of superstructure only applies for freight vehicles. For track sections on a bridge, a constant correction is added to rolling noise sound power.
- Traction noise strongly depends on operating conditions. CNOSSOS-EU considers only (1) constant speed, which serves as an approximation for acceleration and deceleration, and (2) idling. Traction noise in the CNOSSOS-EU sense is actually more than traction noise *stricto sensu*, since it is the superposition of noise from the powertrain, gear transmission, electrical generators, electrical sources like converters, HVAC systems, intermittent noises from compressors, valves and other equipment.
- Aerodynamic noise is only potentially relevant at speeds above 200 km/h. This noise component increases rapidly with vehicle speed.
- Impact noise is caused by crossings, switches, rail joints or other discontinuities. CNOSSOS-EU integrates impact noise in the calculation of rolling noise by a correction term. Impact noise depends on the size and shape of the discontinuities and on the number of impacts per unit length.

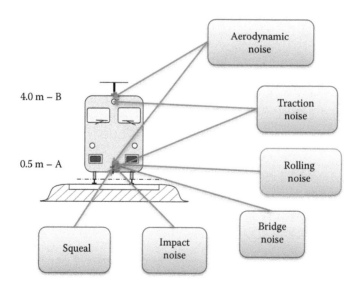

FIGURE 2.6 Railway noise – equivalent source heights, contributions from the different noise generation mechanisms.

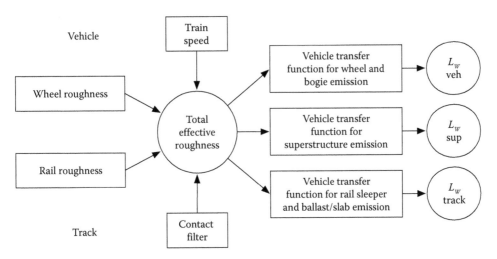

FIGURE 2.7   Roughness Rolling Noise Components and Required Input Parameters.

- Squeal noise has strong tonal components. It is generated in curves or when braking. CNOSSOS-EU considers only *curve squeal*, which will therefore only occur at particular locations along the infrastructure. CNOSSOS-EU takes this mechanism into account by adding a penalty to sound power for a radius of curvature below 500 m.

The CNOSSOS-EU railway emission model appears to be relatively sophisticated, with many input parameters required. In many MS the key parameter of rail roughness may not be readily available. Collecting it for all major infrastructures at the scale of a country could be challenging. Fortunately, default values are provided in CNOSSOS-EU. Concerning railbound vehicles, the annexes of CNOSSOS-EU contain emission parameters for a few vehicle types. But they are far from reflecting the diversity of the rolling stock across the EU. Some guidance is available about the acquisition of emission parameters for a specific vehicle or train [15], but the measurement protocol is expensive and time-consuming.

2.2.4.3 Industry Sources

CNOSSOS-EU acknowledges that there is considerable variability among industrial sites in terms of size, number of sources and source characteristics. Therefore, only general requirements are stated in the method. The general principles are:

1. To replace physical noise sources by equivalent ones. Point sources, source-lines or surface distributions are allowed.
2. When the largest dimension of a source is lower than half of the distance between source and receiver, the source can be modelled as a single point source.
3. For sources close to the ground, the equivalent source height deserves special attention, since ground effect (see A_{ground}) strongly depends on it.

From the intrinsic sound power of the source, CNOSSOS-EU introduces two corrections to obtain the sound power to be used in simulations:

$$L'_W = L_W + C_W + \Delta L_{W,dir} \tag{2.6}$$

The first correction C_W is specified first when moving vehicles are involved, and second to take into account the proportion of working hours T over the reference period T_{ref} (12 hours for day, 4 hours for evening, 8 hours for night).

$$C_W = 10\log_{10}\frac{T}{T_{ref}} \tag{2.7}$$

$\Delta L_{W,dir}$ is directivity. By definition it is expressed as a function of direction. For industrial sources, CNOSSOS-EU considers two possibilities for the modelling of sound power and directivity. First, these characteristics may be estimated in a hemi-anechoïc field, and an accurate modelling of the obstacles surrounding the sources is necessary in order to evaluate reflections and diffractions during propagation. Second, they may be estimated in a specific setting. In this case *equivalent* sound power and directivity are obtained and the surrounding obstacles must be excluded from propagation calculations, since they are already included in sound power.

Tabulated values of L_W are given in CNOSSOS-EU for only four industrial sources, assuming zero directivity. They must be seen only as examples that will be difficult to transpose even to similar facilities. Dedicated measurements are probably the only suitable way to model an industrial site in a satisfying manner. Unfortunately, the typical budget for the SNM of an agglomeration does not allow for detailed investigations of individual industrial plants. Since industrial plants covered by the END often fall under the 2008/1/EC directive concerning IPPC*, there are potential opportunities to reuse existing models.

2.2.5 Propagation

The previous sections dealt with sound power levels for different kinds of environmental noise sources. The next one considers how to pass from L_W to a facade noise level L_p.

The propagation method for terrestrial sources in CNOSSOS-EU is a simplified version of NMPB2008, a standardized method developed in France [16,17]. For the sake of conciseness, only a general overview of the method is provided. The interested reader shall refer to [1,16] for more details.

We assume that propagation paths have been identified, and we will consider the procedure for any path between two points, M and N. Before discussing the different attenuation terms involved, a few words on ground and atmosphere are necessary.

2.2.5.1 Ground Description

Digital terrain models are not used directly in CNOSSOS-EU. The ground profile is simplified to a so-called mean ground plane (Figure 2.8). A specific procedure describes how to calculate in a robust way the mean ground plane from a series of connected ground geometrical segments. Heights and distances are recomputed with respect to the mean ground plane.

Ground absorption is represented in CNOSSOS-EU by an adimensional frequency-independent factor named G. It can take discrete values between 0 for a reflecting ground like dense asphalt and 1 for an absorbent one like a forest floor. Even if G varies between 0 and 1, which makes it look like an absorption factor, G must be seen as a normalized airflow resistivity.

Each ground segment is assigned a value for G. For the purpose of acoustic calculations, the mean ground plane is associated to G_{path}, which is the average of G values along a path between M and N. When source and receiver are close to each other a correction is applied to G_{path}. This correction is necessary for the common case of propagation over an absorbent ground beside a reflecting road. Otherwise, the reflection on the ground would be underestimated.

G does not apply to other obstacles like facades and walls, for which the usual diffuse field absorption coefficients are used in octave bands.

---

* Integrated Pollution Prevention and Control.

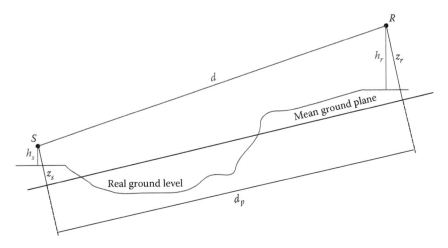

FIGURE 2.8 Mean ground plane for the calculation of acoustic attenuation.

2.2.5.2 Atmospheric Conditions and Long-Term Sound Levels

For ranges above 100 m, temperature and wind speed gradients influence sound propagation. They can be combined in an effective sound speed gradient. If we assume a monotonous gradient that is not range-dependent, there are three possible propagation conditions. Figure 2.9 shows the paths of sound rays from a point source in the three different cases.

Upward refraction conditions lead to shadow zones and lower sound levels than non-refracting conditions. They are difficult to simulate with ray tracing [18]. Downward refraction conditions (also called favourable conditions) lead in general to higher sound levels. Non-refracting conditions (also called homogeneous conditions) rarely occur in the field but they are easy to simulate, and a noise level in homogeneous conditions can be used as an upper boundary of noise levels in upward refraction conditions. Therefore, CNOSSOS-EU considers only two standard atmospheres and computes one attenuation A_H in homogeneous conditions and one A_F in favourable conditions for each path. Obtaining the related sound levels is straightforward:

$$\begin{cases} L_H = L_W - A_H \\ L_F = L_W - A_F \end{cases} \tag{2.8}$$

The site- and orientation-dependent probability p of occurrence of favourable conditions allows one to compute the long-term sound level. CNOSSOS-EU does not give any indication of how to obtain these occurrences. The reader may refer to References 16 or 19 for indications on this topic.

$$L_{LT} = 10\log_{10}(p \cdot 10^{L_F/10} + (1-p) \cdot 10^{L_H/10}) \tag{2.9}$$

The components of attenuations A_H and A_F are outlined in the next section.

2.2.5.3 Acoustic Attenuation

In CNOSSOS-EU the attenuation can be split into three components:

$$A = A_{div} + A_{atm} + A_{bnd} \tag{2.10}$$

Where

- $A_{div} = 20\log_{10}d + 11$, where d is the Euclidean distance, accounts for geometrical spreading. It does not depend on frequency. A_{div} is linked to the conservation of energy. As a spherical wave-front propagates, its amplitude must decrease.

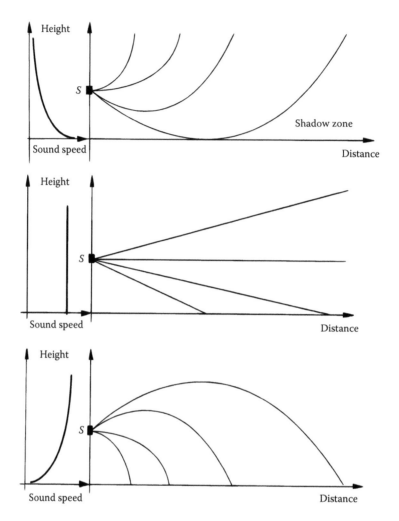

FIGURE 2.9 The three possible propagation conditions (top: upward refracting; middle: homogeneous; bottom: downward refracting), depending on the shape of a monotonous range-independent sound speed gradient shown on the left of each diagram.

- A_{atm} is the atmospheric absorption, expressed in dB/km. It reflects dissipation of acoustic waves by their interaction with air molecules. The calculation of A_{atm} complies with ISO 9613-1 [20] and depends in general on temperature, static pressure and humidity. For road and rail sources a reference atmosphere is assumed (15°C, 101325 Pa, 70% humidity). For industrial sources other values may be used. A_{atm} behaves as a low-pass filter.
- A_{bnd} is the attenuation relating to the boundary characteristics and the only attenuation term that depends on sound speed gradient. The boundary consists of the ground in the broad sense and occasional manmade obstacles like buildings or noise barriers.

The description of the details of A_{bnd} is beyond the scope of these pages, and only general indications are provided below. Refer to CNOSSOS-EU for further information.

The entry point in the calculation of A_{bnd} is the decision as to whether a diffraction occurs along the path or not. If it does, then two mean ground planes (Cf Figure 2.8) will be calculated – one on each side of the edge of diffraction – and a specific set of formulae named A_{dif} will apply; otherwise, another set named A_{ground} will apply and only one mean ground plane will be used. The decision is based on the difference

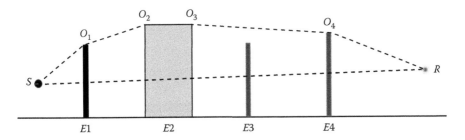

FIGURE 2.10 Convex hull around the obstacles between S and R.

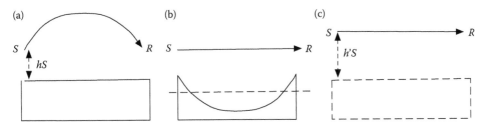

FIGURE 2.11 Equivalence between (a) curved rays (favorable conditions) + straight ground and (b) straight rays (homogeneous conditions) + curved ground. In (b) the dotted line is the mean ground plane. In (c) straight rays straight ground configuration approximating (a).

in path length between the length of the convex hull from source to receiver surrounding all obstacles and the Euclidean distance (see Figure 2.10). Diffraction may occur even though source and receiver are in direct sight. For downward refracting conditions a particular definition of the convex hull must be used. The reader shall refer to the dedicated part of the ISO 17534 series for more detail.

2.2.5.3.1 A_{ground}

The main formula for ground effect is defined for homogeneous conditions. For favourable conditions the equivalence between curved rays over a flat ground and straight rays over a curved ground is used (Figure 2.11).

The very same formula that is used for homogeneous conditions can be used, except that positive height corrections are introduced for both the source and the receiver.

2.2.5.3.2 A_{dif}

The main formula in A_{dif} is a Maekawa-like [21] expression which depends on Fresnel's number $N = 2\delta/\lambda$ where δ is the path length difference and λ is the wavelength. Ground effect occurs along with diffractions, and specific components of A_{dif} account for that.

CNOSSOS-EU provides a specific set of formulas for diffractions around vertical edges, like the sides of a noise barrier or the junction of adjacent facades of a building. For SNM purposes, however, diffractions around vertical edges can usually be ignored when simulating source-lines.

Not all combinations of diffractions on vertical and horizontal edges are addressed in CNOSSOS-EU.

2.2.5.3.3 *Reflections on Obstacles*

Reflections on obstacles like noise barriers or building facades are dealt with by image sources (Figure 2.12). In SNM the so-called *order of reflection* n taken into account is usually not greater than 2. This means that paths containing more than n reflections are ignored. If one assumes a reflection on a surface whose diffuse

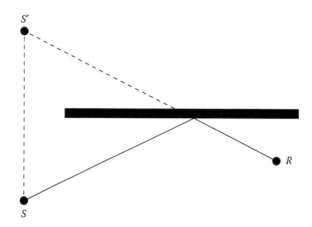

FIGURE 2.12   Image source S′ of a physical source S with respect to a reflecting surface.

field absorption coefficient is α, the sound power level of the source is modified accordingly:

$$L'_W = L_W + 10\log_{10}(1 - \alpha) \tag{2.11}$$

CNOSSOS-EU also proposes an alternative procedure for reflections. It is based on the so-called retro-diffraction introduced in [17]. Arguably it is too sophisticated for the purpose of SNM. Retrodiffraction is mostly relevant when high orders of reflection are taken into account, and it will not be described further here.

In the case of a multiple horizontal diffraction, reflections are not allowed between two edges of diffraction. Reflections on the ground are not handled as reflections in CNOSSOS-EU. They are taken into account implicitly in A_{bnd}.

2.2.6 Conclusion

From the point of view of terrestrial noise sources, CNOSSOS-EU appears to be a major step toward consistency and comparability of the data delivered by the MS to the EU, thanks to the introduction of a unified prediction method.

CNOSSOS-EU is, however, not completely self-contained. The representativeness of the tabulated data it provides is limited. Guidance would also be needed when it comes to collecting or producing complementary input data, particularly for road pavements, rail rolling stock and occurrences of favourable conditions.

References

1. European Parliament and Council. 2015. Commission Directive (EU) 2015/996 of 19 May 2015 establishing common noise assessment methods according to Directive 2002/49/EC. http://eur-lex.europa.eu/legal-content/EN/TXT/?uri=CELEX:32015L0996
2. European Parliament and of the Council. 2002. Directive 2002/49/EC of the European Parliament and of the Council of 25 June 2002 relating to the assessment and management of environmental noise. http://eur-lex.europa.eu/legal-content/en/TXT/?uri=CELEX:32002L0049
3. ECAC. 2016. Doc 29 – Report on Standard Method of Computing Noise Contours around Civil Airports. https://www.ecac-ceac.org/documents/10189/51566/02.+Doc29+4th+Edition+Volume+2.pdf/d9164c10-339a-4650-ba87-6540cd68e4a9

4. Krokstad, A., Strom, S. and S. Sørsdal. 1968. Calculating the acoustical room response by the use of a ray tracing technique. *Journal of Sound Vibration* 8:118–125.

5. Allen, J.B. and D.A. Berkley. 1979. Image method for efficiently simulating small-room acoustics. *Journal of the Acoustical Society of America* 65(4), 943–950.

6. ISO/TC43/SC1/WG56. 2015. ISO 17534-1 Acoustics – Software for the calculation of sound outdoors – Part 1: Quality requirements and quality assurance.

7. ISO/TC43/SC1/WG56. 2014. ISO/TR 17534-2 Acoustics – Software for the calculation of sound outdoors – Part 2: General recommendations for test cases and quality assurance interface.

8. Jonasson, H. 2007. Acoustical source modelling of road vehicles. *Acta Acustica United with Acustica* 93(2), 173–184.

9. Dittrich, M. 2007. The Imagine source model for railway noise prediction. *Acta Acustica United with Acustica* 93(2), 185–200.

10. Besnard, F., Hamet, J.F., Lelong, J., Le Duc, E., Guizard, E., Fürst, N., Doisy, S. and G. Dutilleux. Road noise prediction 1-Calculating sound emissions from road traffic.

11. ISO/TC43/SC1/WG33. 1997. ISO 11819-1 Acoustics – Measurement of the influence of road surfaces on traffic noise – Part 1: Statistical Pass-By method.

12. Anfosso-Lédée, F., Dutilleux, G. and M. Conter. 2016. Compatibility of the ROSANNE noise characterization procedure for road surfaces with CNOSSOS-EU model. InterNoise 2016, Hamburg.

13. Shilton, S., Anfosso-Lédée, F. and H. van Leeuwen. 2015. Conversion of existing road source database to use CNOSSOS-EU. Euronoise 2015, Maastricht. 469–474.

14. Dutilleux, G., Soldano, B. and W.H. Lee. 2016. Adaptation de la directive 2015/996/CE (CNOSSOS-EU) au modèle français de prévision de bruit : revêtements routiers. Cerema report. 48p.

15. CROW. 2006. Rail traffic – Technical regulations for methods of measuring emission.

16. Besnard, F., Dutilleux, G., Defrance, J., Bérengier, M., Ecotière, D., Gauvreau, B., Junker, F. and D. Van Maercke. Road noise prediction 2-Noise propagation computation method including meteorological effects (NMPB2008).

17. AFNOR. 2011. NF S 31-133 Acoustics – Outdoor noise – Calculation of sound levels.

18. Salomons, E.M. 2001. *Computational atmospheric acoustics*. Kluwer.

19. Maijala, P., Kokkonen, J. and O. Kontkanen. 2016. CNOSSOS-EU Sensitivity to Meteorological and to Some Road Initial Value Changes. Inter Noise 2016, Hamburg, 1367–1378.

20. ISO/TC43/SC1. 1993. ISO 9613-1 Acoustics – Attenuation of sound during propagation outdoors – Part 1: Calculation of the absorption of sound by the atmosphere.

21. Maekawa, Z. 1968. Noise reduction by screens. *Applied Acoustics* 1(3), 157–173.

2.3 Noise Measures for Action Plans

It is well known that the implementation of Directive 2002/49/EC [1] has created among the European Member States the responsibility to improve the environmental health of citizens; in this case, it focuses on noise pollution. In addition, the adoption of Directive 2015/996 [2] will help to improve the quality of strategic noise maps which, in turn, will allow a better diagnosis of the problem. Using a medical analogy, the usefulness of a diagnosis is not understood without the prescribing of a treatment.

On the other hand, the implementation of mitigation measures includes public investments that form part of the environmental health policies to prevent the effects of noise pollution. In some cases, we will be talking about large investments in public works.

Along with all the steps ranging from diagnosis to treatment, there is a real risk of inefficiencies. To avoid them, the noise action plan must be based on an appropriate methodology and supported by a systematic and objective decision system. Roughly speaking, six steps are needed to carry out the action plans as shown in Figure 2.13.

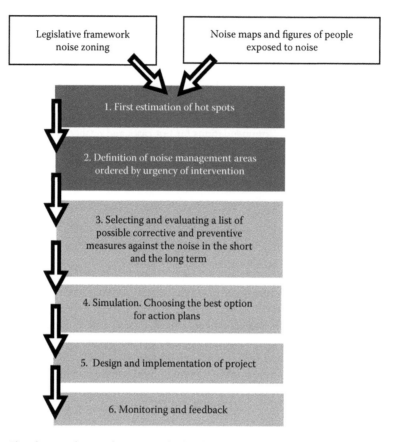

FIGURE 2.13   Flow diagram showing the steps involved in the development of an action plan. The diagnosis phase is shown in dark gray, the treatment phase in light gray.

Step 1 talks us about 'where' in the city to focus our concern (the hot spots) and the urgency of the intervention ('when'). This step is part of the diagnosis phase, and it is obtained directly from the conclusions of the noise maps, taking into account the legislation in force in each country. But what is a 'hot spot'? It can be defined as a 'high concentration of people highly annoyed by noise inside sensitive buildings'. This definition is important, because the proposed noise mitigation measures should consider what the problem to solve is. Actually, what we obtain after step 1 is a set of candidates to be considered noise management areas ordered according to the gravity of the problem. Normally, a geographic information system (GIS) tool is used to analyze a great amount of spatial data.

Step 2 supplies additional information to complete a broad picture of the scope and understanding of the noise problems and the possibilities of mitigation. Sometimes this reevaluation of the diagnosis is carried out by 'on-site' study visits. With the new knowledge, it is determined which of the candidates will be considered consolidated noise management areas to work on.

Step 3 is the beginning of the treatment, and, continuing with the medical analogy, to prescribe the best remedy we need a *vade mecum*. Therefore, step 3 implies an expert system that links the available information (output of step 2) with the most promising noise mitigation list (the expert experience) to give an answer about which are the potential best solutions (Figure 2.14). This catalog has to be a useful working document for the noise technician, so what information is contained in it and which requirements should be satisfied is very important.

We need to know 'what happens if' (step 4) is applied to each of these mitigation candidates on the real hot spot. To obtain this information, we have noise prediction software and other 'ad hoc' simulation tools

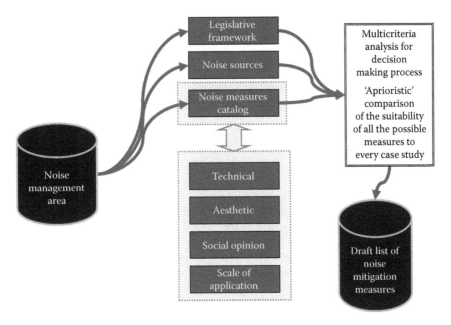

FIGURE 2.14 This scheme describes step 3 of Figure 2.13 as a tool composed of an expert system that uses the catalog in order to choose the most promising noise measures that will be candidates to be part of the action plan. The parameters for comparison are those that will be treated in Table 2.3.

in order to quantify the real results. The best noise mitigation solution provided by the decision system says what to do and how to apply it, and predicts the benefits and side effects and how much the solution will cost. Finally, step 5 refers to the design and execution of the noise project, while step 6 focus on monitoring the effectiveness of the measures implemented and predicting future trends.

2.3.1 Noise Mitigation Measures: Types

Designing noise mitigation measures for urban areas is always a complex task. The implementation period of the action plan can be defined for the short term or the long term, depending on its complexity. Solutions can be adopted from different angles: technical, planning, educational, increasing public awareness, economical, etc., alone or in combination. If we focus exclusively on technical measures, the intervention could be applied to the noise source, in the propagation path, at the receiver and through urban planning.

From the point of view of municipalities, when noise legislation is implemented and begins to apply, the administration must face the current pressing problems of the city and invest the funds in solving them. These corrective measures will try to avoid, or at least minimize, the noise in sensitive noise areas and hot spots. But over time, increasingly fewer public resources are invested in correction and increasingly more in prevention. Noise preventive measures are designed and integrated into the complete urban plan under the requirements of the sustainable development model. Preventive measures set the basis for environmentally compatible city growth. Infrastructure projects, new residential developments, hospitals, schools, etc., could be designed under certain noise constraints. They usually are the most cost-effective measures, and most of them will be part of plans for the middle to long term. Apparently, the most popular mitigation measures in the EU agglomerations are [3] those related to urban planning (23% of the total action plans are related to agglomerations).

Within the corrective measures, reducing the emissions of noise at the source may have been considered the most valuable way to proceed, as noise pollution is reduced everywhere. In an urban environment, most noise is caused by transportation infrastructures (road traffic, railways, airports, ports, etc.), industrial facilities and recreational and leisure activities. Intervening in the propagation of noise could normally involve the construction of barriers around the noise source, which is not very efficient and not always viable. Abating the noise at the receiver must always be taken as the last option, as in this case pollution is only reduced at the individual level and when the person is inside the building, so it is the most inefficient option.

2.3.2 Noise Mitigation Measures: Criteria to Assess the Appropriateness of the Solution

The catalog of mitigation measures should not only seek to define and describe every measure, but also needs to include data that can be used analytically. It is therefore important to quantify the experiences in the implementation of the measures that have been carried out in the cities around the EU and abroad. Obviously, a catalog is the starting point to decide which noise measure is most appropriate for resolving every problem. Now we are going to try to give meaning to the adjective 'appropriate'. In other words, which criteria are going to be included to compare different mitigation measures, and how can we weigh the importance of each of these criteria? The response to the first question is included in Table 2.3. An introductory answer to the second question will be outlined in the next section.

At this point, it is interesting to explain that there are some exclusionary parameters (pass/fail requirements) that by themselves would advise against going ahead with the study of the measure analyzed.

- Feasibility. Is it technically possible to construct (implement) a noise mitigation measure with the proposed technical specifications? It must be taken into account that noise mitigation measures have to be designed, constructed, installed and/or implemented in compliance with the country's legal framework.
- Acoustic effectiveness requirements. Technical specifications used to contain the noise reduction requirements. In that sense, the before-and-after comparison that demonstrates a substantial reduction in the number of people exposed to noise and other figures that show noise reduction (noise indicators) should be analyzed.
- Safety requirements. The materials and the installation itself must guarantee local safety requirements and general protection against accidents, flame, fuel, smoke, etc. Also, the design must avoid distractions, be well protected and not exacerbate the negative effects of adverse weather and reflexes, etc.

TABLE 2.3 Parameters to Be Considered in the Assessment of the Appropriateness of Noise Measures

Type of Criteria	Parameters Involved in Assessing the Suitability of each Mitigation Measure
Technical	Feasibility
	Cost
	Safety
	Acoustic effectiveness
	Benefit/cost ratio
	Synergies
	Negative side effects
	Level of complementarities between noise measures
Others	Aesthetics
	Public opinion
	Scale of implementation

- Acceptance of public opinion. To go ahead with a project it is necessary to have a social understanding and positive evaluation of what the proposal for the community is. Any social opposition or resistance to the proposed noise mitigation measures must be considered and integrated to reach consensus, or the project should be abandoned. We have taken into account that informing and consulting the public on the development of a noise action plan is compulsory for its administration.
- The economic feasibility of the project is as important as the technical viability. It should consider the construction costs, the maintenance costs and durability aspects (including acoustic and physical properties).

Other parameters that may be part of the decision may be based on certain 'fuzzy' logic about which is the best noise prevention measure. All of them look for rating the efficiency of the noise measures and other quality values, such as aesthetics and landscape values.

- Simplicity and ease of installation will be paramount.
- Benefit/cost ratio. A monetary approach to assessing 'return on investment' using the relationship between the investment in noise plans and their benefits.
- Synergies. Taking advantage of renewal and replacement plans (transport, urban planning, etc.) and urban policies.
- Negative side effects. Analyzing the possible adverse effects, especially on urban mobility and sustainability, and being aware of the need to prevent the transfer of noise problems to other areas of the city.
- Level of complementarities between noise measures. The joint introduction of two different noise mitigation measures sometimes does not imply that the final efficacy will be the sum of the efficacies of these measures separately.
- Aesthetics. Analyzing the level of integration in the landscape and other visual effects. Avoiding interference with architectural heritage and landscape. Other things to be considered are art design and anti-graffiti measures.
- Scale of implementation. Analyzing in which cases a measure or combination of measures has an effect at the macro level (the entire city, a district, a neighbor, etc.). Specifying when a design appears to be effective only at the micro level.

2.3.3 Noise Mitigation Measures: Requirements for the Development of the Catalog

This section deals with the requirements that must be fulfilled to be a useful working document for the noise technician, presenting an example. The catalog must have a structure that makes it manageable and provides easy access to information. The content should include a broad description of measures and any previous experience that serves to quantify the parameters treated in Table 2.3. In this sense, there is a considerable amount of easily accessible information from the projects funded by EU which can serve as a basis for developing the catalog. Also, the know-how acquired can be transferred to the catalog. The document should be organized in a logical way. For example, the section 'reduction of noise at the source: road' could be easily divided into four or more subsections that contain related mitigation measures: 1. reduction of road traffic density by traffic management; 2. control measures for noisy vehicles; 3. traffic calming – noise control measures which can moderate speed and uneven traffic flow; 4. quiet surfaces and street maintenance.

Every mitigation measure has been described and evaluated on the basis of literature and other contrastable experiences. Looking for an analytical approach, every case study, example or experience has been evaluated alone and not in conjunction with others (as are carried out by local authorities most of the time). We must assess the adaptability of solutions for every case, taking into account the peculiarities of every country, its legislation and the date on which the mitigation proposal was considered positive.

Let's take a look at Tables 2.4 and 2.5, which include an example of a very common measure in some cities. The measure consists of calming traffic in certain areas of the city where it is forbidden to travel more than 30 km/h. In this example, it is shown how to incorporate the knowledge and experience of noise experts.

In some circumstances, a simplification of the information contained in the catalog is required. This simplification can be carried out in two ways. Some of the parameters contained in Table 2.3 could be codified in graphical mode in order to allow quick access to the most important features of each measure. In Figure 2.15, we propose five levels for the evaluation of mitigation measures.

Sometimes it is very difficult to collect all the required information suggested in Table 2.3. In these situations, it is imposed using information that is easier to find. As can be seen in Table 2.6, the most

---

TABLE 2.4 Example of Noise Mitigation Measure Description

There is a necessity to change the relationship between road traffic and the residents of our cities. Designated 30 km/h areas imply the reduction of speed limits in combination with other types of engineering measures (see other mitigation measures: 'Re-designing of street layout', 'Speed reducers' and 'Enforcement of speed limit with dissuasive measures like radar, cameras and police'). This measure allows, among other things:

- Improving the safety of these areas, including the percentage of survival after a pedestrian run-over
- Making vehicle traffic compatible with other non-motorized modes, particularly interesting for cycling
- Improving the attractiveness of commercial areas

As an extension of designated 30 km/h areas, there are the concepts of home zones and playing streets. Both are defined by a residential area specially designed for enhancing pedestrian use and in which there is infrastructure that limits speed and discourages the use of these routes for non-local traffic. There are no separations for different street users. Playing streets are supposed to be in a home zone and are designed especially for children playing during specific hours of the day, so there is traffic suppression during these hours.

TABLE 2.5 Example of Experiences Regarding Acoustic Efficacy

The expected effect of reduction in speed from 50 to 30 km/h in certain zones of the city is a reduction of the equivalent noise level from 2 to 4 dB for car traffic and up to 2 dB in the case of heavy traffic. The noise reduction is higher for the maximum noise levels, in which reductions up to 7 dB are expected.

References to real cases:

Graz, Austria [4]. Noise reductions of 0.9–1.9 dB and 0.9–2.5 dB on the maximum level.

Vienna, Austria [4]. Reduction of 3–4 dB can be achieved by implementing a speed reduction to 30 km/h. In other words, even if you are driving in the most 'loudish' way (according to the engine of the vehicle) at a speed of 30 km/h, you will not be louder than you would be if you were driving at a speed of 50 km/h.

Berlin. Annecy. [5]. Reductions of 2–3 dBs were reported (taking into account the traffic ban for vehicles with more than 3.5 Tn).

Amsterdam [6]. Reduction of 2–3 dBs. Speed reduction from 50 km/h to 35 km/h on all roads inside the circular road with a traffic intensity of >10,000 vehicles per 24 h.

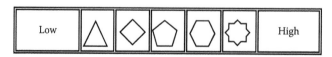

FIGURE 2.15 Simple iconic codification in which valuations are divided on a scale of five possibilities from the simplest to the most complex (low/high, cheap/expensive).

TABLE 2.6　Quick Access Information to the Most Important Parameters of the Example of Mitigation Measure Described along Tables 2.4 and 2.5

Action	Cost	Effectiveness	Complexity	Scale
Designated 30 km/h areas	△	⬠	△	Macro
Playing areas	△	✦	△	Local

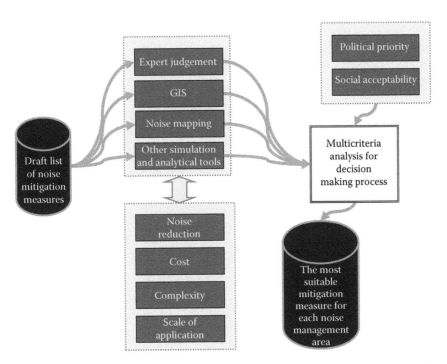

FIGURE 2.16　This scheme describes step 4 of Figure 2.13 as a system composed of an expert system that uses the output of analytical tools applied in a real scenario in order to choose the best solution for a noise action plan in each specific noise management area. The parameters for comparison are those that will be treated in Table 2.6 with the incorporation of the political opinion of the city council.

important indicators for the evaluation are the cost, effectiveness and complexity of the solution. The feasibility of this noise mitigation plan is now evaluated based on the complexity of implementation (construction). The scale factor indicates whether the measures will be applicable to the entire municipality (macro measures) or only for parts of it (local measures).

We stress once more that the catalog should be built for use by local authorities to design noise mitigation plans. Therefore, those responsible should be able to select from the catalog the most interesting candidates for each proposed noise problem.

Thereby, the next task (step 4 of Figure 2.13) should include a decision-making process such as that proposed in Figure 2.16. It establishes how to compare the different proposed noise mitigation measures in every specific case study. So we have a real assessment of the pros and cons that allow us to decide what the best action plan for each hot spot is. Figure 2.16 includes, in addition to the parameters of Table 2.6, the political priorities of the city council.

References

1. Directive 2002/49/EC of the European Parliament and of the Council of 25 June 2002 relating to the assessment and management of environmental noise.
2. Commission Directive (EU) 2015/996 of 19 May 2015 establishing common noise assessment methods according to Directive 2002/49/EC of the European Parliament and of the Council.
3. European Environment Agency. EEA Report No 10/2014. Noise in Europe 2014. http://www.eea.europa.eu/
4. Danish Road Institute Report 137. Traffic management and noise reducing pavements – Recommendations on additional noise reducing measures (2004). http://www.vejdirektoratet.dk/
5. Sustainable Mobility Initiatives for Local Environment Guidelines for road traffic noise abatement. Recommendations for Local Authorities (2004). http://ec.europa.eu/environment/life/project/Projects/index.cfm?fuseaction=search.dspPage&n_proj_id=1869
6. Quiet City Transport. Deliverable D 5.14 Amsterdam – effectiveness road traffic noise reduction measures. (2007). http://www.qcity.org/

3

Exposure to Noise at the Workplace

Lenzuni Paolo
Italian National Workers'
Compensation Authority

Tønnes Ognedal
Sinus

3.1 Introduction

Despite being recognised as an occupational hazard for more than a century, noise at the workplace is still a lingering problem well into the 21st century. Repeated workplace exposure to high noise levels lead to Noise-Induced Hearing Loss (NIHL), which is one of the most common work-related pathologies.

Estimates of the number of subjects exposed to high noise levels in the workplace are very uncertain, but they span a range from millions (9 million workers in the United States, NIOSH 1998) to tens of millions (40 million workers in Europe, EASHW 2005). Depending on the definitions of exposure and impairment, NIOSH estimates that there are between 5 and 30 million workers in the U.S. who are exposed to noise levels at work that put them at risk of hearing loss (CDC website). Quoting a previous study by the Dublin Foundation, the *Non-binding guide to good practice for the application of Directive 2003/10/EC, 'Noise at work'*, drafted by the European Commission in 2007 (hereafter EC 2007), shows that a rising fraction of European workers is exposed to noise for more than 25% of their working time, this fraction having reached 30% in the five-year term 2000–2005.

For several decades, different countries have developed national legislation for the protection of workers against various risk factors. The 1959 and 1961 'Factories Acts' in the United Kingdom, the 1955 'General disposition for occupational hygiene' in Italy, and the 1975 Workplace Ordinance in Germany are just three examples. In the mid-1980s the first piece of European legislation dealing with occupational exposure to noise was drafted as Directive 1986/188/EEC. This document created a common European basis for the assessment and evaluation of occupational exposure to noise, and many of its fundamental elements are still cornerstones of current EU legislation. Directive 1986/188/EEC has been subsequently replaced by Directive 2003/10/EC, the current reference legislation on occupational exposure to noise in all EU countries. While some of the original elements dating from 1986 have been kept in place, first and foremost the two descriptors of exposure, and some others, such as the threshold values, have undergone only moderate adjustments, the 2003 Directive is undeniably much wider in its scope and much more focused on the actions which must be taken to prevent (first) and to handle (later) possible risks.

Directive 2003/10/EC lies at the root of any current procedure for risk assessment in occupational exposure to noise. It seems natural, therefore, to use it as a script and replicate its structure here, so the sections of this chapter mirror the corresponding articles of the Directive.

We shall first discuss descriptors of exposure in Section 3.2 and thresholds (limits) in Section 3.3, which the Directive handles in articles 2 and 3, respectively.

In article 4, the Directive first clarifies the role of measurements in the broader context of risk assessment. But it is section 4.3 of the article that really represents the core of the document, as it includes a long list of elements that must be taken into consideration in order to create a comprehensive and truly effective risk assessment document. These elements shall all be discussed in Section 3.4 of this chapter.

Finally, methods for the reduction of exposure shall be discussed in Section 3.5 and hearing protective devices in Section 3.6.

3.2 Descriptors of Exposure

3.2.1 Noise Exposure Level

The first of the two quantities, which we shall refer to as 'descriptors of exposure', that are identified by Directive 2003/10/EC (article 2b) for the assessment of occupational exposure to noise is the 'daily noise exposure level' $L_{EX,8h}$

$$L_{EX,8h} = 10 \times \log \left[\frac{\int_{T_{e\text{-}day}} (p_A^2(t)/p_0^2)dt}{T_0} \right] \tag{3.1}$$

In Equation 3.1

- $p_A(t)$ is the time-varying A-weighted sound pressure,
- $p_0 = 20 \ \mu\text{Pa}$ is the reference pressure,
- $T_{e\text{-}day}$ is the duration of the working day,
- $T_0 = 8$ hours is the reference duration of the working day.

This quantity, which was already included in Directive 1986/188/EEC, albeit with a different name ($L_{ep,d}$), is directly proportional to the sound energy density that the worker is exposed to during his/her working day. In other words, the exposure level $L_{EX,8h}$ calculated in Equation 3.1 is such that an 8-hour exposure to this level would have the same sound energy density of the actual exposure of duration T_e to a time-varying sound pressure $p_A(t)$. Equation 3.1 also shows that for the same sound pressure p_A, $L_{EX,8h}$ increases as the exposure time T_e increases. Comparison of Equation 3.1 with Equation 3.4, which gives the continuous equivalent (mean) sound pressure level, shows that long shifts (exceeding $T_0 = 8$ hours) result in an exposure level that is larger than the continuous equivalent sound pressure level of the shift itself.

The mathematical definition of $L_{EX,8h}$ (3.1) assumes the validity of the so-called 'equal energy principle', which states that equal amounts of sound energy produce equal amounts of hearing impairment. Accordingly, exposure can be simply determined by the time integral of the energy density of the local sound pressure field that the worker is exposed to, irrespective of the detailed time evolution.

The equal energy principle implies the well-known 3-dB exchange rate rule between noise level and exposure time (an exposure to noise level 3 dB higher is equivalent to an exposure twice as long). Discussing the validity of the equal energy principle in the assessment of occupational exposure to noise is beyond the scope of this work. In synthesis, over timescales ranging from minutes to days the equal energy principle has been proven to hold up fairly well to sound pressure levels of order 110–115 dB. This is a long-established result (e.g. Suter 1993) that has been more recently reassessed and confirmed, provided that noise has at most a moderately impulsive nature (Qiu, Davis & Hamernik 2007). Over timescales of order seconds, there is considerable evidence that a sequence of short bursts of noise, usually referred to as impulsive noise, is more harmful to hearing than continuous noise of identical energy. Over extended timescales (months to years), processes involving recovery of hearing become heavily involved and the equal energy principle might indeed overestimate the resulting hearing impairment. A critical analysis of the equal energy principle, including a historical perspective, and an extensive discussion of the possible extra damage due to impulsive noise is presented in NIOSH 1998 and references therein.

For the purposes of interest in this section, it appears safe to assume that, apart from impulsive noises, which shall be separately discussed in Section 3.4, the equal energy principle can be applied to the daily/weekly exposure to noise of all workers.

In general, a worker is exposed to noise levels which vary on timescales of days, months, and years in unpredictable ways. Therefore, a realistic estimate of the lifetime integrated exposure would require a long-term analysis and measurements that take into account the long-term variability of the exposure. However, based on the previous discussion of the equal energy principle, an estimate based on a simple integration of the sound energy is unlikely to give a meaningful prediction of the hearing impairment potential of the exposure. But the reason why long-term measurements of noise exposure are not carried out are not technical, but practical: Directive 2003/10/EC is clearly uninterested in a good estimate of lifetime exposure, favoring a short-term estimate. The latter represents a simple diagnostic tool that allows a fast determination of the 'risk category' that a worker should be assigned to. This way, appropriate actions can be taken so that hearing risk is eliminated, or at least minimised, before any hypothetical damage takes place. This is the line of thought that has led to the adoption of a descriptor with a one-day reference time.

Under conditions in which 'the daily noise exposure varies markedly from one working day to the next' but are otherwise unconstrained, the use of a weekly averaged value

$$\overline{L_{EX,8h}} = 10 \times \log\left[\frac{\sum_{j=1}^{N} 10^{0.1 \times L_{EX,8h,j}}}{5}\right] = 10 \times \log\left[\frac{\int\limits_{\text{week}} (p_A^2(t)/p_0^2)dt}{T_0}\right] \tag{3.2}$$

is also allowed (article 2c). In Equation 3.2, $T_0 = 40$ hours is now the reference duration of the working week. The weekly exposure level (3.2) is not really an average, since the summation is extended over N days but the result is then divided by 5. Accordingly, the weekly exposure level resulting from six- or seven-day working weeks is larger than the simple mean of individual daily noise exposure.

In order to avoid unnecessary confusion, the symbol $\overline{L_{EX,8h}}$ used in Directive 2003/10/EC is often replaced by the symbol $L_{EX,week}$. $L_{EX,week}$ is the conceptual equivalent of $L_{EX,8h}$, extended to a time baseline of 40 hours.

The noise exposure level calculated using Equations 3.1 and 3.2 depends on which particular day (week) has been selected for measurements to be carried out. Because the Directive's main target is to identify the appropriate protective actions to be taken, it has been suggested to select a day (week) characterised by noise levels which are large without being exceptionally large. This provision has actually been included in some national legislation (e.g. the Italian Law 81/2008, art. 189.3).

Equations 3.1 and 3.2 show that evaluation of the noise exposure level requires that the A-weighted sound pressure is integrated over the total exposure time T_e accumulated over one day (one week). The strategy where the measurement time coincides with the exposure time is certainly feasible but, understandably, rarely pursued in practice. This 'full-day strategy' shall be discussed in Section 3.2.3. Much more widespread is the use of an alternative approach known as 'task-based strategy', which shall be discussed in Section 3.2.2.

3.2.2 Task-Based Strategy

The calculation of the daily exposure level using the task-based strategy is extensively discussed in ISO 9612, the reference international standard on occupational exposure to noise. In the task-based strategy, the working day selected for the analysis (called nominal day by the standard) is first divided into M independent tasks. ISO 9612 explicitly requests that each individual task is identified such that its noise 'is likely to be repeatable', and that 'all relevant noise contributions are included'.

Each individual task is then characterised through its continuous A-weighted equivalent sound pressure level $L_{p,Aeq,Tm}$ and the associated exposure time T_m. The noise exposure level can then be computed as

$$L_{EX,8h} = 10 \times \log\left[\frac{\sum_{m=1}^{M} 10^{0.1 \times L_{p,Aeq,Tm}} T_m}{T_0}\right] \tag{3.3}$$

(ISO 9612, equation 9). Sections 3.2.2.1 and 3.2.2.2 shall discuss how estimates of the equivalent continuous sound pressure level $L_{p,Aeq,Tm}$ and the associated exposure time T_m can be found.

3.2.2.1 Continuous Equivalent A-Weighted Sound Pressure Level $L_{p,Aeq,Tm}$

Consistent with the definition of the noise exposure level, the equivalent continuous A-weighted sound pressure level

$$L_{p,Aeq,Tm} = 10 \times \log\left[\frac{\int_{Tm} (p_A^2(t)/p_0^2)dt}{T_e}\right] \tag{3.4}$$

is a value proportional to the integral of the sound field energy density that the worker is exposed to while performing a given task. Just like $L_{EX,8h}$, $L_{p,A,eqT,m}$ is calculated using the equal energy principle. This implies that doubling or halving the exposure time is equivalent to increasing/decreasing the sound pressure level by 3 dB.

ISO 9612 goes into extensive detail on how to carry out measurements that lead to a reliable estimate of the continuous sound equivalent level of a given task. The following is a list of the most relevant points, including a reference to the appropriate section of ISO 9612:

1. *Type of instrumentation (section 5.1 and section 12.1):* Both personal sound exposure meters and sound level meters can be used. Personal sound exposure meters are better suited when making long duration measurements for a mobile worker engaged in complex or unpredictable tasks or carrying out a large number of discrete tasks. Under such circumstances, the use of a sound level meter would introduce unnecessary uncertainties and should be avoided. In all other cases, the use of a sound level meter is a viable option.

2. *Field calibration (section 12.2):* Before each series of measurements and at the start of each daily series of measurements, a field calibration with appropriate instrumentation compliant with IEC 60942:2003, class 1, must be performed. At the end of each series of measurements and at the end of each daily series of measurements, a field calibration shall also be performed. If the reading at any frequency at the end of a series of measurements differs from the reading at that frequency at the beginning of the series by more than 0.5 dB, the results of the series of measurements shall be discarded.

3. *Presence/absence of the worker (section 12.3):* Whenever a task is carried out in a fixed or mildly variable position, measurements are preferentially taken in the absence of the worker. This, however, rarely occurs in practice because following a worker in his/her working activity provides a much more reliable and detailed picture of his/her actual noise exposure.

4. *Microphone position and orientation (section 12.3 and section 12.4):* When using a sound exposure meter, the microphone should be positioned on the shoulder at a distance of at least 0.1 m from the entrance of the external ear canal at the side of the most exposed ear and approximately 0.04 m

above the shoulder. When using a sound level meter, in the absence of the worker the microphone should be placed at the center plane of the worker's head, on a line with the eyes, with its axis parallel to the worker's vision; if the worker is present, it should instead be positioned at a distance between 0.1 m and 0.4 m from the entrance of the external ear canal and at the side of the most exposed ear.

5. *Measurement duration (section 9.3):* The duration of each measurement must be long enough that the measured level is representative of the average equivalent continuous sound pressure level for the actual task. If the duration of the task is shorter than 5 minutes, the duration of each measurement must be equal to the duration of the task. For longer tasks, the duration of each measurement shall be at least 5 minutes. The duration of each measurement may, however, be reduced if the level is found to be constant or repeatable. ISO 9612 does not include a quantitative definition of a constant or repeatable noise. ANSI/ASA S3.20 (2015) includes the following definition of stationary or steady noise: 'Noise with negligibly small fluctuations of level within the period of observation'. This is also non-quantitative. A quantitative criterion to identify constant noise was provided by an earlier version of ISO 1999 (1990) by requiring that the difference between the largest and the smallest of slow-weighted sound pressure levels is below 5 dB. Unfortunately, this criterion has been eliminated in the current version of the standard ISO 1999 (2013), so this matter is still hung.

6. *Number of measurements (section 9.3):* For each task, at least three measurements must be carried out. It is recommended to measure at different times during the task.

The equivalent continuous A-weighted sound pressure is calculated as the energetic (not the arithmetic) mean of the $I \geq 3$ measurements carried out

$$L_{p,Aeq,Tm} = 10 \times \log\left[\frac{\sum_{i=1}^{I} 10^{0.1 \times L_{p,Aeq,Tm,i}}}{I}\right] \tag{3.5}$$

The associated standard uncertainty on the mean is calculated using ISO 9612 equation (C.6)

$$u\left(L_{p,Aeq,Tm}\right) = \sqrt{\frac{\sum_{i=1}^{I} \left(L_{p,Aeq,Tm,i} - \overline{L_{p,Aeq,Tm}}\right)^2}{I(I-1)}} \tag{3.6}$$

where $\left(\overline{L_{p,Aeq,Tm}}\right)$ is the arithmetic mean of the I values of $L_{p,A,eqT,m}$. Equations 3.5 and 3.6 are not mutually consistent, since Equation 3.5 calculates an 'energetic' mean (that is, a mean of levels based on the mean of squared sound pressures, or sound energy densities), while Equation 3.6 calculates a standard deviation based on the arithmetic mean. The calculation of the uncertainty due to sampling resulting from the use of Equation 3.6 along the path given by the Guide to the Uncertainty of Measurements (GUM 2003), limited to first order, leads to an underestimate compared to the exact results (Thiery & Ognedal 2008). As long as the measurement variability is limited, this underestimate is also limited.

Indeed, the three measurements recommended by ISO 9612 allow a reliable computation of $L_{p,A,eqT,m}$, provided the standard deviation is below 1.5 dB, which is equivalent to requesting that the difference between the largest and the smallest (hereafter max-min difference) is less than 3 dB. The enforcement of this 3/3 rule (3 measurements, 3 dB max-min difference) makes sure that the error due to the coarse treatment given by ISO 9612 of sampling uncertainties in small-sized samples is less than 1 dB.

It should also be pointed out that ISO 9612 allows the three measurements required to acoustically characterise a given task to be carried out, not only on an individual worker, but also on 'different workers within a group'. In this latter case it will be extremely hard to keep the standard deviation below 1.5 dB. Assuming a more realistic value of 3 dB, this translates into a max-min difference of 7 dB, and in order to keep the error below 1 dB, as in the single-worker case, a total of $N = 6$ measurements are needed (Thiery & Ognedal 2008). So when different workers within a group are sampled for the estimate of a given continuous sound equivalent level, a 6/7 rule appears to be better suited.

Other non-acoustical elements play a role in the noise assessment procedure and should be recorded when acoustic measurements are carried out (EC 2007):

- The worker or workers to whom the measurement is applicable
- The work activity during the measurement
- The measurement location
- The measurement duration
- The background noise level
- Any hearing protection worn by the worker

In case the exposure consists of a series of short-duration events, two possible strategies are possible:

1. If events repeat themselves at more or less regular intervals, the equivalent continuous sound pressure level can be measured over a period that includes several events, and exposure can be computed along the lines discussed above.
2. If events occur along a very irregular temporal pattern, the sound energy level (SEL) of a few events can be measured. The sound energy level of an event is defined as

$$SEL = 10 \times \log\left[\int_{Tm} \left(p_A^2(t)/p_0^2\right) dt\right] \qquad (3.7)$$

and its physical meaning is that SEL is the continuous equivalent sound pressure level that would be measured if all the acoustic energy of the event were compressed into a 1-second window.

The sound energy levels of different events can then be averaged to give a mean SEL and the noise exposure level generated by N events can be calculated as

$$L_{EX,8h} = \overline{SEL} + 10 \times \log\left[\frac{N}{T_0}\right] \qquad (3.8)$$

3.2.2.2 Exposure Time T_m

Equation 3.3 requires that the duration T_m of a task (i.e. the exposure time) must also be determined along with the noise equivalent level $L_{p,A,eqT,m}$. ISO 9612 presents different options in this respect, including:

1. Interviewing the workers and the supervisor
2. Observing and measuring durations during noise measurements
3. Gathering information regarding operation of typical noise sources (e.g. work processes, machines, activities at the workplace and in its surroundings)

Regardless of the specific method used to collect information, it is important that multiple determinations are obtained, so that the mean value and the associated uncertainty can be calculated on a statistical basis. If J determinations $T_{m,j}$ are available, than the task duration is best estimated using the arithmetic mean

$$T_m = \frac{\sum_{j=1}^{J} T_{m,j}}{J} \qquad (3.9)$$

(ISO 9612, section 9.2), while the associated standard uncertainty is

$$u(T_m) = \sqrt{\frac{\sum_{j=1}^{J} \left(T_{m,j} - T_m\right)^2}{J(J-1)}} \qquad (3.10)$$

(ISO 9612, section C.2). Note that there is no coupling between the determinations of $L_{p,Aeq,Tm}$ and T_m, so the two numbers I and J are usually different. In case the available information on the task duration is limited to an upper and a lower limit (i.e. a range T_{min}; T_{max}), ISO 9612 recommends that the uncertainty is estimated as $u_m = 0.5 (T_{max} - T_{min})$. This appears to be an unnecessarily crude approximation. If nothing more than the upper and the lower limit of the range is known, then a rectangular distribution is a fair assumption, which leads to $u_m = 1/(2\sqrt{3}) (T_{max} - T_{min})$. A better estimate can be found by recognising the fact that the available information presumably leads to some kind of central concentration of the probability density function, which can therefore be approximated using a trapezoidal rather than a simple rectangular distribution. By defining the trapezoidal probability density function such that β is the minor to major base ratio, the uncertainty is

$$u(T_m) = \frac{1/2(T_{max} - T_{min})\sqrt{(1 + \beta^2)}}{\sqrt{6}} \tag{3.11}$$

Because the degree of central concentration is unknown, a fair way to proceed is to take the mean overall values of β, which leads to $u(T_m) \approx (T_{max} - T_{min})/4$.

3.2.3 Job-Based Strategy

The name of this strategy indicates that it is related to a specific job and not the tasks that make up the job. The job-based strategy is useful when it is hard to describe the job by a set of typical tasks. It may also be used for verification in parallel with task-based measurements; for instance if it is desired to do a separate check of the noise exposure by simply leaving a noise-dose meter on one person or more.

A precondition of the application of the job-based strategy is the identification of a homogeneous group of workers.

ISO 9612 specifies the total duration minimum of measurements, which depends on the number of workers in the homogeneous group. Based on the minimum total duration, the number and duration of individual measurements is determined. All the measurements have identical duration and are taken on two or more workers in the homogeneous group. Measurements are taken randomly throughout the day, but in order to avoid underestimation of the exposure, one should ensure that typical high noise periods are included.

In principle, the job-based strategy is a straightforward and effective way to calculate the exposure of a group of workers who share the same job. In practice, however, its application is hampered by the very long measurement time implied. Indeed, Table 1 of ISO 9612 shows that even for more moderately sized groups (say $N = 6 - 10$), total measurement times range between 5.5 and 7.5 hours. This implies that the job-based strategy is best pursued by analysing data taken using noise dose-meters. One may then pick out a sufficient number of measurements from the time-level history afterwards. This is especially useful when the job-based strategy is used to verify the task-based measurements.

Finally, the noise exposure for the day is calculated taking the energetic mean of the noise levels measured in individual noise measurements, as indicated in ISO 9612 Equation 3.11

$$L_{p,Aeq,Te} = 10 \times \log\left[\frac{\sum_{n=1}^{N} 10^{0.1 \times Lp,Aeq,T,n}}{N}\right] \tag{3.12}$$

The uncertainty is directly influenced by the variation in the level of the separate periods. Thus, a day with a relatively constant noise level has less uncertainty than a day with large variations in noise levels. This is also quite natural, as large uncertainty indicates that days are more different. The procedure for the calculation of the uncertainty is detailed in ISO 9612, section C.3.

3.2.4 Full-Day Strategy

The full-day strategy is conceptually naïve and leads to accurate results, but it can be very time-consuming. ISO 9612 mandates that at least three measurements of the equivalent continuous A-weighted sound pressure level $L_{p,Aeq,Te}$ are carried out over the total working day. Two additional measurements are required if the range spanned by the initial three measurements is greater than 3 dB. The energy average of the three (five) measurements is then calculated and the result is simply scaled to the nominal duration of the working day of 8 hours to give the daily noise exposure

$$L_{EX,8h} = 10 \times \log\left(\frac{\sum_{j=1}^{N} 10^{0.1 \times L_{p,Aeq,Tej}}}{N}\right) + 10 \times \log\left(\frac{T_e}{8\,\text{hours}}\right) \tag{3.13}$$

The calculation of the uncertainty of $L_{EX,8h}$ calculated using the full-day strategy is detailed in ISO 9612, section C.4. It stems directly from Equation 3.10 that there is no contribution from exposure times. However, the calculation of contribution deriving from the propagation of the uncertainty on the equivalent level is more complex than in than case of task-based strategy. The procedure is not fully analytic as in the previous case, and requires instead the use of tabulated values.

3.2.5 Peak Sound Pressure Level

Many decades of studies have shown that the vast majority of noise-induced hearing loss (NIHL) is due to repeated exposure to high noise levels over many years. This justifies the much larger attention that has been paid to the measurement, assessment, and reduction of long-term exposure to 'continuous' noise, compared to very short, very-high-energy sound pressure peaks. This latter element is nevertheless taken care of in article 2a of Directive 2003/10/EC, using a second descriptor of exposure for the assessment of occupational exposure to noise, called 'peak sound pressure' p_{peak}. The quantity p_{peak} is defined as the maximum value of the C-weighted instantaneous noise pressure. C-weighting has a much flatter profile than A-weighting and is considered to be better suited to quantify the damage potential of large peaks of sound pressure which may be dominated by low-, medium-, or high-frequency components.

It is unclear why in Directive 2003/10/EC the choice was made to express this quantity in terms of the physical unit (Pa) rather than as a 'level', similar to what was done for the noise exposure level. The equivalence between peak sound pressures and C-weighted peak sound pressure levels was established by Directive 2003/10/EC itself in a note. While it is certainly not a major issue, this may have undesired consequences because the same peak sound pressure, quantified using Pa or dB, leads to values which have widely different rounding uncertainties.

ISO 9612 always deals with C-weighted peak sound pressure levels (symbol $L_{p,Cpeak}$). Measurements of $L_{p,Cpeak}$ are only recommended 'if relevant', that is, if the task is such that it includes highly impulsive signals. This is conceptually correct, since the vast majority of tasks have peak levels well below 130 dB. However, this is practically irrelevant, since contemporary instrumentation provides simultaneous measurements of peak sound pressure levels and equivalent continuous sound pressure levels (as well as many other quantities), so no additional burden is created by the systematic measurement of $L_{p,Cpeak}$.

ISO 9612 does not include a procedure to calculate the uncertainty on the C-weighted peak sound pressure level. In its section C.1 it states that 'in most cases, the uncertainty for the peak sound pressure level can be expected to be greater than the uncertainty for the A-weighted equivalent continuous sound pressure level', but in its section C.5 this becomes 'the uncertainty for $L_{p,C,peak}$ can be considerably higher (than the uncertainty on $L_{p,A,eqT,m}$)'.

Some progress on this topic can be made by splitting the uncertainty into separate contributions due to instrumentation and (incomplete) sampling. Uncertainty due to instrumentation can be quantified by using the tolerances for tests related to peak levels indicated in IEC 1672-1. The uncertainty due to incomplete sampling has been recently estimated assuming the statistical distribution of peak levels (Lenzuni 2015). It is interesting to note that for the peak sound pressure level, incomplete sampling leads not only to a finite uncertainty on the estimated value, but to a net underestimate of the value itself. This happens because the descriptor included in Directive 2003/10/EC is the single largest sound pressure peak that occurs during the working day, which is most likely not caught by fractional samplings such as those which are commonly performed. Both the uncertainty and the underestimate of the peak sound pressure level decrease as the sampled fraction of the working day increases and vanish when the working day (or the task where the largest peak occurs, if this can be unambiguously identified) is fully sampled.

3.3 Thresholds

Article 3 of Directive 2003/10/EC introduces three thresholds for each of the two descriptors $L_{EX,8h}$ and $L_{p,Cpeak}$. Table 3.1 shows the threshold denominations and their values for the noise exposure level and the peak level, respectively.

The two lower thresholds for $L_{EX,8h}$ are identical to the thresholds set by Directive 1986/188/EEC about 30 years ago. These two values both have clear, unambiguous meaning:

1. The lower exposure action value set at 80 dB(A) is the level below which even long-term exposures will not induce any measurable NIHL for the vast majority of the exposed population. Many analyses carried out during the 1970s, 1980s, and 1990s agree on mean hearing threshold shifts below 5 dB after 40 years of occupational exposure to levels around 80 dB(A) (NIOSH 1998).
2. The higher exposure action value set at 85 dB(A) is the level above which a sizeable hearing loss will occur in a vast majority of the exposed population.

Both the lower and the upper action values have been set based on the assumption that typical 8-hour working days will be followed by typical 16-hour periods of acoustic rest. The existence of extended acoustic rest periods interposed between successive exposure periods is paramount to the numerical identification of thresholds. In the absence of such rest periods, noise-induced hearing loss would develop much more quickly. Precise indications of the time evolution of NIHL as a function of the noise/rest duty cycle are not available. However, it is advisable that workers whose work schedules include repeated extended work shifts are assigned to a category of particularly sensitive workers and kept under strict medical surveillance for any signs of incipient NIHL.

The exposure limit value, set by the 2003/10/EC Directive at 87 dB(A), is down 3 dB from the previous value of 90 dB(A) set by the 1986/188/EEC Directive. This 3 dB reduction would be sizeable enough to have a significant impact on the workers' hearing protection. However, it becomes insignificant in light of the Directive's indication (of opposite sign) that compliance with the exposure limit value should be checked taking into account the attenuation provided by any hearing protector device worn.

The three thresholds for $L_{pC,peak}$ are all identical to the thresholds set by Directive 1986/188/EEC. Their meaning is, however, much less clear-cut in terms of hearing damage. Initial criteria set to limit the exposure to peaks were balancing duration, number, and peak sound pressure levels of the events in a day (Ward 1968). Such criteria were considered too difficult to grasp for the average user and too complicated to enforce. Instead, a simple one-number rule was chosen, setting a limit at 140 dB. According to NIOSH (1998), a 'ceiling' value of 140 dB is based on a work by Kryter et al. (1966), and was considered at the time 'little more than a guess'. However, no evidence has emerged since then that is strong enough to induce regulators to either modify this value or to introduce a more complex and biomechanically realistic criterion. The lower and upper action values of 135 and 137 dB(C) do not have any real scientific basis.

TABLE 3.1   Threshold Values for $L_{EX,8h}$ and $L_{p,Cpeak}$

Name	Threshold for $L_{EX,8h}$	Threshold for $L_{p,Cpeak}$
Lower exposure action value	80 dB(A)	135 dB(C)
Higher exposure action value	85 dB(A)	137 dB(C)
Exposure limit value	87 dB(A)	140 dB(C)

Compliance of calculated noise exposure values with both the lower and the higher exposure action values must be established using measurements which do not take into account the effect of any personal hearing protector device (HPD). This is in line with the Directive's approach that the worker's protection against noise should fundamentally be handled by improving environmental conditions, not by the use of HPDs. On the other hand, compliance with the exposure limit value must be checked, considering that 'the determination of the worker's effective exposure shall take account of the attenuation provided by the individual hearing protectors worn by the worker' (article 3.2). This has a twofold consequence:

1. With respect to the noise exposure level, because the effectiveness of hearing protectors is routinely assessed by means of the experimenter-fit method (ISO 4869-1, ISO 4869-2), which usually provides insertion loss values ranging from 25 to 35 dB (see Section 3.6), a situation of non-compliance with the 87 dB(A) exposure limit value is indeed hardly conceivable.
2. With respect to the peak sound pressure level, this makes compliance with the limit value extremely hard to check, given the large uncertainties on the attenuation provided by HPDs to the peak SPL.

Article 3.3 allows the daily noise exposure level to be replaced by the weekly noise exposure level if the former 'varies markedly' over the week. As discussed in Section 3.2.1, neither the directive nor ISO 9612 (or any other technical standard, for that matter) includes a quantitative criterion to decide if the fluctuation is 'marked' enough. This criterion might well be unnecessary: the noise exposure assessment is certainly more reliable when carried out on a week-long temporal baseline rather than on a day-long temporal baseline. Whenever enough information is available, the calculation of $L_{EX,W}$ provides an excellent contribution to a correct assessment of the exposure. The use of $L_{EX,8h}$ is acceptable if care is taken to select the nominal day such that it gives a picture of an exposure that is rarely exceeded. This establishes a fair trade-off between the largest value that $L_{EX,8h}$ will take on compared to $L_{EX,W}$ and the reduced amount of time/energy/money required for its calculation.

ISO 9612 does not include an explicit equation for the calculation of the weekly noise exposure level (which appears instead in ISO 1999, section 3.1), and accordingly does not include any procedure to calculate the associated uncertainty.

3.4 Elements That Must Be Taken into Consideration in the Process of Risk Assessment

Before discussing the various elements that concur to create a solid noise risk assessment document, it must be made clear that assessing and measuring are two distinct concepts. Measurements are certainly needed in order to correctly quantify the exposure. However, the process of risk assessment extends well beyond measuring and, indeed, measuring may even be unnecessary when the absence of significant noise sources is obvious.

Once this point has been clarified, the many individual points discussed below give a comprehensive description of the risk assessment process that the Directive is focused on. These elements shall be dealt with following the same order in which they appear in Directive 2003/10/EC, article 4.

1. *The level, type, and duration of exposure, including any exposure to impulsive noise:* With regard to the noise level and duration, the mathematical formulation used to compute $L_{EX,8h}$ shows how these factors are already factored in. How any hypothetical impulsive nature of noise should be handled is instead a long-standing but still unsettled issue.

   The simple fact is that no consensus has yet been reached on this topic at the international level, so neither Directive 2003/10/EC nor ISO 9612 (see Section 3.2.1) provide methods to take into account the impulsive nature of noise. Indeed, the definition of impulsive noise is by no means agreed upon in technical documents.

According to superseded ISO 11204 (1996), appendix C, a noise is impulsive if

$$L_{pAIeq} - L_{pAeq} \geq 3 \qquad (3.14)$$

where L_{pAIeq} is the A-weighted continuous equivalent sound pressure level determined using impulse time weighting, while L_{pAIeq} is the corresponding A-weighted continuous equivalent sound pressure level. Unfortunately, Annex C has been altogether eliminated from the current revision of ISO 11204 (2010), where any reference to the concept of impulsive noise has disappeared.

ISO 1999 (1990), while not including a quantitative definition of impulse noise, did include the following statement: 'Some users may, however, wish to consider … impulsive/impact noise about as harmful as a steady … noise that is approximately 5 dB higher in level'. This sentence has been eliminated from the current revision of ISO 1999 (2013).

According to EN 458, impulse noise is a result of 'sudden change of pressure that can consist of a unique single event or form either a series of impulses with pauses between'. This definition, again, is not quantitative.

The absence of a definition of impulsive noise that is globally agreed upon has also generated an unwanted side effect: many practitioners confuse impulse noise with noise characterised by large peak sound pressure values. While it is true that large peak sound pressure values imply the existence of impulsive noise, the opposite is certainly not.

Moving on to more scientific ground, NIOSH (1998) provides an excellent synthesis of the many experimental studies which support the existence of extra hearing damage associated with impulsive noise, as well as the few studies which do not find any such evidence. The sheer size of the two samples suggests that some effect is at least likely.

Additional evidence supporting the claim that impulse noise is more harmful to hearing than steady noise of identical energy has been provided in the last 15 years in a series of studies carried out by a joint U.S.-China workgroup (Hamernik et al. 2003; Goley et al. 2011; Davis et al. 2012; Xie et al. 2016). Such studies have shown that noise signals whose pressure distribution in time is strongly non-normal, and as such include a fair amount of impulsive noise (the 'large' tails of the distribution), lead to higher hearing loss both in animals and in humans. In particular, the extra damage has been found to be well correlated to the kurtosis (i.e. the fourth moment) of the pressure distribution. This is a promising pathway that might eventually lead to establishing a quantitative, continuously distributed penalty that takes into account the larger hearing damage potential of highly impulsive noise compared to non-impulsive noise of identical energy. Such penalty would, of course, have to depend on practically measureable quantities. Incidentally, this would also make any discussion on the definition of impulse noise moot, since the penalty would be applicable to any noise with values ranging from zero up to some maximum number.

2. *The exposure limit values and the exposure action values laid down in Article 3 of this Directive:* The concept of taking into consideration the exposure limit values and the exposure action values is likely more clearly rephrased as comparing the results of measurements to the exposure limit values and the exposure action values. This is precisely the same idea behind article 4.2: '*to decide whether, in a given case, the values fixed in Article 3 have been exceeded*'.

The EC 2007 guide explicitly states that 'When it is possible, rather than certain, that an exposure action value or limit value is exceeded, action shall be taken on the assumption that it is exceeded'. Thus the guide clarifies that the comparison between the result of a measurement, with its uncertainty, and a legally mandated threshold devoid of any uncertainty is an inherently statistical problem. ISO 9612 does not explicitly tackle the issue of compliance of measurements/ estimates with thresholds. However, it defines the expanded uncertainty U as the quantity that, when added to the result, gives the upper limit of the monolateral confidence interval with a 95% confidence level (section C.1). This implicitly indicates that the expanded uncertainty $U(L_{EX,8h}$ must enter the process that leads to checking compliance of the calculated value of the noise exposure level with thresholds. Otherwise, a bilateral confidence interval would have been better suited. Different national technical documents or standards (INRS 2009 in France, UNI 11326-2 2015 in Italy) have made this concept explicit, so that it is now possible to claim that compliance can only be established if the inequality

$$L_{EX,8h} + U\big(L_{EX,8h}\big) < \text{Threshold} \qquad (3.15)$$

is fulfilled.

3. *Any effects concerning the health and safety of workers belonging to particularly sensitive risk groups:* The identification of workers belonging to particularly sensitive risk groups is not a technical issue, rather a clinical issue that falls under the jurisdiction of medical surveillance action. It is the medical doctor's responsibility to determine whether a worker displays some kind of peculiar sensitivity. This information should then be passed on to the employer, so that he/she takes appropriate actions to ensure that this worker is given additional and customised protection from noise.

4. *As far as technically achievable, any effects on workers' health and safety resulting from interactions between noise and work-related ototoxic substances, and between noise and vibrations:* An extensive literature exists supporting synergistic effects between exposure to noise on one side and some chemical substances (known as ototoxic) or vibration on the other. Ototoxic substances include solvents (ethanol, n-exane, styrene, toluene, xyrene), metals (arsenic, lead, mercury), and other substances (e.g. hydrogen cyanide). Interactions of noise and ototoxic substances have been extensively discussed in a publication by the European Agency for Safety at the workplace (EU-OSHA 2009), with several more recent updates (e.g. Nies 2012). The same issue has been the topic of Appendix D of the OSHA Technical Manual on Noise (OSHA 2013).

   Quantitative relationships indicating possible penalties to be introduced when simultaneous exposure to noise and ototoxic substances occur do not exist. Following the general principle of precaution, it has been suggested that action values be lowered by 5 dB for combined exposure to noise and ototoxic substances (EC 2007). Given that this issue is at least as ill-defined as that regarding impulsive noise, it appears that the introduction of any penalty is premature.

   A similar situation exists for the combined exposure to noise and both hand-arm vibration and whole-body vibration. Here again, scientific studies (see Pettersson et al. 2014 for a recent review) suggest the existence of some synergistic effect, but quantification will have to await a much more solid experimental body of evidence.

5. *Any indirect effects on workers' health and safety resulting from interactions between noise and warning signals or other sounds that need to be observed in order to reduce the risk of accidents:* The interaction between noise and warning signals is not dealt with by technical documents which handle the occupational exposure to noise. This is because they focus on the effect of noise on the hearing system, which may ultimately result in noise-induced hearing loss. This is also the viewpoint that Directive 2003/10/EC usually adopts.

   Here a different perspective is assumed which considers implications for the worker's safety, due to the fact that noise at the workplace may hide acoustic signals used to warn people against possible risks, a phenomenon called sound masking.

   Sound masking can become dangerous in cases where an employee has to be warned about potential hazards (e.g. machines or their parts in motion) or has to perform verbal orders. As a result of sound masking, an employee may not be able to hear or recognise warning signals, which may result in an accident (EC 2007).

From an acoustical point of view this problem might be tackled by tuning the warning signals in such a way that their audibility is only marginally affected by workplace noise. However, this problem is more commonly dealt with by introducing warning lights.

6. *Information on noise emission provided by manufacturers of work equipment in accordance with the relevant Community directives:* This same concept is also replicated elsewhere in Directive 2003/10/EC: 'The risks arising from exposure to noise shall be eliminated at their source or reduced to a minimum ... by taking into account in particular ... (b) the choice of appropriate work equipment, taking account of the work to be done, emitting the least possible noise, ...'. The concept of 'emitting the least possible noise' is best quantified using the sound power level L_W of a device. All devices commercialised in the EU which fall under the so-called Machinery Directive (2006/42/EC) must indicate, among other acoustically relevant quantities, their A-weighted sound power level (L_{WA}). More in detail, section 1.7.4.2 (*u*) of the Machinery Directive requests that the manufacturer provide the A-weighted sound power level emitted by the machinery whenever the A-weighted sound pressure level at a workstation exceeds 80 dB(A). The sound power level may be measured following a specific noise code illustrated in a tailored technical standard, if available. This is the case for handheld non-electric power tools where a noise code is provided to measure the sound pressure level. More often it shall be measured according to a general-purpose standard such as ISO 3744 or ISO 3746, which provide a procedure for measuring the sound power level that is applicable to any noise source.

As clarified by the Guide to the application of the Machinery Directive, section 1.7.4.2 (*u*) also requires the uncertainties on the measured values to be specified in the noise emission declaration. Guidance on determining the uncertainty associated with the measurement of L_{WA} should be given in the relevant test codes. If ISO 3744 or ISO 3746 are used, then the uncertainty $u(L_{WA})$ is not calculated following an analytic procedure, but an approximate fixed number is given.

Additionally, if machinery falls under the scope of the Outdoor Equipment Directive 2000/14/EC, the measured sound power level, L_{WA} and the associated uncertainty should be replaced by a single quantity called *guaranteed sound power level*. The guaranteed sound power level is given by the sum of the sound power level calculated according to the method set out in Annex III of the Directive, plus the value of the uncertainties due to production variation and measurement procedures.

Sound power in itself gives no precise indication of the sound pressure level that the worker is actually exposed to. The conversion of sound power into sound pressure is a complex problem that involves the geometry and properties of materials in the environment where sound propagates. However, replacing a piece of equipment (tool/vehicle/machinery) with a quieter one will give a benefit which is very closely related to the sound power level differential, provided the relative position of the worker and the point of greatest emission of sound in the environment do not change too much.

7. *The existence of alternative work equipment designed to reduce the noise emission:* There are two approaches to finding alternative, less noisy equipment.
   a. There are often alternative methods for the same process, as described in Section 3.5.1.
   b. The same type of equipment may have different types of noise levels. When purchasing large machinery, it is recommended to set a noise limit based on previous experience of achievable values. When buying smaller units, like handheld tools, noise data for tools from various producers or vendors should be part of the evaluation. In this evaluation, one has to be aware that manufacturer data often are misleading, as many of the tools seem to be measured without any load. Noise databases for handheld tools with practical working noise values are available on the Internet, and specific tools may be found. If not, special

attention should be paid if manufacturer data deviates much from the typical noise data of the equipment group.

8. *The extension of exposure to noise beyond normal working hours under the employer's responsibility:* Because of the way the noise exposure level $L_{EX,8h}$ is calculated (see Equation 3.1), exposures of any duration can be take into account without problems. Of course, extensions of the work shift well beyond 8 hours cause the daily period of acoustic rest to shrink below 16 hours, which creates a conflict with the assumption underlying the values selected for the lower and upper action values (see Section 3.3). As long as such extensions are not the rule and do not characterise a large fraction of working life, this is more a problem of principle, and has no practical implications.

9. *Appropriate information obtained following health surveillance, including published information, as far as possible:* Health surveillance performed in a workplace should aim at obtaining valuable information about the possible impact of various processes on hearing impairment. Together with results collected through measurements, this information can be used to draw a more comprehensive and detailed picture of where the most serious risks lie, as well as to increase awareness among the personnel. The latter will ultimately lead to better understanding of the importance of noise reduction, low noise operations, and the use of hearing protection.

10. *The availability of hearing protectors with adequate attenuation characteristics:* Hearing protector devices (HPDs) have traditionally played a very strong role in any actions taken to keep the exposure to noise under control. The combination of low cost and large attenuation values would seem to make HPDs the primary tool in any noise reduction strategy. This will be discussed in Section 3.6, which is entirely dedicated to HPDs.

Finally, two additional elements of interest which appear in article 4 deserve some attention:

The methods and apparatus used shall be adapted to the prevailing conditions particularly in the light of the characteristics of the noise to be measured, the length of exposure, ambient factors and the characteristics of the measuring apparatus. These methods and this apparatus shall make it possible to determine the parameters defined in Article 2 and to decide whether, in a given case, the values fixed in Article 3 have been exceeded. The methods used may include sampling, which shall be representative of the personal exposure of a worker.

This statement implies that different methods can be used, provided that they are able to give reliable determinations of the descriptors of noise exposure and that enough information is collected to decide if the exposure/action limit values are exceeded. Different methods include both different instrumentation and different measuring strategies. The different strategies proposed by ISO 9612 have been reviewed in Section 3.2.

When applying this Article, the assessment of the measurement results shall take into account the measurement inaccuracies determined in accordance with metrological practice.

The idea that measurement inaccuracies (uncertainties) must be part of the process leading to check compliance with thresholds as well as the identification of appropriate actions to be taken is very important.

It is interesting to note the radical difference in the approach compared to that of its predecessor, Directive 1986/188/EEC. The latter document included a short and, to some extent, visionary paragraph on uncertainty, which correctly mentioned the implication of both instrumental and procedural uncertainty on decision-making when it comes to assessing compliance or non-compliance with noise thresholds (indicated as 'limits'). Directive 1986/188/EEC opted for the three-way rule that allows a decision (either compliance or non-compliance with threshold values) or no decision be made. This approach is at variance with more contemporary documents on this topic, but it is not faulty in itself. The final sentence of the paragraph, however, 'Measurements of the highest accuracy enable a decision to be taken in all cases', remains somewhat mysterious, given that the existence of a finite amount of uncertainty inevitably implies the existence of a grey area.

Directive 2003/10/EC does not spend a word on how to first determine and then use the uncertainty in the context of noise risk assessment. This task is left to technical documents, and ISO 9612 does indeed include an in-depth analysis of uncertainty. As already discussed, ISO 9612 lacks a section on compliance with limits, but this issue has been tackled by other documents and technical standards (see item b of this section), so the path leading from the statement of principle (2003/10/EC) to the practical implementation is now fully laid out.

3.5 Actions Aimed at Avoiding or Reducing Exposure

Article 5 is where Directive 2003/10/EC makes it clear that risk assessment is of limited significance, unless it is a source of information for technical/organisational actions aimed at lowering the exposure to noise. A list of possible pathways leading to the desired reduced exposure of workers is provided in article 5.1. Each item shall be labeled as either of technical (T) or organisational (O) nature and commented upon.

It is important to point out that the vibration reduction that can be expected by pursuing different approaches is extremely variable. Large variations can be expected even for the same type of action, underlining the intrinsic complexity of the situation.

Proper evaluation of noise-reducing measures requires fundamental understanding of the various quantities used as input data as well as the physics of noise reduction. The following examples may, however, be used as a general guide for achievable noise reduction of various actions. Actions on or close to the source are normally preferred compared to actions further away (and especially compared to hearing protection).

3.5.1 Actions

1. *Other working methods that require less exposure to noise:* (O) This is a more radical solution to the noise problem, where the source of the problem itself (the noisy working method) is eliminated. This option is rarely a pursued, since it only becomes available when alternative technology becomes available at comparable costs. There may be several alternative ways to perform a specific task. As an example, there may be several ways to clean a metal surface: sandblasting creates A-weighted levels around 100–115 dB(A), while vacuum blasting reduces the A-weighted level to 80–90 dB(A). High-pressure water-jetting is around 100–110 dB(A), while sand water-jetting is more than 10 dB less noisy. If a grinder is used, A-weighted levels are typically 95–100 dB(A) at working distance.

2. *The choice of appropriate work equipment, taking account of the work to be done, emitting the least possible noise, including the possibility of making available to workers work equipment subject to Community provisions with the aim or effect of limiting exposure to noise:* (O) This sentence regards the possibility of replacing existing equipment with more quiet machinery. This topic has been previously tackled in Section 3.4(f). Noise requirements should be set when purchasing larger equipment. For smaller equipment, like handheld tools, one can check vendor data or noise databases on the Internet (ref. Section 3.4 item g). The variations in noise level for different makes are typically 5–10 dB, while for some groups of tools the variation may be even larger from the most silent to the noisiest. Whereas the noise levels stated for grinders are typically 95–100 dB(A), they may actually have levels from 85 dB(A) and up to 105 dB(A) or even higher. The situation becomes even more complex when a tool is used with various types of 'workbits'. A low-noise grinding plate can reduce the noise level by 5–8 dB compared to normal grinding plates. The fact that low-noise plates are normally less efficient, more expensive, and wear out faster also enters into the evaluation process.

3. *The design and layout of workplaces and work stations:* (T) This approach is best pursued using simulation software, which allows noise levels resulting from alternative layouts to be predicted and compared. An extensive discussion of the main concepts supporting the layout of low-noise workplaces is given in standard ISO 11690, parts 1, 2, and 3 (ISO 11690-1,1996; ISO 11690-2, 1996; ISO/TR 11690-3, 1997). For the average noise exposure in the workplace, it is often better to group noisy processes together than to spread them around. By putting two noise sources with A-weighted contributions of 90 dB in the same room, one may have one room with 93 dB and one quiet room instead of two rooms with 90 dB. If operations are only done for part of the day, a similar principle applies to the simultaneity of operations.

4. *Adequate information and training to instruct workers to use work equipment correctly in order to reduce their exposure to noise to a minimum:* (O) Choose the least noisy equipment for the job and use it in a quiet way. For instance, many processes are less noisy when less force is applied to the tool, and the reduced force does not necessarily reduce efficiency correspondingly.

5. *Noise reduction by technical means:*

   a. *Reducing airborne noise, e.g. by shields, enclosures, sound-absorbent coverings:*

      I. Enclosures: Using enclosures is a very efficient way to reduce noise emitted from a source, provided that they can be built in. Fully covering enclosures, which are internally dampened with absorption, may easily reduce the level more than 20 dB. However, there is often a need to have openings for ventilation, access to service points, or to the process. This may affect the geometry of the enclosure, which is often no more than a screen on the noisiest parts of the machine, with associated lower acoustic effectiveness. Enclosures are often built of double metal elements with absorption in the cavity, perforated on the inside and unperforated on the outside.

      II. Screens: In theory, screens can provide noise reduction above 15 dB, but in many cases practical limitations reduce this value considerably. The screening effect depends on the weight of the screen, the number of gaps and reflections from the ceiling above, and/or potentially nearby walls. With good and airtight design, a reduction of 8–10 dB is achievable. Screens may be built of the same material as the enclosure, but there are also tarpaulin solutions available that provide a lot of flexibility with respect to access and service. Such tarpaulin screens may have absorption integrated on one or both sides.

      III. Absorption panels: Absorbing surfaces are important in the reduction of the general noise level in the room. In the nearfield of a source, however, the direct sound transmission dominates the noise level. Absorption is therefore of minor importance for the noise level close to the source in large rooms. Still, an industry hall sounds much better with sufficient absorption.

      The distance at which the direct sound level is more or less equal to the diffuse field level is called the hall radius. It is typically between 3 and 10 meters in large halls. When the amount of absorption is doubled, the noise level theoretically drops 3 dB. Often it drops more, due to increased damping over distance.

      IV. Personal enclosures, automated operations, and remote-controlled operations: A large reduction in noise exposure can obviously be obtained by removing the subject from the source. This has been done for many years in some industrial sectors. Re-design of the workplace so that the subject can operate from within a noise-isolated cabin is an extremely effective and definitive solution. Windows or screens to monitor the process may be required.

      Finally, robot technology allows personnel to operate noisy equipment from a distance, so that noise exposure is reduced.

   b. *Reducing structure-borne noise, e.g. by damping or isolation:* There are several methods to reduce noise emission from structure noise or structure-borne noise transmission. Vibration isolators, elastic padding, viscoelastic layers, and mufflers are some of the many examples. These can be used either close to the source, in the transmission path, or at the receiver.

6. *Appropriate maintenance programs for work equipment, the workplace, and workplace systems.*
7. *Organisation of work to reduce noise:*
   a. Limitation of the duration and intensity of the exposure
   b. Appropriate work schedules with adequate rest periods

3.5.2 Synthesis

Articles 5.2 and 5.3 of the Directive include some of the actions that are requested when different thresholds are exceeded. The existence of three thresholds implies that the worker's exposure can be assigned to one of four categories, each one characterised by specific actions to put into practice for appropriately handling the risk. In this perspective, it is useful to summarise the overall picture that emerges, including a citation of the specific article of the Directive where the issue is dealt with in more detail:

1. $L_{EX,8h} < 80$ dB(A) or $L_{p,Cpeak} < 135$ dB(C)
   - No action is needed
2. 80 dB(A) $< L_{EX,8h} < 85$ dB(A) or 135 dB(C) $< L_{p,Cpeak} < 137$ dB(C)
   - Workers are entitled to receive adequate information about risks (art. 8)
   - Hearing protective devices must be made available (art. 6.1)
3. $L_{EX,8h} > 85$ dB(A) or $L_{p,Cpeak} > 137$ dB(C)
   - Workers are entitled to receive adequate information about risks (art. 8).
   - Hearing protective devices must be worn (art. 6.1).
   - Workers have the right to have their hearing checked by a doctor or by another suitably qualified person (art. 10.2).
   - Areas of high sound pressure level must be marked with appropriate signs (art. 5.4). Because the Directive explicitly indicates 'workplaces where workers are likely to be exposed to noise exceeding the upper exposure action values', high sound pressure level must be read as 85 dB(A).
   - Areas of high sound pressure level must be delimited and access to them restricted where this is technically feasible and the risk of exposure so justified.
   - The employer shall draw up and apply a program of measures of a technical nature and/or of organisation of work aimed at reducing as far as reasonably practicable the exposure of workers to noise (art. 5.2).
4. $L_{EX,8h} > 87$ dB(A) or $L_{p,Cpeak} > 140$ dB(C) taking into account the attenuation provided by HPDs, immediate action must be taken to reduce the exposure to below the exposure limit values; the reasons why overexposure has occurred must be identified and existing protection and prevention measures must be amended in order to avoid any recurrence (sections 7.2 and 7.3).

   This fourth category is essentially of academic interest. As discussed in Section 3.3, it is extremely unlikely that $L_{EX,8h}$ cannot be kept below 87 dB(A) once the attenuation provided by an HPD is taken into account, as this would imply noise exposure levels exceeding 110 dB(A). The same applies to exposure to high-level impulses, where a hypothetical non-compliance would require peak levels above 160–165 dB(C), which can only be envisaged in connection to the use of large-caliber weapons or explosives.

A comprehensive comparison between the provisions which appear in the two Directives (2003/10/EC and 1986/188/EEC) is provided in EC 2007.

3.6 Personal Protective Equipment – Hearing Protector Devices

HPDs are dealt with by EN 458, a European standard which handles many different issues concerning HPDs, covering acoustical as well as non-acoustical aspects.

3.6.1 What's In

From an acoustical standpoint, the contents of EN 458 include:

1. A short review of the different available types of HPD, with directions on how to select the appropriate HPD to best handle specific exposures. Passive HPDs include: ear-muffs, helmet-mounted ear-muffs, acoustic helmets, ear-plugs, pre-shaped ear-plugs, individual custom-moulded ear-plugs, user-formable ear-plugs, and banded ear-plugs. Active HPDs include: level-dependent hearing protectors, flat frequency response hearing protectors, active noise reduction (ANR) protectors, and hearing protectors with communication facilities.

2. Methods for calculating the sound attenuation of a passive hearing protector with respect to the equivalent continuous A-weighted sound pressure level, including the octave-band method (OBM), the high-medium-low (HML) method, and the single number rating (SNR) method, all discussed in Annex A.

3. Some indications of the performance expected from the combination of ear-muffs and ear-plugs (also known as double protection). EN 458 recommends that dual protection is used when the noise exposure level L_{EX} exceeds 105 dB(A), especially if substantial energy is concentrated below 500 Hz. A more conservative approach has been adopted by NIOSH, which, in its 1998 Revised Criteria for Occupational Noise Exposure, recommends implementation of double protection for L_{EX} above 100 dB(A). The attenuation provided by the simultaneous use of ear-plugs and ear-muffs is always less (often much less) than the sum of the individual attenuation values. This happens because the plugs and muffs interact mechanically with each other, and thus do not behave as two completely independent attenuators. Indeed, small changes in attenuation are predicted when different ear-muffs are used with the same ear-plugs, but for a given ear-muff the choice of ear-plug is critical, particularly at frequencies below 2000 Hz (Berger 2001). The classical study by Berger (1983) shows that the overall attenuation varies from near 0 to about 15 dB over the better single device. Larger attenuations can be expected at frequencies below or near 500 Hz. As frequencies approach 1000 Hz, the added value of double protection is limited by the presence of bone conduction, which introduces an additional pathway that bypasses the HPD to directly stimulate the middle and inner ears of the wearer. This effect is most pronounced at 2000 Hz, when it limits the achievable attenuation to about 39 dB. Above 2000 Hz the achievable attenuation is entirely set by bone conduction. Despite some experimental activity (e.g. Behar and Kunov 1999), no algorithm has yet been developed that provides a sufficiently accurate prediction of double protection.

4. A method for establishing the correct noise level to be achieved when HPDs are used in order to avoid acoustic risks associated with under-protection, as well as safety issues which may arise from over-protection. These are again discussed in Annex A of EN 458. Use of any of the methods for calculating the attenuation provided by the HPD (OBM, HML, SNR) allows an effective level $L'_{p,Aeq}$ to be computed. This is called 'level effective to the ear', but it is a somewhat misleading expression, as it seems to imply that this is what would be measured by positioning a microphone at the ear, downstream from the HPD. Because of the methods employed in the lab, attenuation is quantified not as a transmission loss, but rather as an insertion loss; accordingly, any 'effective sound pressure level at the ear' is actually the external sound pressure level (at 15–40 cm from the ear) that would give the same pressure at the ear as the actual noise when attenuated by the HPD. The effective level to the ear can be assessed based on a table shown in EN 458, here replicated as Table 3.2.

   This table does not provide any absolute reference, since all thresholds are given relative to the value L'_{NR} supposedly specified by national regulations. EN 458 does, however, clarify in Section 3.6.2 that the level when attenuation is taken into account must be 'ideally between 75 and 70 dB(A)', so a reasonable conclusion is that $L'_{NR} = 80$ dB(A).

TABLE 3.2 Assessment of the Effective Level to the Ear

Level Effective to the Ear ($L'_{p,A,eq}$ in dB)	Protecting Rating
Greater than L'_{NR}	Insufficient
Between L'_{NR} and $L'_{NR}-5$	Acceptable
Between $L'_{NR}-5$ and $L'_{NR}-10$	Good
Between $L'_{NR}-10$ and $L'_{NR}-15$	Acceptable
Less than $L'_{NR}-15$	Risk of over-protection[a]

[a] Sound attenuation may be too high and speech intelligibility could be hindered; check for communication and acoustical isolation.

5. Methods for assessing the sound attenuation of a hearing protector for impulsive sounds.

A first important distinction is made between impulses where $L_{pC,peak}$ is below or above 140 dB(C).

If $L_{pC,peak} < 140$ dB(C), it is recommended that a procedure be used which basically ignores the impulsive nature of sound. Indeed, the recommended procedure is the same as the one given in the standard's Annex A, adopted to quantify the HPD performance regarding the equivalent continuous sound pressure level (section 5.3.2).

If on the opposite $L_{pC,peak} > 140$ dB, one should use a specific procedure which takes into account the impulsive nature of sound. This procedure is outlined in the standard's Annex B and is aimed at calculating the 'effective' peak sound pressure level L'_{peak} by means of the following steps:

a. L_{peak} of the noise is measured.

b. The noise type (1, 2, or 3) is determined based on the source characteristics.

c. The HPD attenuation to the peak sound pressure level, called d_m, is determined as a function of the noise type identified in the previous step.

d. The effective peak sound pressure level at the ear $L'_{peak} = L_{peak} - d_m$ is calculated.

If L'_{peak} is below the national peak regulation level $L_{NR,pk}$, then the hearing protector is considered adequate. The national peak regulation level $L_{NR,pk}$ is defined as the peak pressure level effective to the ear according to national regulations. By analogy with the effective level L'_{NR} specified by national regulations for continuous noise, set equal to the lower action value of 80 dB(A), $L_{NR,pk}$ can be equal to 135 dB(C).

An additional important point made by the standard is that if $L_{peak} > 140$ dB, the equivalent sound pressure level L_{Aeq} is presumably largely determined by impulses. Accordingly, the performance of the HPD should be quantified using the procedure developed for impulses, even with respect to L_{Aeq}. Hearing protector performance is determined using the quantity d_m, and it is considered adequate if $L'_{Aeq} = L_{Aeq} - d_m$ is below the national regulation level L_{NR}.

A second important point is the following: there is no indication in EN 458 regarding the frequency weighting of the peak level, so L_{peak} should be assumed to be unweighted. EN 458 also states that adequacy of the HPD is assessed by comparing L'_{peak} with the national regulation level $L_{NR,pk}$. Since $L_{NR,pk}$ is a C-weighted value, in order to make this comparison possible, L_{peak} must also be measured in dB(C).

3.6.2 What's Out

The most relevant issue that is not discussed in EN 458 is 'real world attenuation' of HPDs. The estimate of real world attenuation values for HPDs is a long-standing issue that has been debated in a vast number of studies (see the list of HPD-related studies by Berger 2011). HPDs commercialised in Europe are required to comply with requirements established by EN 458. In its Annex A, EN 458 explicitly states that attenuation values must be measured according to EN ISO 24869-1. This latter standard adopts a method for the quantification of attenuation based on differential thresholds reported by tested subjects. This method, which therefore belongs to the family of methods called real-ear attenuation at threshold (REAT), is known as 'experimenter fit' and provides estimates of attenuation close to the ideal performance. These values are never achieved in actual workplaces, where workers have neither the time nor the education nor the motivation to go through a careful routine for accurately positioning the HPD on/in their ears each and every time. In order to get more realistic estimates of attenuation to be used in the context of noise exposure reduction, NIOSH (1998) has proposed the use of 'derated' values, which can be obtained by multiplying nominal values by 25% for earmuffs, 50% for formable earplugs, and 75% for all other earplugs. The NIOSH method was originally meant to be applied to NRR values, which are qualitatively similar to SNR values with some quantitative difference. In consideration of the small conceptual distance between the two descriptors, their extension to SNR has been proposed. Other institutions have followed less articulated approaches. As an example, the United States Department of the Navy (2013) recommends that labelled NRR values are first cut by 7 dB to make them adequate for A-weighted levels, and then reduced by 50%. Indeed, a standardised procedure exists (ISO/TS 4869-5, 2006) that specifies a method for measuring noise reduction using 'typical groups of users in real-world occupational settings, who may lack the training and motivation to wear hearing protectors in an optimum manner'. This method has also been included in the more recent U.S. standard ANSI/ASA S12-6 (2016) as method B (as opposed to method A, the traditional experimenter fit). Application of the two methods to several HPD's has been discussed by Murphy et al. (2009). However, adoption of this method has been limited, and the general consensus is that attenuation will continue to be quantified using the traditional experimenter fit in the foreseeable future (Berger 2017).

References

Legislation

Directive 1986/188/EEC Council Directive of 12 May 1986 on the protection of workers from the risks related to exposure to noise at work, O.J. No L 137 of 24.05.1986, page 28.

Directive 2000/14/EC of the European Parliament and the Council of 8 May 2000 on the approximation of the laws of the Member States relating to the noise emission in the environment by equipment for use outdoors. O. J. No L 162 of 03.07.2000, page 1.

Directive 2003/10/EC of the European Parliament and of the Council of 6 February 2003, on the minimum health and safety requirements regarding the exposure of workers to the risks arising from physical agents (noise), O.J. No L 42 of 15.02.2003, page 38.

Directive 2006/42/EC of the European Parliament and of the Council of 17 May 2006 on machinery, and amending Directive 95/16/EC, O.J. No L 157 of 9.6.2006, page 24.

International Technical Standards

EN 458 (2016) Hearing protectors – Recommendations for selection, use, care and maintenance. Guidance document. European Standard Committee, Brussels, Belgium.

IEC 1672-1 (2002) Electroacoustics – Sound level meters – Part 1: Specifications. International Electrotechnical Commission, Geneva, Switzerland.

ISO 1999 (2013) Acoustics – Estimation of noise-induced hearing loss, International Organization for Standardization, Geneva, Switzerland.

ISO 3744 (2010) Acoustics – Determination of sound power levels and sound energy levels of noise sources using sound pressure – Engineering methods for an essentially free field over a reflecting plane, International Organization for Standardization, Geneva, Switzerland.

ISO 3746 (2010) Acoustics – Determination of sound power levels and sound energy levels of noise sources using sound pressure – Survey method using an enveloping measurement surface over a reflecting plane, International Organization for Standardization, Geneva, Switzerland.

ISO 4869-1 (1990) Acoustics – Hearing protectors – Part 1: Subjective method for the measurement of sound attenuation, International Organization for Standardization, Geneva, Switzerland.

ISO 4869-2 (1994) Acoustics – Hearing protectors – Part 2: Estimation of effective A-weighted sound pressure levels when hearing protectors are worn, International Organization for Standardization, Geneva, Switzerland.

ISO/TS 4869-5 (2006) Acoustics – Hearing protectors – Part 5: Method for estimation of noise reduction using fitting by inexperienced test subjects, International Organization for Standardization, Geneva, Switzerland.

ISO 9612 (2009) Acoustics – Determination of occupational noise exposure. Engineering method. International Organization for Standardization, Geneva, Switzerland.

ISO 11204 (1996) Acoustics – Noise emitted by machinery and equipment. Measurement of emission sound pressure levels at a work station and at other specified positions. Method requiring environmental corrections, International Organization for Standardization, Geneva, Switzerland.

ISO 11204 (2010) Acoustics – Noise emitted by machinery and equipment – Determination of emission sound pressure levels at a work station and at other specified positions applying accurate environmental corrections.

ISO 11690-1 (1996) Acoustics – Recommended practice for the design of low-noise workplaces containing machinery – Part 1: Noise control strategies.

ISO 11690-2 (1996) Acoustics – Recommended practice for the design of low-noise workplaces containing machinery – Part 2: Noise control measures.

ISO/TR 11690-3 (1997) Acoustics – Recommended practice for the design of low-noise workplaces containing machinery – Part 3: Sound propagation and noise prediction in workrooms.

National Technical Standards or Technical Documents

ANSI/ASA S12.6 (2016) Methods for Measuring the Real-Ear Attenuation of Hearing Protectors, Acoustical Society of America, Melville, NY.

ANSI/ASA S3.20 (2015) American National Standard Bioacoustical Terminology, Acoustical Society of America, Melville, NY.

EC (2007) Non-binding guide to good practice for the application of Directive 2003/10/EC 'Noise at work' http://ec.europa.eu/social/BlobServlet?docId=4388&langId=en

EU-OSHA (2009) Combined exposure to Noise and Ototoxic Substances, https://osha.europa.eu/it/tools-and-publications/publications/literature_reviews/combined-exposure-to-noise-and-ototoxic-substances.

INRS (2009) Evaluer et mesurer l'exposition professionelle au bruit, in French, http://www.inrs.fr/media.html?refINRS=ED%206035

NIOSH (1998) Criteria for a recommended standard: Occupational noise exposure; revised criteria. National Institute for Occupational Safety and Health, U. S. Dept. HHS, Report DHHS (NIOSH) 98-126, Cincinnati, OH.

OSHA (2013) OSHA Technical Manual – Noise, https://www.osha.gov/dts/osta/otm/new_noise/index.pdf

UNI/TS 11326-2 (2015) Acustica – Valutazione dell'incertezza nelle misurazioni e nei calcoli di acustica – Parte 2: Confronto con valori limite di specifica, in Italian.

United States Department of the Navy, (2013) Bureau of Medicine and Surgery, BUMED Notice 6260.

Scientific Literature

Behar, A., Kunov H. (1999). Insertion loss from using double protection, *Applied Acoustics* 57(4), 375–385.

Berger, E. H. (1983). Laboratory attenuation of earmuffs and earplugs both singly and in combination, *Am. Ind Hyg Assoc J.* 44(5), 321–329.

Berger, E. H. (2001). Extra Protection: Wearing Earmuffs and Earplugs in Combination.

Berger, E. H. (2011). Bibliography on hearing protection, hearing conservation and aural care, hygiene and physiology 1831–2010, E-A-R 82-6/HP Version 28.1, 3 January 2011, http://www.docsford.com/document/4470863

Berger, E. H. (2017). The History of REAT since 1957, and the new ANSI S12.6, http://c.ymcdn.com/sites/www.hearingconservation.org/resource/resmgr/2017_conference/Friday_Podium/Berger_-_T16-22_NHCA_REAT_hi.pdf

Davis, R. I., Qiu, W., Heyer, N. J., Zhao, Y., Yang, Q., Li, N. et al. (2012). The use of the kurtosis metric in the evaluation of occupational hearing loss in workers in China: Implications for hearing risk assessment. *Noise & Health*, 14(61), 330–342.

Goley, G., Song, W., Kim, J. (2011). Kurtosis corrected sound pressure level as a metric for risk assessment of occupational noises. *J Acoust Soc Am*, 129(3), 1475–1481.

Hamernik, R. P., Qiu, W., Davis, B. (2003). The effects of the amplitude distribution of equal energy exposures on noise-induced hearing loss: The kurtosis metric. *J Acoust Soc Am*, 114(1), 386–395.

Kryter, K. D., Ward, W. D., Miller, J. D., Eldredge, D. H. (1966). Hazardous exposure to intermittent and steady-state noise. *J Acoust Soc Am*, 39, 451–464.

Lenzuni, P. (2015). Application of the extreme value distribution to estimate the uncertainty of peak sound pressure levels at the workplace, *Ann Occup Hyg*, 59(6), 775–787.

Murphy, W. J., Byrne, D. C., Gauger, D., Ahroon, W. A., Berger, E. H., Gerges, S. N. Y., McKinley, R., Witt, B., Krieg, E. F. (2009). Results of the National Institute for Occupational Safety and Health – U.S. Environmental Protection Agency Interlaboratory Comparison of American National Standards Institute S12.6-1997 Methods A and B, *J Acoust Soc Am*, 125(5), 3262–3277.

Nies, E. (2012). Ototoxic Substances at the workplace: A brief update, *Arh Hig Rada Toksikol* 63, 147–152.

Nataletti, P., Sisto, R. (2010). Proceedings of the International Workshop 'Synergistic exposure to noise, vibrations and ototoxic substances' Rome, Italy – Università Urbaniana 30th September 2010.

Pettersson, H., Burström, L., Hagberg M., Lundström, R., Nilsson, T. (2014). Risk of hearing loss among workers with vibration-induced white fingers. *Am J Ind Med* 2014; 57(12), 1311–1318.

Qiu, W., Davis, B., Hamernik, R. P. (2007). Hearing loss from interrupted, intermittent and time-varying Gaussian noise exposures: The applicability of the equal energy hypothesis. *J Acoust Soc Am*, 121(3), 1613–1620.

Suter, A. H. (1993). The relationship of the exchange rate to noise-induced hearing loss. *Noise News International*, 1(3), 131–157.

Thiery, L., Ognedal, T. (2008). Note about the statistical background of the methods used in ISO/DIS 9612 to estimate the uncertainty of occupational noise exposure measurements. *Acta Acustica United with Acustica*, 94, 331–334.

Ward, W. D. ed. (1968). Proposed damage risk criterion for impulse noise (gunfire). Washington, D.C: Committee of Hearing. Bioacoustics and Biomechanics, National Academy of Science, National Research Council.

Xie, H. W., Qiu, W., Heyer, N. J., Zhang, M. B., Zhang, P., Zhao, Y. et al. (2016). The use of the kurtosis-adjusted cumulative noise exposure metric in evaluating the hearing loss risk for complex noise. *Ear & Hearing*, 37(3), 312–323.

4

Exposure to Vibration at the Workplace

Lenzuni Paolo
Italian National Workers' Compensation Authority

Tønnes Ognedal
Sinus

4.1 Introduction

Although it has been studied for many decades, it is fair to say that occupational exposure to vibration is still one of the least understood among the physical risk factors. There are three main reasons why this is the case: a) the identification of a simple yet effective descriptor of exposure (the conceptual analogue to L_{EX} for noise) has turned out to be quite complicated because of the presence of many co-factors (e.g. body mass and body mass index for whole-body vibration, grip force, push force for hand-transmitted vibration) along with the two core elements 'intensity of vibration' and duration of exposure; b) apart from the well-known, but admittedly rare Raynaud syndrome (also known as 'white finger syndrome'), exposure to vibration is often ill-correlated with pathological effects. Both objective factors (e.g. posture) and individual susceptibility play a large role, and the resulting dose-effect relationship is quite vague; c) while the cochlea usually can be assumed to be the only target in the case of exposure to noise, there are several possible targets for exposure to vibration, which can be categorised as belonging to the muscle-skeletal system, the neurological system or the vascular system. Data about individual effects have been insufficient to determine individual dose-effect relationships. The single relationship established in the standard, which was meant to be applied to all effects, is clearly inadequate to provide a good picture for any of them individually. Most work has traditionally focused on damage to the vascular system, so presumably this is the effect that the standards are most fitted to deal with.

The number of subjects exposed to high levels of vibration is definitely smaller than the number of those exposed to noise: a study by Palmer et al. (2000) estimates that about 2% of the working population in the United Kingdom is exposed to levels which may imply significant risk.

Similar to what happened with exposure to noise, initial provisions on exposure to vibration were included in general-purpose legislation on safety and health at the workplace. The 1959 and 1961 'Factories Acts' in the United Kingdom, the 1955 'General disposition for occupational hygiene' in Italy and the 1975 Workplace ordinance in Germany are just three examples.

However, it was not until 1993, when the 'Commission proposal for a Council Directive on the minimum health and safety requirements regarding the exposure of workers to the risk arising from physical agents' was issued, that occupational exposure to vibration was aligned to other risk factors, both in terms of public awareness and in terms of legislative tools for risk assessment, evaluation and reduction.

While not specifically focused on exposure to vibration, it nevertheless had a basic structure that has been transposed, unaltered, to all contemporary EC directives and an annex dedicated to vibration, which included recommendations for measurement procedures, risk assessment and action levels.

However, it took 9 years to see this proposal materialise in a EC Directive entirely dedicated to occupational exposure to vibration. With respect to the 1993 proposal (EC 1993), Directive 2002/44/EC has a similar approach but is much more comprehensive, especially with respect to the actions which must be taken to prevent (first) and handle (later) possible risks.

Directive 2002/44/EC is the reference document for risk assessment in occupational exposure to vibration. Three other documents exist that supplement the EU directive by providing additional elements or clarifying points which would otherwise have ambiguous interpretation. These documents are the *EU Guide to Good Practice on Hand-Arm Vibration* (EC 2006), the *EU Guide to Good Practice on Whole-Body Vibration* (EC 2008) and *Workplace Exposure to Vibration in Europe: An Expert Review by the European Agency for Safety and Health at Work* (EU-OSHA 2008). A very comprehensive and detailed summary of the current understanding of human response to vibration can be found in Professor Griffin's *Handbook of Human Vibration* (Griffin 2012). This volume includes a huge amount of information of a technical, medical and normative nature, and represents the optimal starting point for anyone interested in a deeper understanding of the topic of human exposure to vibration.

Similar to what was done for exposure to noise, the structure of this chapter will mirror the formal structure of Directive 2002/44/EC. Most references shall be made to technical standards or to EC good practice documents. Reference to scientific papers shall be made on topics about which knowledge is still limited and issues that are still matters for debate. We will first distinguish in Section 4.2 between hand-arm (or hand-transmitted) vibration and whole-body vibration; exposure limits and action values will be discussed in Section 4.3. Risk assessment will be tackled in Section 4.4, where particular emphasis will be placed on the various elements that must be considered in order to create a comprehensive and truly effective risk assessment document, listed by the 2002/44/EC Directive in its Section 4.4. Finally, methods for the reduction of exposure shall be discussed in Section 4.5, and anti-vibration gloves in Section 4.6.

4.2 Types of Vibration

4.2.1 Hand-Arm Vibration

The first of the two types of vibration that Directive 2002/44/EC deals with is hand-arm vibration. In this case, vibration is initially transmitted from the tool handle to the hand (palm and fingers) and subsequently transmitted up along the forearm and the upper arm. The positive contact between the source of vibration (the tool) and the receptor (the hand-arm system) is one of the main distinctions between exposure to noise and exposure to vibration. The former is largely independent of the subject's behaviour, apart from the possible orientation of the head relative to the sound source, and shows only a moderate dependence on the subject's individual characteristics for a wide range of frequencies and amplitudes. On the contrary, the strength of the grip, the hand-tool contact area and the angles of flexion at the wrist, elbow and shoulder all have significant effects on the quantity (intensity) and quality (frequency spectrum) of hand-arm vibration transmitted into and eventually absorbed by the human body. The subject's physical characteristics also play a significant role.

Vascular, neurological and musculoskeletal disorders have all been found to be associated with exposure to hand-arm vibration. Vascular and neurological disorders are mostly localised in the hand (for a recent review see Nilsson et al. 2017). Musculoskeletal disorders occur throughout the hand-arm system.

While commonly perceived as less incapacitating than disorders traditionally associated with whole-body vibration, the inability to perform work that requires fine motor skills should also be considered, since it has an impact on both the subject's work performance (Gerhardsson and Hagberg 2014) and his or her self-confidence.

4.2.2 Whole-Body Vibration

The second type of vibration that Directive 2002/44/EC deals with is whole-body vibration. This type of vibration is usually transmitted from a vehicle to the lower back of a seated subject, or more occasionally from a platform to the feet of a standing subject. Vibration is then transmitted to the trunk, and, if intensity is strong enough, can propagate to the head and upper limbs (Kelsey et al. 1984, Lan et al. 2016).

Exposure to whole-body vibration shows less inter-individual variability compared to hand-arm vibration, as contact between the exposed subject and the vibrating structure is mostly governed by gravity. However, individual characteristics, such as posture and body mass, have a sizeable influence on the actual interaction of the vibration with the human body. Most involved disorders are of a musculoskeletal nature, but neurological disorders have also been claimed (Prisby et al. 2008).

4.3 Descriptors of Exposure, Measurements and Threshold Values

Directive 2002/44/EC mandates in Annex A.1 and Annex B.1 that both hand-arm and whole-body vibration are assessed using the same descriptor A(8), called daily exposure value. However, substantial conceptual differences exist between the two analytical expressions that allow the calculation of daily exposure values for hand-arm and for whole-body vibration. Significant differences also exist with respect to measurement strategies, instrumentation, action and limit values.

Two separate subsections have accordingly been created, with hand-arm vibration being the topic of Section 4.3.1 and whole-body vibration the topic of Section 4.3.2.

4.3.1 Hand-Arm Vibration

4.3.1.1 Daily Vibration Exposure Value

Apart from setting exposure limit values and exposure action values, which shall both be discussed in Section 4.3.1.11, Directive 2002/44/EC delegates all technical issues to ISO standard 5349-1 (2001), the reference international standard on occupational exposure to hand-arm vibration.

In principle, vibration can be quantified using one of three basic kinematical quantities: displacement, velocity or acceleration. In practice, it is acceleration that is usually measured. Because acceleration is a vector, an accurate description of vibration requires three separate measurements of the acceleration's axial components. No precise understanding of the effect of individual axial components of vibration on the human hand and arm has emerged yet, so ISO 5349-1 merges the information on individual axial components to calculate a single tri-axial descriptor

$$A(8) = \sqrt{\frac{\int\limits_{Te-day} a_{hwx}^2(t) + a_{hwy}^2(t) + a_{hwz}^2(t)dt}{T_0}} \tag{4.1}$$

In Equation 4.1

- a_{hwk} is the time-varying, frequency-weighted acceleration along k axis ($k = x, y, z$);

- $T_{e\text{-}day}$ is the duration of the working day;
- $T_0 = 8$ hours is the reference duration of the working day.

Apart from the absence of logarithms, due to the choice of adopting physical units rather than 'levels' for the descriptors, there is perfect formal analogy between the quantity used to quantify the daily exposure to noise (L_{EX}) and vibration A(8).

Because identical formalism is adopted, exposure to vibration, just like exposure to noise, assumes the validity of the so-called 'equal energy principle', which states that equal amounts of sound energy produce equal amounts of damage, irrespective of the detailed time evolution.

4.3.1.2 Reference Frames

ISO 5349-1 allows two different choices for the coordinate system:

- A 'basicentric' coordinate system, where orthogonal axes are defined with respect to the tool handle or appliance. Here the y axis lies along the handle, while x and z lie in the cross-sectional plane of the handle (see Figure 1 of ISO 5349-1).
- A 'biodynamic' coordinate system, where orthogonal axes are defined with respect to the hand. ISO 5349-1 provides the following description of how the coordinate system should be set: The z_h-axis (i.e. hand axis) is defined as the longitudinal axis of the third metacarpal bone and is oriented positively towards the distal end of the finger. The x_h-axis passes through the origin, is perpendicular to the z_h-axis and is positive in the forward direction when the hand is in the normal anatomical position (palm facing forward). The y_h-axis is perpendicular to the other two axes and is positive in the direction toward the fifth finger (thumb).

Despite going into detail on the biodynamic coordinate system, ISO 5349-1 does not try to enforce it in practice, and seems to accept the fact that the basicentric coordinate system is usually used. This choice may be dictated by practical considerations, since users are unlikely to have the technical ability and the time to correctly identify the biodynamical system. Using a basicentric coordinate system might also result in better reproducibility. However, it fails to provide a true tri-axial description of the hand motion, which would be of critical importance should an appropriate axis-by-axis evaluation and assessment scheme eventually be developed. As discussed in Section 4.3.1.1, this is currently not the case. Equation 4.1 shows that A(8) is the module of the vector whose components are the three frequency-weighted accelerations a_{hwx}, a_{hwy} and a_{hwz}, normalised to an 8-hour nominal exposure, which implies that the choice of the reference frame is irrelevant in the context of the calculation of daily exposure.

Using the vector module as the descriptor of hand-arm vibration also has the advantage of requiring only one frequency weighting to be applied to all three axial accelerations. The selected weighting curve is called W_h and is shown in Figure 1 of ISO 5349-1 in graphical form and Table A.2 in tabular form. W_h is a single-mode curve, has a shallow maximum near 12.5 Hz and remains within 2 dB of the largest value between 6.3 and 25 Hz. Despite the fact that it falls off more slowly at high frequency than at low frequency, it is still steep enough that the weight is down to 1/25 around 400 Hz, which is a de facto upper limit for the range of human effects of hand-arm vibration. As for exposure to noise, the frequency weighting curve is applied to measured data using a specific algorithm, which can be approximated by the simple analytical expression

$$a_{hw} = \sqrt{\sum_i (W_{hi} a_{hwi})^2} \qquad (4.2)$$

where a_{hwi} are individual one-third octave-band accelerations.

The selection of a specific reference frame is of paramount importance in other areas of hand-arm vibration, such as evaluation of the vibration transmissibility of gloves at the palm of the hand. The test code included in the reference document on this topic, ISO 10819, requires a 1-D shaker to provide the desired vibration stimulus, oriented along the subject's forearm. Thus it (implicitly) supports the adoption of the biodynamic reference frame. Anti-vibration gloves shall be discussed in more detail in Section 4.6.

4.3.1.3 Measurement Strategy

ISO 5349-1 does not formally define any measurement strategy comparable to the task-based or job-based strategies of ISO 9612 for exposure to noise. However, this document provides a specific algorithm (Equation 4.3, Section 4.5.3) for cases where the daily exposure is split into I independent tasks, which establishes a de facto task-based strategy for the evaluation of exposure to vibration. If a_{hwki} are the acceleration axial values associated with the i-th task, then

$$a_{hvi} = \sqrt{a_{hwxi}^2 + a_{hwyi}^2 + a_{hwzi}^2} \qquad (4.3)$$

is the vibration total value (the acceleration vector module) of the i-th task ($i = 1$ to I). Equation 4.1 can then be rewritten in discrete form as

$$A(8) = \sqrt{\frac{\sum_{i=1}^{I} a_{hvi}^2 \times T_i}{T_0}} \qquad (4.4)$$

by summing up the contributions from all the I tasks. In Equation 4.4, T_i is the exposure time to the i-th task.

4.3.1.4 Weekly Exposure Value

There is no indication in Directive 2002/44/EC that vibration exposure values averaged over periods longer than one day (e.g. a weekly vibration exposure value) can be used in the context of risk assessment. It is possible that this more conservative approach is due to the poorly established dose-response relationship (s). In particular, the injury potential of intermittent exposure, and the applicability of the equal energy principle of periods longer than one day, are still up for debate. Indeed, ISO 5349-1 goes as far as stating that the equal energy principle has been adopted basically in the absence of better ideas (Annex C.1). ISO 5349-2 is definitely more open-minded on this topic. Its note to Section 4.5.5 and Annex B provide some (limited) guidance on simple algorithms for the calculation of averages over longer periods. It should be noted, however, that the apparent conflict between the two documents vanishes when they are put into the right perspective. The daily exposure value that Directive 2002/44/EC focuses on provides an estimate of the exposure to be used for risk assessment and for implementing methods to reduce and possibly eliminate risks associated with the exposure to vibration. The average over longer periods that ISO 5349-2 discusses is instead useful 'when developing vibration control plans'.

4.3.1.5 Number and Duration of Measurements

The measurement strategy is not dealt with in ISO 5349-1, but it is extensively discussed in its technical counterpart, ISO 5349-2.

As with noise, both long-duration and short-duration measurements are possible. In principle, the result that is obtained from long-term measurement is usually more reliable, as it results from consideration of all work phases, including rest periods, with their actual durations. In practice, however, it is more energy-consuming and costly, and it is definitely more cumbersome, as it forces the worker to have his or her tool wired, and the technician to maintain physical contact with the worker, for extended time periods.

The short-term measurement strategy (ISO 5349-2, Section 4.5.3.c) is therefore much more popular. Being conceptually identical to the task-based strategy for noise assessment, this approach requires that for any task an estimate of the vibration magnitude and an estimate of the exposure time are provided through some kind of sampling.

With regard to acceleration measurements, Section 4.5.4.1 of ISO 5349-2 seems to proceed along two pathways simultaneously:

On one hand, this document provides a set of minimal requirements: at least three measurements should be carried out for each task; the duration of individual measurements must be greater than 8 seconds; and the minimum accumulated measurement time is 1 minute.

At the same time, it recommends that 'many more than three measurements' be taken and that the total sampling time be longer than 1 minute. While offering these two options may be acceptable in a legislative context (see, e.g. the co-existence of lower action values and higher action values), it is somewhat unfortunate in technical standards, where simple and unambiguous procedures should be set out.

It is important to note that the variability shown by three measurements performed consecutively in the same work situation is expected to be much lower than the actual variation over a working day, during which the same tool may be used for different purposes and may act on different materials. It is therefore important that the measurements be taken so as to include as many relevant operating modes of the tool as possible, in order to more precisely characterise the exposure to vibration.

4.3.1.6 Additional Recommendations on How to Measure

Besides the number and duration of measurements, ISO 5349-2 provides hints on how to carry out measurements that lead to a reliable estimate of the acceleration associated with a given task. The following is a list of the most relevant items, including references to the appropriate section of ISO 5349-2:

1. *Accelerometer response*: accelerometers must be able to withstand very high acceleration at very high frequencies while maintaining enough sensitivity to provide accurate measurements of the relatively low acceleration in the relevant range for risk assessment (tens to hundreds of Hz). This is more easily achieved by the interposition between the accelerometer and the tool handle of a mechanical device that filters out very high accelerations.
2. *Accelerometer resonance frequency and mass*: the ideal accelerometer should have a very high resonance frequency (Section 4.6.1.2), which would introduce negligible distortion into the transduced signal, and very small mass (Section 4.6.1.5), in order not to alter the vibration characteristics of the investigated system.
3. *Accelerometer position*: in principle, measurements should be made as near as possible to the entry point of the vibration into the body. This is because the vibration of a complex structure can exhibit substantial variation in different locations, due to the existence of complex vibrational modes that imply rotational as well as translational movements. However, as clearly stated in ISO 5349-2, Section 4.6.1.3, positioning the accelerometer near the middle of the gripping zone will generally interfere with the normal grip used by the operator, which rules this option out unless some kind of ergonomic adaptor is used. Such adaptors that fit under the hand are an attractive option, because they have been designed so that they can also be used inside gloves and can therefore be used to measure glove transmissibility. Care must be taken, however, to make sure such adaptors do not have any structural resonance which would introduce unwanted artifacts into the measured vibration.
4. *Accelerometer mounting*: another critical issue, extensively discussed in ISO 5349-2, Section 4.6.1.4. The bottom line is that accelerometers should be rigidly attached to the vibrating surface. Any non-rigid mounting would imply some vibration of the accelerometer with respect to the tool surface, with associated distortion of the original signal.

4.3.1.7 Mean Acceleration

Each acceleration axial value for a specific task is calculated as the time-weighted root mean square of the acceleration values found in the J individual measurements carried out

$$a_{hw} = \sqrt{\frac{\sum_{j=1}^{J} a_{hwj}^2 \times T_j}{T}} \qquad (4.5)$$

where *T* is the total measurement time. In principle this formulation is attractive, as it provides the same result that would be found in a single measurement of duration *T*. In practice, however, this makes the user's life more complicated for two reasons:

1. Care must be taken to keep track not only of acceleration values but also of measurement times. This additional information is not requested in noise exposure evaluation procedures, where the root mean square of the measured pressure values is not time-weighted (see ISO 9612, Equation 4.7). It is therefore hardly justifiable in the context of vibration exposure, where knowledge is certainly more incomplete and numerical subtleties should be kept at a minimum.
2. The calculation of associated uncertainty becomes much more involved. While it is true that Directive 2002/44/EC makes no explicit mention of the concept of uncertainty in the context of risk assessment, measurement uncertainty is nevertheless a valuable quantity to compare different estimates and to quantify the measurement's reliability. The existence of a detailed and elegant formulation for the estimate of the uncertainty of the noise equivalent continuous sound pressure level and noise exposure level (ISO 9612, Annex C) would represent an excellent basis, which could be transposed to the field of vibration with minimal effort, should the formal expression for averaging individual measurements be the same. Thus it is unfortunate that its calculation is made unnecessarily messy.

4.3.1.8 Uncertainty

Directive 2002/44/EC does not indicate any role for uncertainty in the risk assessment process. However, very general rules on establishing compliance with legal limits require that uncertainty should always be taken into account. As previously discussed, limited effort has been made in ISO 5349-1 and ISO 5349-2 to allow a simple analytic calculation of uncertainty. This appears to be due to the existence, due to factors such as instrumental performance, incomplete sampling and exposure duration, which have been accommodated in the formalism developed for exposure to noise, of vibration-specific contributions to the uncertainty which are difficult or impossible to quantify, such as

- Calibration;
- Electrical interference;
- Mounting of accelerometers;
- Mass of accelerometers;
- Location of accelerometers.

Finally, elements of significant uncertainty are brought about by unpredictable changes in

- Normal operation of the power tool and changes to hand posture and applied forces brought about by the measurement process (i.e. mounting of accelerometers and associated cables);
- The operator's method of working, as a response to being the subject of the measurement;
- The maintenance of power tools and accessories (e.g. replacing the worn wheel of a grinder with a new one may dramatically change the vibration transmitted to the operator);
- Posture and applied forces;
- Characteristics of the materials being processed.

Having ruled out the possibility of an analytical calculation, ISO 5349-2 just quotes a range (20% to 40%) for global uncertainty on the daily exposure value. Thus it would appear that a fair estimate of the relative uncertainty, to be used when assessing compliance of the calculated daily vibration exposure value with both the limit and the action values given in Directive 2002/44/EC, is $\pm 30\%$. However, other sources (e.g. EC 2006) indicate an asymmetric range for the uncertainty ('as much as 20% above the true value to 40% below'). Because it is the upper limit of the confidence interval that is relevant when assessing compliance to a legal limit (e.g. DIN 45660-2), this would imply that the reference value should be 20% rather than 30%.

A comprehensive discussion of uncertainty in vibration, including both estimates derived from measurements and estimates derived from laboratory tests, is presented in the German standard DIN 45660-2. Annex A of this document includes a very detailed analysis of uncertainty for exposure to hand-arm vibration. An extended set of contributions is discussed, with both mandatory terms given by sampling, instrumentation, choice of measuring points, coupling and optional terms for inter-subject variability and application of results to comparable workplaces that can be added if required by the use made of measurements.

4.3.1.9 Exposure Time T_m

Equation 4.5 requires that the duration T_i of a task (i.e. the exposure time) must also be determined along with the acceleration. ISO 5349-2 presents different options in this respect, including:

1. Observing and measuring durations during noise measurements;
2. Gathering information regarding operation of typical sources (e.g. work processes, machines, activities at the workplace and in its surroundings);
3. Interviewing the workers and the supervisor.

Measurements of the actual exposure time can be made over a complete work cycle, and several techniques are possible, including the use of a stopwatch, the use of a dedicated data logger linked to power tool usage and the analysis of video recordings. Estimates based on interviews with workers may be less reliable because of the confusion between operation time (time when the tool actually vibrates) and contact time (time when the tool is handled by the worker, including pauses in tool operation). The acceleration measured during operation time is higher than the acceleration measured during contact time. This is, in principle, taken care of by adjusting the exposure time, which is obviously longer in association with trigger time than in association with working time. However, this is easier said than done, as the working process varies from situation to situation and from person to person, so that the duration of the pauses and the actual trigger time will vary correspondingly. This is another factor in uncertainty evaluation.

Once the real nature of the information has been explicated and any ambiguity has been removed, this estimate can be used in association with the appropriate measurement of the acceleration and a correct determination of the exposure shall be obtained.

There is no hint on how to average different estimates of the measurement time, so it can be implicitly assumed that the task duration is best estimated using the arithmetic mean

$$T_m = \frac{\sum_{j=1}^{J} T_{m,j}}{J} \tag{4.6}$$

similar to the ISO 9612 recommendation for exposure to noise.

4.3.1.10 Shocks and Peak Values

Despite the fact that no mention is made in Directive 2002/44/EC of 'peak' or 'ceiling' limit values in the context of the occupational exposure to vibration, this is a topic that deserves some attention.

Several tools generate strong vibration impulses or bursts whose actual injury potential should be carefully scrutinised. However, the vast majority of the scientific evidence on which ISO 5349-1 is based deals with exposure to periodic and random, or non-periodic, continuous vibration. As such, this standard is intrinsically unsuitable for application to the exposure to multiple single shocks or transient vibration (e.g. fastening tools). In the absence of customised methods developed for these cases, ISO 5349-1 allows that methods indicated therein may be applied to repeated shocks and transient vibration, but this should be done with caution and indicated in the information to be reported. ISO 5349-2 also states that such methods 'may not be adequate and underestimate the severity of shock exposure'. Under similar

TABLE 4.1 Threshold Values

Name	Threshold
Daily exposure action value	$2.5\,\mathrm{ms}^{-2}$
Daily exposure limit value	$5\,\mathrm{ms}^{-2}$

circumstances, it is unclear to what extent the methods of measurement and the limits of acceptability outlined in these technical standards can be reliable for application to multiple shocks or transient vibration, and how the concept of caution should be transposed into practice.

4.3.1.11 Thresholds

Article 3 of Directive 2002/44/EC introduces two thresholds, here displayed in Table 4.1. The exposure action value is the threshold above which employers are required to control the hand-arm vibration risks of their workforce. As such, it should represent a threshold below which damage due to hand-arm vibration is rare. ISO 5349-1 states that symptoms of the hand-arm vibration syndrome are rare in persons exposed with an 8-hour energy-equivalent vibration total value, A(8), at a surface in contact with the hand, of less than $2\,\mathrm{ms}^{-2}$ and unreported for A(8) values of less than $1\,\mathrm{ms}^{-2}$. However, some recent documents argue that a level around $2.5\,\mathrm{ms}^{-2}$ is still not a safe level of exposure.

The exposure limit value is the threshold above which workers must not be exposed. It has been criticised for being a difficult challenge for some sectors of industry (HSE 2010). However, based on the dose-response relationship presented in ISO 5349-1, enforcing this limit should keep the probability of developing the most severe form of vibration-induced disability, Raynaud syndrome, below 0.1 after 5 years of continuous exposure at this level. This should ensure that any early sign of finger blanching can be detected by appropriate medical surveillance, thus minimising the risk of a more severe disability.

These action values and exposure limit values set by 2002/44/EC are identical to the action level and maximum level which appeared in the 1993 Commission proposal for a Council Directive on the minimum health and safety requirements regarding the exposure of workers to the risk arising from physical agents (EC 1993). Interestingly, the 1993 proposal also included two additional threshold values that were subsequently abandoned:

1. A $1\,\mathrm{ms}^{-2}$ target value for A(8);
2. A $20\,\mathrm{ms}^{-2}$ short-term value not to be exceeded during short exposures (a few minutes).

The former has been abandoned, presumably being perceived as utopic in practice. The latter has re-emerged to be included in the current Italian legislation on occupational exposure to vibration.

4.3.2 Whole-Body Vibration

4.3.2.1 Daily Vibration Exposure Value

Directive 2002/44/EC delegates all technical issues to ISO standard 2631-1 (1997), the reference international standard on occupational exposure to whole-body vibration.

Similar to hand-arm vibration, whole-body uses acceleration to quantitatively describe vibration magnitude. Individual measurements along the three orthogonal axes of a Cartesian reference frame are required. As with hand-arm vibration, no precise understanding of the effect of individual axial components on the human body has emerged yet. Where whole-body and hand-arm methods of risk assessment diverge is in the choice of the spatial dimensionality of the descriptor of exposure. While ISO 5349-1 is firm in advocating the use of a tri-axial descriptor, ISO 2631-1 recommends that exposure be evaluated by means of the largest of three quantities $A_l(8)$

$$A(8) = \max \left[A_l(8)\right] = \max \left[k_l \frac{a_{wl}}{T_0^{1/2}}\right] \tag{4.7}$$

When vibration in two or more axes is comparable, ISO 2631-1 suggests that the vector sum may be used to estimate health risk. However, given that the typical conversion factor from single axis to tri-axial acceleration has been estimated to be around 1.4 (Nelson 1997), this leads to substantial ambiguity in establishing compliance with limits. Common practice has largely ignored this statement by sticking to single-axis descriptors for the assessment of risks.

In Equation 4.7

- a_{wl} is the time-averaged frequency-weighted acceleration along axis l ($l = x, y, z$)
- $T_0 = 8$ hours is the reference duration of the working day
- k_l is the axial weight of axis l. For health-related effects, the following values are adopted: $k_x = 1.4$, $k_y = 1.4$ and $k_z = 1$. Other sets of values are recommended by ISO 2631-1 to be applied for the assessment of comfort and perception ($k_x = 1$, $k_y = 1$ and $k_z = 1$). Directive 2002/44/EC does not consider issues related to comfort and perception, so these alternative axial weighting factors will not be considered further in this work.

As previously discussed in the context of Hand-Arm vibration, there is a strong formal analogy between noise and vibration, A(8) being the conceptual analogue of L_{EX} and a_w the conceptual analogue of the equivalent continuous equivalent level L_{Aeq}.

Because identical formalism is adopted, exposure to vibration, just like exposure to noise, assumes the validity of the so-called 'equal energy principle', which states that equal amounts of vibrational energy produce equal amounts of damage, irrespective of the detailed time evolution of the exposure. It should be noted, however, that, unlike ISO 5349-1 for hand-arm vibration, ISO 2631-1 does not allow the equal energy principle to be applied all the way up to very high vibration for very short times, restricting the field of application above 10 minutes. Accordingly, exposure characterised by acceleration values in excess of about 3.5 ms^{-2} do not comply with the action value, regardless of the exposure time; similarly, exposure characterised by acceleration values in excess of about 5.5 ms^{-2} do not comply with the exposure limit value, regardless of the exposure time.

4.3.2.2 Reference Frames

ISO 2631-1 recommends that a basicentric coordinate system is used, whose orthogonal axes are defined as follows:

- X in the fore-aft direction (longitudinal direction)
- Y in the left-right direction (lateral direction)
- Z in the up-down direction (vertical direction)

Figure 1 of ISO 2631-1 shows that the same coordinate system applies to both the more common seated position of the exposed subject and the less common standing position.

Because whole-body vibration is evaluated and assessed along individual directions, it makes sense that frequency weighting is carried out by using different curves along different axes. Indeed, two frequency weightings are applied by ISO 2631-1 and shown in its Figure 2, W_k for the vertical Z direction and W_d for the horizontal X and the Y direction. Both frequency weighting curves share the same bell shape, with low values at very low and very high frequencies and a broad maximum at intermediate frequencies. The main difference between the two is the definition of 'intermediate' frequencies: the W_d curve remains within 2 dB of the maximum between 0.5 Hz and 2 Hz, more than two octaves below the W_k curve, which remains within 2 dB of the maximum between 3.15 Hz and 12.5 Hz. Like the W_h curve for hand-arm vibration, W_d and W_k also drop off more slowly at high frequency than at low frequency. A rough upper limit for the range of interest (here set at a weight of 1/25) is 50 Hz for horizontal vibration and 150 Hz for vertical vibration. A second significant difference is that while the W_d curve has a single mode, the W_k curve has two modes (the secondary mode lies at lower frequencies than the quoted main mode), so the overall shape is much broader than that of W_d.

Measurements on the seat backrest are also recommended (ISO 2631-1, Section 4.7.2.2) but are not included in the actual algorithm for the evaluation of A(8) due to the paucity of scientific support.

It is interesting to note that in the assessment of comfort and perception, ISO 2631-1 recommends a tri-axial descriptor. Because axial weighting factors for comfort assessment are all equal to one in this context, this implies that the descriptor of exposure is the acceleration vector module. Effects on comfort are outside the scope of Directive 2002/44/EC and will not be discussed further.

4.3.2.3 Measuring Strategy

ISO 2631-1 does not formally define any measurement strategy comparable to the task-based or job-based strategies of ISO 9612 for noise. However, it provides a specific equation (Equation 4.3, Section 4.5.3) for the case where the daily exposure is split into I independent tasks, which establishes a de facto task-based strategy for the evaluation of exposure to vibration. If a_{wli} are the acceleration axial values for the i-th task along axis l, then Equation 4.7 can be rewritten in discrete form as

$$A(8) = \max\left[k_l \sqrt{\frac{\sum_{i=1}^{I} a_{wli}^2 \times T_i}{T_0}} \right] \qquad (4.8)$$

In Equation 4.8 T_i is the exposure time to the i-th task, while the other quantities have already been defined.

4.3.2.4 Weekly Exposure Value

As previously discussed in Section 4.3.1.4, there is no indication in Directive 2002/44/EC that a vibration exposure value averaged over periods longer than one day (e.g. a weekly vibration exposure value) can be used in the context of risk assessment. EN 14253 is explicit in requesting a daily average, since 'it is not acceptable to determine A(8) by averaging over several days on which different levels of variation exposure have occurred'. Accordingly, the daily exposure value is currently the only acceptable descriptor.

4.3.2.5 Number and Duration of Measurements

Measurement duration is not specified in ISO 2631-1. Some indications are given with respect to the minimum duration required to taper statistical fluctuations of a random signal down to a limiting frequency, which leads to a duration of about 4 minutes for a good accuracy down to the one-third octave band centred at 0.5 Hz. This would seem to be a good argument for setting a minimum duration of a few minutes. However, ISO 2631-1 goes on to state that the measurement duration is 'usually much longer', but with no real quantitative indication. More useful elements are given in the European standard EN 14253, where a minimum duration of 3 minutes and a total measurement time of at least 15 minutes are recommended. The 2008 *Good Practice Guide* (EC 2008) also recommends a total measurement time of 20 minutes.

The number of measurements is not dealt with in ISO 2631-1. Nor is any indication provided by the *EC Guide to Good Practice* (EC 2008) or by EN 14253, which merely states that 'the number of sample measurements made shall be sufficient to show that the average value obtained is representative of the vibration occurring throughout the day'. Under these circumstances, two options appear viable:

1. Considering that Directive 2002/44/EC does not make reference to any possible role of uncertainty in vibration risk assessment, only one measurement (possibly extended to 15 minutes) is carried out

for any one exposure condition. This does not allow any statistical analysis and prevents any quantification of uncertainty, but does not violate any rule set by either legislative or technical standards.

2. If quantification of uncertainty is desired, then at least three measurements must be carried out, and some simple statistical analysis can be carried out along the lines established by ISO 9612 for exposure to noise. In this latter case, three individual 5-minute measurements should be considered adequate.

4.3.2.6 Additional Recommendations on How to Measure

The document that provides the most detailed indications on the procedure to follow in order to make reliable measurements is EN 14253. The following is a list of the most relevant items:

1. *Accelerometer response, resonance frequency and mass*: Indications are not as quantitative as those given for measurements of hand-arm vibrations. Only a general request is made of compliance with the requirements of ISO 8041.
2. *Accelerometer position*: As for hand-arm vibration, the basic idea is to measure as close as possible to the entry point of the vibration into the body. This is usually an achievable condition in subjects exposed while sitting on a vehicle seat, by positioning the accelerometer between the seated subject and the seat, near the bottom end of the spine.
3. *Accelerometer mounting (Section 4.6.1.3)*: While measurements of hand-arm vibration are usually carried out by positioning the accelerometer on a rigid surface, this is unfeasible in measurements of whole-body vibration, where the rigid frame of the seat cannot be exposed. Neither would it make sense to take measurements upstream of the damping mechanisms that operate in a suspended seat. In general there is no need to attach the disc on the seat, since the weight of the operator is adequate to hold it down in the range of frequencies being measured. Under particularly severe conditions, in which the driver might make unwanted lateral movements on the seat, the disc may be held in place with adhesive tape (EN 14253, Section 6.1.3.2).

4.3.2.7 Mean Acceleration

Each acceleration axial value for a specific task is calculated as the time-weighted root mean square of the values found in the J individual measurements carried out (EN 14253, Equation 4.3)

$$a_w = \sqrt{\frac{\sum_{j=1}^{J} a_{wj}^2 \times t_j}{T}} \tag{4.9}$$

where t_j is the duration of the j-th measurement and T is the total measurement duration. As already discussed in the section on hand-arm vibration, in principle this formulation is correct, as it provides the same result that would be found in a single measurement of duration T. In practice, however, the calculation of associated uncertainty becomes much more complicated. The German standard DIN 45660-2, which outlines the full path for the calculation of uncertainty in exposure to vibration, does not calculate the average acceleration according to the time-weighted root mean square formalism shown above, but using a simple arithmetic mean. The actual difference between the two estimates is usually insignificant compared to the large magnitude of the uncertainties due to many sources (see the following paragraph). It is nevertheless annoying that the most authoritative technical documents on exposure to vibration do not agree on such a fundamental issue as the algorithm to compute the mean of a series of measurements.

4.3.2.8 Uncertainty

The picture of uncertainty quantification is even more fuzzy for whole-body vibration then it is for hand-arm vibration. Most technical documents (EC 2008, EN 14253) just include a generic list of elements that may contribute to the uncertainty, such as instrument accuracy, calibration, electrical interference, mounting of accelerometers, location and orientation of accelerometers, changes from the normal operation of the machine and body posture brought about by the measurement process (e.g. presence of accelerometers and associated cables), changes in the operator's method of working as a result of being the subject of measurement, changes in the condition of the machine and equipment (e.g. changes in tyre pressure with temperature) and changes in the characteristics of the travelling surface. No tentative estimate of a typical uncertainty value or range is provided.

A much more in-depth analysis of uncertainty in whole-body exposure to vibration can be found in the German standard DIN 45660-2. As indicated in the section on hand-arm vibration, this document includes a detailed procedure for an analytical estimate of the uncertainty. This is certainly a powerful and flexible tool, but since it requires substantial effort to implement it, it might discourage most potential users, with the end result of excluding uncertainty from the risk assessment procedure for whole-body vibration.

4.3.2.9 Exposure Time T_m

Equation 4.8 requires that the duration T_i of a task (i.e. the exposure time) must also be determined along with the acceleration. ISO 2631-1 provides no clue on how to make a reliable estimate of the exposure time. In order to fill this void, EN 14253 simply transposes to whole-body vibration the methods recommended by ISO 5349-2 for hand-arm vibration:

1. Observing and measuring durations during noise measurements;
2. Gathering information regarding operation of typical sources (e.g. work processes, machines, activities at the workplace and in its surroundings);
3. Interviewing the workers and the supervisor.

There is no clear indication of how to average different estimates of the measurement time, so it can be implicitly assumed that the task duration is best estimated using the arithmetic mean

$$T_m = \frac{\sum_{j=1}^{J} T_{m,j}}{J} \tag{4.10}$$

similar to what is done for noise and for hand-arm vibration.

4.3.2.10 Shocks and Peak Values

Exposure to repeated acceleration bursts and shocks is one of the few components of vibration that is clearly at a more advanced stage in whole-body vibration than it is in hand-arm vibration, in terms of both scientific knowledge and inclusion in technical documents.

Following the recognition that consistent scientific evidence supporting the validity of the equal energy principle is very thin, an alternative descriptor of exposure to whole-body vibration, called Vibration Dose Value, or VDV,

$$VDV = \left(\int_{Te} a_{hwx}^4(t) + a_{hwy}^4(t) + a_{hwz}^4(t)dt \right) \tag{4.11}$$

has been developed (Griffin 2012), and was eventually incorporated into Directive 2002/44/EC.

Equation 4.11 shows that VDV is based on a combination of acceleration and time ($a^4 \times t$) different from the combination of A(8) ($a^2 \times t$). The larger exponent of acceleration implies a higher sensitivity to short acceleration bursts, so VDV appears to have been developed specifically for use in cases where significant evidence of impulsive vibration exists. However, Directive 2002/44/EC does not indicate that VDV should be preferentially or exclusively used under certain conditions. It just provides two alternative metrics for the action value and limit value, A(8) and VDV. This is unfortunate since, as indicated in EN 14253, the use of VDV instead of A(8) generally results in a different evaluation of the health risk.

Luckily, ISO 2631-1 eliminates any possible ambiguity on this matter, as it recommends that VDV be used alongside A(8) for the assessment of impulsive vibration, where impulsivity is quantified by a ratio of peak-to-r.m.s. acceleration (known as the 'crest factor') larger than 9.

Additionally, it should be noted that, unlike A(8), VDV is not a mean acceleration, as it incorporates the exposure duration by performing an integral over time. Therefore, as clearly pointed out by the *Good Practice Guide* (EC 2008), if the VDV is measured over a period that is shorter than the full working day, then the resulting measurement must be scaled up.

4.3.2.11 Thresholds

Article 3 of Directive 2002/44/EC introduces two thresholds, here displayed in Table 4.2.

The exposure action value for daily vibration exposure is the threshold above which employers are required to control whole-body vibration risk to their workforce. The exposure limit value is the threshold above which workers must not be exposed.

Unlike the situation discussed for hand-arm vibration, the daily exposure limit value is not identical to, and indeed is much lower than, the 'maximum level' set by the EC 1993 proposal at $0.7 \, \text{ms}^{-2}$. This very large change may reflect a better understanding of the human response to vibration or may be induced by some pressure applied to moderate costs, which otherwise would have been very large for some economy sectors (or possibly both).

As previously discussed, the EC 1993 proposal also included two more threshold values that were lost in time:

1. A $0.25 \, \text{ms}^{-2}$ target value for A(8);
2. A $1.25 \, \text{ms}^{-2}$ maximum value not to be exceeded during any 1-hour interval.

The former has been abandoned, presumably because it was perceived as being utopic in practice. The latter has instead re-emerged to be included in the current Italian legislation on occupational exposure to vibration scaled up to $1.5 \, \text{ms}^{-2}$.

A final proof of the magmatic nature of action and exposure limit values for whole-body vibration is provided by the comparison of values included in Directive 2002/EC/44 with those given in ISO 2631-1.

The values given by Directive 2002/44/EC are summarised in Table 4.2. As previously noted, ISO 2631-1 does not include limit values in textual or tabular form. However, a lower value of about $0.45 \, \text{ms}^{-2}$ and an upper value of about $0.8 \, \text{ms}^{-2}$ can be inferred from its Figure B.1 for an 8-hour exposure. Thus, despite being explicit on the fact that all the exposure descriptors should be 'determined in accordance with Chapters 5, 6 and 7, Annex A and Annex B to ISO Standard 2631-1(1997)', there is no agreement on such a fundamental issue as the numerical values of threshold values.

TABLE 4.2 Threshold Values

Name	Threshold	
	A(8)	VDV
Daily exposure action value	$0.5 \, \text{ms}^{-2}$	$9 \, \text{ms}^{-1.75}$
Daily exposure limit value	$1.15 \, \text{ms}^{-2}$	$21 \, \text{ms}^{-1.75}$

4.4 Elements That Must Be Taken into Consideration in the Process of Risk Assessment

Articles 4.1 and 4.2 discuss two themes which are at the heart of vibration risk assessment.

Article 4.1 clarifies that, similar to what happens for the exposure to noise, 'the employer shall assess and, if necessary, measure the levels of mechanical vibration to which workers are exposed'. There are many circumstances in which a quantitative estimate of the vibration magnitude may be unnecessary, when the absence of significant sources of vibration is obvious. Only when the presence of significant vibration cannot be ruled out must some estimate of the vibration magnitude be collected.

Article 4.2 indicates three different ways to estimate the vibration magnitude (improperly called 'level'):

1. Direct measurements;
2. References to existing information on the probable magnitude of the vibration associated with the equipment used in the specific conditions of use;
3. Information provided by the manufacturer of the equipment.

Option 1 has been extensively discussed in Section 4.3.1 for hand-arm vibration and in Section 4.3.2 for whole-body vibration.

Acceleration values obtained through option 2 are much less reliable, since they have been found under operating conditions that differ from the one under investigation, both with respect to the specific action carried out with the tool/vehicle and with respect to the operator. Such conditions are often poorly laid out, so that their proximity to the conditions of interest is usually unknown. Use of these values should be restricted to cases where the objective is to decide whether the exposure action value or the exposure limit value is likely to be exceeded, but no accurate estimate of the exposure is needed.

Option 3 involves the use of results collected for use in compliance with the Machine Directive that mandates declaration of acceleration values in its Section 3.6.3.1. Since they have all been collected under the same rigidly set standardised test, declared emission values allow purchasers to perform an objective comparison among machines of the same kind. This comparison is a valuable tool that can lead to the selection of low-emission vibrating tools. The use of such values in the context of the evaluation of occupational exposure to vibration should, however, be limited to cases where options 1 and 2 are not viable. Should this option be pursued, declared factors should be corrected using the multiplicative factors indicated in the European standard CEN/TR 15350. Multiplicative factors range between 1 and 2.

In the rest of this section, the many elements that contribute to an accurate risk assessment shall be presented and discussed following the same order in which they appear in Directive 2002/44/EC, Article 4.4.

a. *The level, type and duration of exposure, including any exposure to intermittent vibration or repeated shocks:* Exposure to intermittent vibration is much more common than exposure to intermittent noise. Indeed, the vast majority of tasks which involve the use of vibrating tools consist of repeated short units (a few seconds to a few minutes) of work. Accordingly, it is possible to speculate that existing limit values have been set based on evidence collected on workers mostly exposed to intermittent vibration. The question that arises is therefore inverted in comparison to exposure to noise: in the latter one may ask whether exposure to intermittent noise might result in less damage compared to exposure to continuous noise of the same total energy. Here we ask whether exposure to continuous vibration might result in higher damage. Mixed evidence has been collected in this sense by the EU-sponsored study VIBRISK. Although some evidence from this study is consistent with intermittent hand-arm vibration having a less severe effect than continuous vibration, this evidence is not yet conclusive (Bovenzi and Griffin 2006). With regard to whole-body vibrations, the topic of intermittent vs. continuous exposure has been investigated by Seidel et al. 1980 and Seidel and Heide 1986.

b. *The exposure limit values and the exposure action values laid down in Article 3 of this Directive:* Directive 2002/44/EC makes no reference to the concept of comparing the results of measurements (or acceleration values found through other sources) to the exposure limit values and the exposure action values. Because no meaningful comparison could be carried out without taking into account the uncertainty on exposure, and uncertainty never shows up in the Directive's text, this is somehow intrinsic to the approach chosen by the Directive.

The *Guide to Good Practice for Hand-Arm Vibration* and the *Guide to Good Practice for Whole-Body Vibration* (EC 2006, EC 2008) also ignore this issue. It is fair to conclude that from a strictly legal viewpoint, taking exposure limit values and action values into consideration is tantamount to checking if the face value of A(8) is above or below such values. If a more rigorous method is needed, then one can proceed along the pathway laid out for exposure to noise, which implies that compliance is established if the inequality

$$A(8) + U(A(8)) < \text{Threshold} \qquad (4.12)$$

is fulfilled. The expanded uncertainty $U(A(8))$ is best calculated using the methods outlined in the German standard DIN 45660-2, as previously discussed in Sections 4.3.1.8 and 4.3.2.8.

c. *Any effects concerning the health and safety of workers at particularly sensitive risk:* As with exposure to noise, the identification of workers belonging to particularly sensitive risk groups is not a technical issue, but a clinical issue that falls under the jurisdiction of the medical surveillance team. It is the medical doctor's responsibility to check whether a worker displays some kind of peculiar sensitivity. This information should then be passed on to the employer, so that he/she can take appropriate action to ensure that this worker is given additional and customised protection from vibration.

d. *Any indirect effects on worker safety resulting from interactions between mechanical vibration and the workplace or other work equipment:* When this item (4d) of the vibration Directive is compared with the corresponding item (4d) in the noise Directive (2003/10/EC), a couple of interesting elements show up: 1) the noise Directive includes the words 'as far as technically achievable', which are absent in the vibration Directive; 2) the noise Directive requests that the effects on health, and not only on safety, are taken into consideration; 3) the noise Directive makes explicit reference to the effects of combined exposure to noise and vibration, while the vibration Directive simply quotes 'interactions between mechanical vibration and the workplace'.

Because the noise Directive (2003/EC/10) was issued several months after the vibration Directive (2002/44/EC), it is possible that it benefits from some additional expertise that led some concepts to be better phrased. In this perspective, combined exposure to noise and both hand-arm vibration and whole-body vibration should be identified as having the primary indirect effect on health. As previously discussed in the chapter on exposure to noise (Chapter 3), scientific studies (see Pettersson et al. 2014 for a recent review) suggest the existence of some synergistic effect, but quantification will have to await a much more solid experimental body of evidence.

With respect to possible effects on worker safety, loss of sensitivity due to neurological disorders is a possible factor that should be looked at carefully. Inability to perform work that requires fine motor skills due to neurological problems may also lie behind accidents and other safety-related issues.

e. *Information provided by the manufacturers of work equipment in accordance with the relevant Community Directives:* Manufacturers of work equipment are required by the Machine Directive (2006/42/EC) to provide quantitative information on the vibration generated by the equipment itself. Test codes are usually built so as to give highly repeatable values in different laboratories. This, however, implies that test conditions are usually artificial and quite far from those encountered in real work life. Additionally, tested equipment is brand new and skillfully operated. The end result is that values provided by manufacturers are usually much lower than values found in field measurements. As specified in item a) of this section, such values should only be used if no other information is available. Whenever possible, ad hoc multiplicative factors should be used to obtain realistic estimates of the actual vibration that workers are exposed to. For hand-arm vibration, multiplicative factors are available in the European standard CEN/TR 15350. Unfortunately, multiplicative factors are unavailable for equipment (vehicles, platforms) involved in whole-body vibration.

f. *The existence of replacement equipment designed to reduce the levels of exposure to mechanical vibration:* There are two approaches to find alternative, low-vibration equipment:
   - There are often alternative methods for the same process, as described in Section 4.5.1.
   - There may be significant variability in the vibration generated by tools that perform the same operation. While of little value in absolute terms (see item e) of this section), manufacturers' self-declared vibration data represent a valuable source of information to compare different tools of the same typology. In this context, care should be taken to take into account quantities such as power and mass, which have significant impact on tool/vehicle acceleration. Additionally, work effectiveness should, in principle, be considered, as it might affect the time needed to carry out a specific task, although this might prove hard to do in practice. Databases for both handheld tools and vehicles, which also include field data, are available on the Internet. Care should be taken to compare 'apples to apples' by making sure that operating conditions are as close as possible for all the tested tools/vehicles under consideration.

g. *The extension of exposure to whole-body vibration beyond normal working hours under the employer's responsibility:* Because of the way the vibration exposure A(8) is calculated (see Equation 4.1), exposures of any duration can be taken into account without problems. Of course, extensions of the working shift well beyond 8 hours cause the daily period of rest to shrink below 16 hours, which might create conflicts with the assumption underlying the values selected for the lower and upper action values (see Section 4.3). As long as such extensions are not the rule and do not characterise a large fraction of the working life, this is more a problem of principle and has no practical implications.

h. *Specific working conditions such as low temperatures:* This topic shall be dealt with in Section 4.5.1i.

i. *Appropriate information obtained from health surveillance, including published information, as far as possible:* Health surveillance performed in a workplace should aim at obtaining valuable information about the possible impacts of various processes on hearing impairment. Together with results collected through measurements, this information can be used to draw a more comprehensive and detailed picture of where the most serious risks lie, as well as to increase awareness among personnel.

4.5 Provisions Aimed at Avoiding or Reducing Exposure

Article 5 is where Directive 2002/44/EC makes it clear that risk assessment is of limited significance unless it is a source of information for technical/organisational actions aimed at lowering exposure to vibration. A list of possible pathways leading to the desired reduced exposure of workers is provided in Article 5.2. Each item shall be labelled as of either a technical (T) or organisational (O) nature and commented upon.

It is important to point out that the vibration reduction that can be expected by pursuing different approaches is extremely variable. Large variations can be expected even for the same type of action, underlining the intrinsic complexity of the situation.

Proper evaluations of vibration-reducing measures requires fundamental understanding of the various quantities used as input data, as well as the physics of vibration reduction. The following discussion may, however, be used as a general guide for achievable vibration reduction of various actions.

4.5.1 Actions

a. *Other working methods that require less exposure to mechanical vibration:* (O) This is the most radical solution to the vibration problem, where the source of the problem itself (the vibrating tool/vehicle) is replaced. It is a very effective but rarely a pursued option, since it only becomes available when alternative technology becomes available at a comparable cost.

b. *The choice of appropriate work equipment of appropriate ergonomic design and, taking account of the work to be done, producing the least possible vibration:* (O) This topic has been previously tackled in Section 4.4f.

c. *The provision of auxiliary equipment that reduces the risk of injuries caused by vibration, such as seats that effectively reduce whole-body vibration and handles which reduce the vibration transmitted to the hand-arm system:* (T) Technical interventions acting at or near the receiver (the exposed subject) are much more popular in vibration-reduction strategies than they are in noise-reduction strategies.

   Development of seats that effectively reduce the transmission of vibration to the seated body has steadily progressed during recent decades. Both passive (see e.g. Paddan and Griffin 2002) and active seats (Gan et al. 2015) are now available for a wide variety of vehicles. While vibration-reducing seats have initially focused on the attenuation of vertical vibration, their potential in the attenuation of tri-axial vibration is now being explored (Ji 2015). Careful design of seats and good decoupling between seat and vehicle is the key to effective vibration reduction in any contemporary vehicle. The effectiveness of suspended sets in reducing the vibration transmitted to the exposed subject can be evaluated with standardized methods outlined in ISO 10326-1 (2016).

   Tool handles designed to reduced vibration have not attracted similar interest, and they lag far behind anti-vibration gloves (see Section 4.6), both in terms of academic studies and also for diffusion in workplaces. Their potential effectiveness has nevertheless been proven (e.g. Oddo et al. 2004). An interesting property of anti-vibration handles is the possibility of fine-tuning their stiffness in order to improve their effectiveness on different tools, which is unachievable with anti-vibration gloves.

d. *Appropriate maintenance programmes for work equipment, the workplace and workplace systems.*

e. *The design and layout of workplaces and workstations:* (T) Because exposure to vibration is mainly (often entirely) direct, the effectiveness of any intervention in the layout of workplaces is much lower than it is for noise reduction. Optimised layouts of workplaces can help to reduce the need to use vehicles, thereby lowering exposure to whole-body vibration. But since many different factors dictate the workplace layout, it is unlikely that significant results will be obtained through this pathway.

f. *Adequate information and training to instruct workers to use work equipment correctly and safely in order to reduce their exposure to mechanical vibration to a minimum:* (O) As clearly indicated in Article 6 of 2002/44/EC, workers must receive adequate information addressing several topics, and in particular:

   i. The measures taken to implement this Directive in order to eliminate or reduce to a minimum the risks from mechanical vibration;
   ii. The exposure limit values and the exposure action values;
   iii. The results of the assessment and measurement of the mechanical vibration carried out in accordance with Article 4 of the Directive and the potential injury arising from the work equipment in use;
   iv. Why and how to detect and report signs of injury;
   v. The circumstances in which workers are entitled to health surveillance;
   vi. Safe working practices to minimise exposure to mechanical vibration.

   The importance of any training process is often underrated by employers and employees alike. Although this topic lies outside the scope of the present section, it should be made clear that a purely technical approach will do little to improve safety and health in the workplace if it is not coupled with an active role played by workers (EC 2006). In particular, risks which have so far received limited attention, such as exposure to vibration, will strongly benefit from involving the exposed subject in the process of risk handling, particularly on issues such as the identification of risks posed by specific tools, early detection of injuries and methods to minimise vibration-related risks while keeping work effectiveness unaltered.

g. *Limitation of the duration and intensity of the exposure:* Because exposure is quantified as the product of vibration intensity (the square of the r.m.s. acceleration) and time, this line of intervention lies at the root of any exposure reduction strategy.

h. *Appropriate work schedules with adequate rest periods:* This topic has been previously discussed in Sections 4.3.1.4 and 4.3.2.4, as well as in Section 4.4a. While it is clear that rest periods must be part of a well-planned work activity from any point of view, the quantification of possible beneficial effects of intermittent exposure in terms of reduced effects of vibration are still a matter of debate.

i. *The provision of clothing to protect exposed workers from cold and damp:* Cold weather or work-related indoor temperatures are well-known aggravating factors that may accelerate or worsen the severity of vibration-related pathologies (see e.g. Carlsson and Dahlin 2014 for a recent paper). This applies both to hand-arm vibration, where Raynaud's phenomenon is strongly temperature-dependent, as well as to whole-body vibration, where low back pain has also been found to be negatively impacted by low temperatures. Warm and waterproof clothing should always be provided to all subjects who operate in cold climates. With respect to hand-arm vibration, anti-vibration gloves (see Section 4.6) represent a particularly attractive solution, as they offer simultaneous protection from injuries (cuts and scratches), vibration-related effects and cold temperatures.

4.5.2 Synthesis

The existence of two thresholds implies that worker exposure can be assigned to one of three categories, each one characterised by specific actions to put into practice to appropriately handle the risk. In this regard, it is useful to summarise the overall picture that emerges, including a citation of the specific article of the Directive where the issue is dealt with in more detail:

1. $A(8) < 2.5 \, \mathrm{ms}^{-2}$ (HAV); $A(8) < 0.5 \, \mathrm{ms}^{-2}$ or $VDV < 9.1 \, \mathrm{ms}^{-1.75}$ (WBV)
   - No action is needed
2. $2.5 \, \mathrm{ms}^{-2} < A(8) < 5 \, \mathrm{ms}^{-2}$ (HAV); $0.5 \, \mathrm{ms}^{-2} < A(8) < 1.15 \, \mathrm{ms}^{-2}$ or $9.1 \, \mathrm{ms}^{-1.75} < VDV < 21 \, \mathrm{ms}^{-1.75}$ (WBV)
   - Workers are entitled to receive adequate information about risks (Art. 4)
   - Workers are entitled to appropriate health surveillance, meaning that a thorough medical examination will be carried out by a medical doctor or by another suitably qualified person (Art. 8.1)
   - The employer shall draw up and apply a program of measures of a technical nature and/or of organisation of work aimed at reducing as far as reasonably practicable the exposure of workers to vibration (Art. 5.2)
3. $A(8) > 5 \, \mathrm{ms}^{-2}$ (HAV); $A(8) > 1.15 \, \mathrm{ms}^{-2}$ or $VDV > 21 \, \mathrm{ms}^{-1,75}$ (WBV)
   - Immediate action must be taken to reduce the exposure to below the exposure limit values; the reasons why overexposure has occurred must be identified, and existing protection and prevention measures must be amended in order to avoid any recurrence (Art. 5.3)

A comprehensive comparison between the provisions which appear in the two Directives (2002/44/EC and 1986/188/EC) is provided by EC 2007.

4.6 Anti-Vibration Gloves

4.6.1 Measurement Method

Several studies have been performed on anti-vibration gloves (Dong et al. 2014, Milosevic and McConville 2012, Pinto et al. 2001). Most of them are performed in accordance with ISO 10819. This standard only describes measurements along one axis, i.e. the axis considered to be of major importance, which is the axis orthogonal to the palm of the hand. This is possibly too simple, since vibrations contributing to the tri-axial descriptor in Equation 4.1 normally occur in two or three axes for many tools. Therefore, it may be claimed that the one-dimensional value obtained with ISO 10819 is of less importance. Following such evaluations, some studies have extended to include measurements in all three axes.

The consequence of measuring only at the palm of the hand and not at the fingers has also been questioned. However, here there is no clear conclusion, partly because the studies seem to disagree on the level and the importance of finger vibrations (Welcome et al. 2014).

4.6.2 Reference Measurements

The studies performed seem to agree more or less with this assessment: Anti-vibration gloves may give some protection at higher frequencies, but at lower frequencies the effect is uncertain or even negative. This can be understood by regarding the tool and gloves as a machine on elastic supports, such as springs. At very low frequencies the spring theoretically has no effect, at resonance amplification will occur and at higher frequencies there will be a damping effect. Unfortunately, many of the worst tools create the most vibration at middle or low frequencies.

Some studies conclude that the effect of anti-vibration gloves is too uncertain to rely on (Hewitt et al. 2015). In practice, this means that you cannot extend your working time by using anti-vibration gloves. In other reports, there are tables that show expected damping for some combinations of tool and glove. The damping has been calculated with vibrations measured on specific tools, and the laboratory measured damping values in three axes. The tables show that damping may be obtained in some situations but also that amplifications may occur.

A Norwegian study (Ognedal et al. 2017) was recently performed as 'site measurements', with more than 20 tools in combinations with more than 10 specific gloves. The site situation was created by using the tool in a practical situation. Several people were operating the tool, running it three times with each set of gloves. Vibrations were measured with and without the various gloves in all three axes. Uncertainty was calculated and taken into account in the evaluations. Several measured values for all combinations were reported. The major important conclusions were:

Average damping values show that gloves in general have a non-negative or even a slight positive effect over time for people working with a combination of tools. For some combinations of glove and tool, considerable damping can be obtained. This applies, not surprisingly, more often to tools working at high frequencies than to tools working at low frequencies.

References

Legislation

Directive 2002/44/EC of the European Parliament and Council, of 25 June 2002, concerning minimum health and safety requirements regarding the exposure of workers to risks arising from physical agents (vibrations), J.O. No L 177 of 06.07.2002, page 13.

Directive 2006/42/EC of the European Parliament and of the Council of 17 May 2006 on machinery, and amending Directive 95/16/EC, O.J. No L 157 of 9.6.2006, page 24.

EC (1993) Commission proposal for a Council Directive on the minimum health and safety requirements regarding the exposure of workers to the risk arising from physical agents, 93/C77/02.

International Technical Standards

CEN/TR 15350 (2013) *Mechanical vibration – Guideline for the assessment of exposure to hand-transmitted vibration using available information including that provided by manufacturers of machinery*. European Standard Committee, Brussels, Belgium.

EN 14253 (2003) *Mechanical vibration – Measurement and evaluation of occupational exposure to whole-body vibration with reference to health – Practical guidance*. European Standard Committee, Brussels, Belgium.

ISO 10326-1 (2016) *Mechanical vibration – Laboratory method for evaluating vehicle seat vibration – Part 1: Basic requirements*. International Standardizing Organization, Geneve, Switzerland.

ISO 10819 (2013) *Mechanical vibration and shock – Hand-arm vibration – Measurement and evaluation of the vibration transmissibility of gloves at the palm of the hand*. International Standardizing Organization, Geneve, Switzerland.

ISO 2631-1 (1997) *Mechanical vibration and shock – Evaluation of human exposure to whole-body vibration – Part 1: General requirements*. International Standardizing Organization, Geneve, Switzerland.

ISO 5349-1 (2001) *Mechanical vibration – Measurement and evaluation of human exposure to hand-transmitted vibration – Part 1: General requirements*. International Standardizing Organization, Geneve, Switzerland.

ISO 5349-2 (2001) *Mechanical vibration – Measurement and evaluation of human exposure to hand-transmitted vibration – Part 2: Practical guidance for measurement at the workplace*. International Standardizing Organization, Geneve, Switzerland.

International Technical Documents

EC (2006) Guide to good practice on hand-arm Vibration, v7.7, http://resource.isvr.soton.ac.uk/HRV/VIBGUIDE/HAV%20Good%20practice%20Guide%20V7.7%20English%20260506.pdf

EC (2007) Non binding guide to good practice for implementing Directive 2002/44/EC (Vibrations at Work) http://ec.europa.eu/social/BlobServlet?docId=3614&langId=en

EC (2008) Guide to good practice on Whole-Body Vibration, v6.7h, http://resource.isvr.soton.ac.uk/HRV/VIBGUIDE/2008_11_08%20WBV_Good_practice_Guide%20v6.7h%20English.pdf

EU-OSHA (2008) European Agency for Safety and Health at Work (2008) Workplace exposure to vibration in Europe: an expert review by the European Agency for Safety and Health at Work https://osha.europa.eu/en/tools-and-publications/publications/reports/8108322_vibration_exposure

National Technical Standards

DIN 45660-2 (2015) Guide for dealing with uncertainty in acoustics and vibration – Part 2: Uncertainty of vibration quantities.

HSE (2010) Health and Safety Executive, Control of Risks from Hand-arm Vibration http://www.hse.gov.uk/vibration/hav/roadshow/bmb2.pdf

Scientific Literature

Bovenzi M., Griffin M.J. (2006) Risks of occupational vibration exposures. Annex 6 to Final Technical Report: Experimental studies of acute effects of hand-transmitted vibration on vascular function. http://www.vibrisks.soton.ac.uk/reports/Annex6%20UTRS%20WP3_1_1%20040107.pdf

Carlsson, I.K., Dahlin, L.B. (2014) Self-reported cold sensitivity in patients with traumatic hand injuries or hand-arm vibration syndrome – an eight year follow up, *BMC Musculoskelet Disord* 15: 83–91.

Gan, Z., Hillis, A., Darling, J. (2015) Adaptive control of an active seat for occupant vibration reduction, *J Sound Vib* 349: 39–55.

Gerhardsson, L., Hagberg, M. (2014) Work ability in vibration-exposed workers, *Occup Med* 64(8): 629–634.

Griffin, M.J. (2012) *Handbook of Human Vibration*. Published: Academic Press, London.

Ji, X. (2015) Evaluation of Suspension Seats Under Multi-Axis Vibration Excitations – A Neural Net Model Approach to Seat Selection, Western University, Ph.D. Thesis.

Kelsey, J.L., Githens, P.B., Walter, S.D., Southwick, W.O., Weil, U., Holford, T.R. et al. (1984) An epidemiological study of acute prolapsed cervical intervertebral disc, *J Bone Joint Surg Am* 66: 907–914.

Lan, F.Y., Liou, Y.W., Huang, K.Y., Guo, H.R., Wang, J.D. (2016) An investigation of a cluster of cervical herniated discs among container truck drivers with occupational exposure to whole-body vibration, *J Occup Health* 58(1): 118–127.

Nelson, C.M. (1997) Hand-transmitted vibration assessment – a comparison of results using single axis and triaxial methods. Proceedings of the UK group meeting on Human Response to Vibration, Southampton, 17–19 September 1997.

Nilsson, T., Wahlström, J., Burström, L. (2017) Hand-arm vibration and the risk of vascular and neurological diseases—A systematic review and meta-analysis, *PLoS ONE* 12(7): e0180795.

Oddo, R., Loyau, T., Boileau, P. E., Champoux, Y. (2004) Design of a suspended handle to attenuate rock drill hand-arm vibration: model development and validation, *J Sound Vib* 275: 623–640.

Paddan, G.S., Griffin, M.J. (2002) Effect of seating on exposures to whole-body vibration in vehicles, *J Sound Vib* 253(1): 215–241.

Palmer, K., Griffin, M., Bendall, H., Pannett, B., Coggon, D. (2000) Prevalence and pattern of occupational exposure to whole body vibration in Great Britain: findings from a national survey, *Occup Environ Med* 57(4): 229–236.

Pettersson, H., Burström, L., Hagberg, M., Lundström, R., Nilsson, T. (2014) Risk of hearing loss among workers with vibration-induced white fingers. *Am J Ind Med* 57(12): 1311.

Prisby, R.D., Lafage-Proust, M.-H., Malaval, L., Belli, A., Vico L. (2008) Effects of whole body vibration on the skeleton and other organ systems in man and animal models: What we know and what we need to know, *Ageing Res Rev* 7(4): 319–329.

Seidel, H., Bastek, R., Brauer, D., Buchholz, Ch., Meister, A., Metz, A.-M., Rothe, R. (1980) On human response to prolonged repeated whole-body vibration, *Ergonomics* 23(3): 191–211.

Seidel, H., Heide, R. (1986) Long-term effects of whole-body vibration: a critical survey of the literature, *Int Arch Occup Environ Health* 58(1): 1–26.

Scientific Literature on Anti-Vibration Gloves

Dong, R.G., Welcome, D.E., Peterson, D.R., Xu X.S., McDowell, T.W., Warren, C., Asaki, T., Kudernatsch, S., Brammer, A. (2014) Tool-specific performance of vibration-reducing gloves for attenuating palm-transmitted vibrations in three orthogonal directions, *Int J Ind Ergon* 44(6): 827–839.

Hewitt, S., Dong, R.G., Welcome, D.E., McDowell, T.W. (2015) Anti-Vibration Gloves? *Ann Occup Hyg* 59(2): 127–141.

Milosevic, M., McConville, K.M. (2012) Evaluation of protective gloves and working techniques for reducing hand-arm vibration exposure in the workplace, *J Occup Health* 54(3): 250–253.

Ognedal, T., Høydal, R., Aumo, L., Nessler, ø.V. (2017) Vibrasjonsdempende hansker (Antivibration gloves), Sinus, Beerenberg, NHO Rapport 108000-0-001 (in norwegian). https://www.nho.no/siteassets/nhos-filer-og-bilder/filer-og-dokumenter/arbeidslivspolitikk/vibrasjonsdempende-hansker.pdf

Pinto, I., Stacchini, N., Bovenzi, M., Paddan G.S., Griffin M.J. (2001) Protection effectiveness of anti-vibration gloves: field evaluation and laboratory performance assessment, 9th International Conference on Hand-Arm Vibration, Nancy, France, 5–8 June 2001.

Welcome, D.E., Dong, R.G., Xu, X.S., Warren, C., McDowell, T.W. (2014) The effects of vibration-reducing gloves on finger vibration, *Int J Ind Ergon* 44(1): 45–59.

Electromagnetic Fields

Giovanni d'Amore
Environmental Protection Agency of Piedmont Region, Italy

Overview of the Basic Concepts of Electromagnetic Field Protection

Giovanni d'Amore
Environmental Protection Agency of Piedmont Region

Sara Adda
Environmental Protection Agency of Piedmont Region

Laura Anglesio
Environmental Protection Agency of Piedmont Region

5.1 Introduction

Human exposure to electromagneticfields (EMFs) comes from many different sources in living and working environments. This exposure is continually changing due to technological developments that lead to new EMF-emitting devices, such as wireless consumer devices, and new infrastructure deployment, such as 4th- and 5th-generation mobile phone networks.

Many laboratory and epidemiological studies have investigated biological effects and health effects from EMF exposure, and the evidence resulting from these studies is the basis for the human exposure restrictions proposed in international guidelines and adopted in many national laws (SCENIHR 2009, EFHRAN 2012).

In recent years, there has been increasing concern about the potential health risk from exposure to power frequency fields and radiofrequency radiation. People and workers are increasingly interested to know EMF exposure levels for risk exposure assessment and for evaluating compliance with limits. To respond to this public concern, the technical institutions of many countries have conducted survey campaigns to assess exposure to magnetic fields of high-voltage electricity lines and electromagnetic fields of telecommunication systems, including mobile base stations and radio and TV broadcasting antennas (EFHRAN 2010, Gajsek et al. 2015, Gajsek et al. 2016, Zhu et al. 2017).

In this chapter, we will discuss the basic concepts of EMF protection, with an overview of the typical emission characteristics of the main sources present in work and residential sites.

5.2 Quantities and Units

The term EMF as used in this chapter covers fields in the frequency range below 300 GHz. EMFs include static fields, extremely low-frequency (ELF) fields having frequencies in the 0 Hz–300 Hz range; intermediate-frequency (IF) fields having frequencies from 300 Hz to 10 MHz; and radiofrequency (RF) fields having frequencies in the 10 MHz–300 GHz range.

Electromagnetic radiation can be represented by electric and magnetic fields that vary sinusoidally with time, according to a sine wave model. The frequency (f) is the number of oscillations per second and is measured in hertz (Hz); the wavelength (λ) is the distance between successive peaks of the sine wave and is related to the frequency according to:

$$\lambda = \frac{c}{f} \tag{5.1}$$

where c is the speed of light, which is the speed with which the wave is propagated in air, equal to $3 \cdot 10^8$ m/sec.

In the area surrounding an EMF source, we can consider three main regions: the reactive near field, the radiating near field or Fresnel region and the far field or Fraunhofer region (see Figure 5.1).

The reactive near field is in the immediate vicinity of the source. The electric (E) and magnetic (H) fields are not in phase with each other, and the angular field distribution is highly dependent upon the distance and direction from the source. In this region, E and H field strengths vary as the inverse cube and inverse square of the distance. Furthermore, the relationship between the E and H fields is very complex and requires measurement of both E and H to determine the power density. The boundary of this region is commonly considered to be a distance of $1/2\pi$ times the wavelength ($\lambda/2\pi$) from the source's surface.

In the radiating near field region (Fresnel region), the radiating fields begin to emerge and the field strength varies with the inverse of the distance from the source, but the angular field distribution continues to vary appreciably with the distance.

The far field region, or Fraunhofer region, is commonly considered to exist at distances greater than $2D^2/\lambda$ meters from the source, where D is the maximum linear dimension of the radiating source. For electrically small radiating sources, whose maximum dimension D is less than $\lambda/2\pi$, the far field region

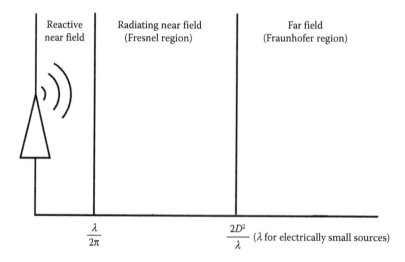

FIGURE 5.1   Propagation regions of the electromagnetic field around a source.

starts at a distance of one wavelength from the source (Stratton 2007). In this region, the electromagnetic field propagation is in accordance with the plane wave model having the following characteristics:

- Field strength varies with the inverse of the distance from the source;
- The E and H vectors and the direction of propagation are mutually perpendicular;
- E, H and power flux density (S) are related by the equations: $E = H \times 377$ and $S = E \times H$. Combining these two equations, we get: $S = H^2 \times 377 = E^2 \div 377$, where 377 is the characteristic impedance of free space expressed in Ohms.

Exposure in the far field region can be described by measuring only E or H field strength, taking into account the above-reported relationships among E, H and S quantities. In the near field, the exposure characterization is more difficult because the field patterns are more complicated and both E and H fields must be measured. In this situation, we cannot consider the power flux density, S, an appropriate quantity for exposure evaluation (CENELEC 2010).

E, H and S quantities are termed 'radiometric quantities' because they specify the physical characteristics of the radiation field. These three quantities are respectively expressed in volt per meter (V/m), ampere per meter (A/m), and watt per square meter (W/m^2). The intensity of a magnetic field can also be expressed by magnetic flux density (B) measured in tesla, T. In air and in non-magnetic materials, the two quantities, H and B, are related by the following simple relationship:

$$B = \mu_0 H \text{ T} \qquad (5.2)$$

where μ_0 is a constant, called the permeability of free space, which has the value of $4\pi10^{-7}$ henry per meter (H/m).

For the description of the coupling mechanisms between fields and the body and the resulting biological and health effects, we have to consider the 'dosimetric quantities'. In the low-frequency region (up to 100 kHz), the dosimetric quantity is the induced current density in tissue, J, measured in ampere per square meter (A/m^2), because the coupling between fields and the body induces currents inside the body which can stimulate the nerve fibers of the central and peripheral nervous systems and the muscles. The current density is related to the internal electric field, induced in tissue, E_i by Ohm's law:

$$J = \sigma E_i \text{ A/m}^2 \qquad (5.3)$$

where σ is the electrical conductivity of the tissue.

Accordingly, we can consider E_i in place of J as a dosimetric quantity. Actually, some effects, such as the induction of retinal phosphenes, are directly related to the exceeding of threshold values of internal electric field strength induced in the central nervous system tissue in the head.

At frequencies above about 10 MHz, exposure to electromagnetic fields can lead to thermal effects consisting in an elevated increase in tissue temperature resulting from significant absorption of energy. The dosimetric quantity in the frequency range 100 kHz–10 GHz is the specific absorption rate (SAR), which is defined by the absorbed power per unit mass at infinitesimal volume of tissue, measured in watts per kilogram (W/kg).

In the frequency range 100 kHz–10 MHz there is an overlapping between stimulation effects and thermal effects, and we can consider all above-reported dosimetric quantities to describe interaction mechanisms. At frequencies above 10 GHz, incident power density, measured in watts per square meter (W/m^2), is more appropriate than SAR as a dosimetric quantity because an electromagnetic field does not penetrate very far into the skin and the energy absorption occurs primarily at the body surface.

The SAR can be calculated as the average over the entire body (global SAR) or over a small mass of contiguous tissue, either 1 g or 10 g (local SAR). Global SAR is related to whole-body heat stress, while local SAR takes into account excessive localised heating of the tissue in critical organs, such as the gonads and eyes. SAR can be expressed in terms of the induced electric field strength or the temperature rise in tissue by:

$$SAR = \frac{\sigma E^2}{2\rho} = c\frac{\Delta T}{\Delta t} \ \text{W/kg} \tag{5.4}$$

where E is the electric field strength in V/m, σ is the dielectric conductivity of tissue in siemens per cm (S/cm), ρ is the density of tissue in g/cm^3, c is the specific heat capacity of tissue in joules per gram per degree Celsius (J/g°C) and ΔT is the temperature rise in degrees Celsius (°C) during the exposure time of Δt in seconds.

5.3 Protection Limits

Various national and international organizations have set limits to protect humans from the adverse health effects of EMFs. The International Commission on Non-Ionising Radiation Protection (ICNIRP), a non-governmental organization formally recognised by the World Health Organisation (WHO), and the International Committee on Electromagnetic Safety (ICES), which is sponsored by the Institute of Electrical and Electronics Engineers (IEEE), have agreed on a set of recommended limits based on well-established acute effects at high fields, such as stimulation effects and thermal effects. Possible other effects at lower fields, such as cancer, are not taken into account because there is lack of evidence to justify setting exposure limits (ICNIRP 1998, IEEE 2005, ICNIRP 2010).

There are also other bodies, in particular the National Radiological Protection Board (NRPB) in the United Kingdom, which have developed international standards on EMF protection. All these standards are based on the same approach and rationale of the ICNIRP guidelines that have formed the basis for national regulations in several countries.

Protection limits in the ICNIRP guidelines are organised into a two-level structure: basic restrictions and reference levels. Basic restrictions are defined in terms of dosimetric quantities: internal electric field, E_i, up to 10 MHz; global SAR and local SAR in the frequency range 100 kHz–10 GHz; and power density for frequencies between 10 GHz and 300 GHz. All SAR limit values are intended as average over any 6-minute period, while local SAR limits are averaged over a mass of 10 g of contiguous tissue. Since the measurement and calculation of dosimetric quantities are complex, reference levels were defined in terms of directly measurable quantities describing the external exposure, the radiometric quantities. Reference levels are derived from basic restrictions considering conditions of maximum coupling between the external fields and the exposed person. Consequently, compliance with reference levels will assure compliance with basic restrictions. On the other hand, if the measured radiometric quantities exceed reference levels it does not necessarily imply that basic restrictions are exceeded.

ICNIRP basic restrictions for workers and for the general public are reported in Table 5.1 (ICNIRP 1998, ICNIRP 2010).

The symbol f reported in the Table 5.1 represents the frequency in Hz.

Reference levels recommended by ICNIRP for workers and for the general public are reported in Table 5.2, where f is the frequency in Hz (ICNIRP 1998, ICNIRP 2009, ICNIRP 2010). In the radiofrequency region, reference levels are frequency dependent because the efficacy of EMF coupling with the human body varies with frequency. The minimum reference levels are fixed in the frequency range (about 60–100 MHz) where the human body has a higher capability to absorb electromagnetic energy. In the frequency range from 100 kHz to 10 MHz, reference levels related to low-frequency effects on the nervous system, as well as high-frequency heating effects, need to be considered, depending on exposure conditions.

As shown in Tables 5.1 and 5.2, basic restrictions and reference levels are frequency dependent, varying across all the considered frequency ranges. In case of simultaneous exposure to multiple frequency fields, the field strength of each frequency component should be compared to the relevant limit. Compliance with exposure limits can be evaluated by defining exposure indexes, which represent the observed exposure

TABLE 5.1   Basic Restrictions for Time-Varying Electric and Magnetic Fields for Frequencies up to 10 GHz

Tissue	Frequency range	Internal electric field V/m	Whole body average SAR W/kg	Local SAR (head and trunk) W/kg	Local SAR (limbs) W/kg
		Occupational Exposure			
CNS[a] Tissues of the head	1–10 Hz	0.5/f	–	–	–
	10 Hz–25 Hz	0.05	–	–	–
	25 Hz–400 Hz	$2 \times 10^{-3}f$	–	–	–
	400 Hz–3 kHz	0.8	–	–	–
	3 kHz–10 MHz	$2.7 \times 10^{-4}f$	–	–	–
All tissues	1 Hz–3 kHz	0.8	–	–	–
	3 kHz–100 kHz	$2.7 \times 10^{-4}f$	–	–	–
	100 kHz–10 MHz	$2.7 \times 10^{-4}f$	0.4	10	20
	10 MHz–10 GHz	–	0.4	10	20
		General Public Exposure			
CNS[a] Tissues of the head	1–10 Hz	0.1/f	–	–	–
	10 Hz–25 Hz	0.01	–	–	–
	25 Hz–1000 Hz	$4 \times 10^{-4}f$	–	–	–
	1000 Hz–3 kHz	0.4	–	–	–
	3 kHz–10 MHz	$1.35 \times 10^{-4}f$	–	–	–
All tissues	1 Hz–3 kHz	0.4	–	–	–
	3 kHz–100 kHz	$1.35 \times 10^{-4}f$	–	–	–
	100 kHz–10 MHz	$1.35 \times 10^{-4}f$	0.08	2	4
	10 MHz–10 GHz	–	0.08	2	4

Source: ICNIRP. 1998. Guidelines for limiting exposure to time-varying electric, magnetic and electromagnetic fields (up to 300 GHz). *Health Physics* 74(4): 494–522. Available from http://www.icnirp.org; ICNIRP. 2010. Guidelines for limiting exposure to time-varying electric and magnetic fields (1 Hz–100 kHz). *Health Physics* 99(6): 818–836. Available from http://www.icnirp.org.

[a] CNS is the central nervous system.

divided by the limit value. Exposure is compliant if the exposure index is less than 1. Considering that exposures are cumulative in their effects, a summation should be performed to assess compliance with limits across the entire frequency spectrum. For electrical stimulation relevant for frequencies up to 10 MHz, electric (E) and magnetic (H) field strength should be added for giving the following exposure indexes EI_E and EI_H:

$$EI_E = \sum_{i=1\,\text{Hz}}^{10\,\text{MHz}} \frac{E_i}{E_{RLi}} \tag{5.5}$$

$$EI_H = \sum_{i=1\,\text{Hz}}^{10\,\text{MHz}} \frac{H_i}{H_{RLi}} \tag{5.6}$$

where E_i and H_i are, respectively, electric and magnetic field strengths at frequency i; E_{RLi} and H_{RLi} are, respectively, electric and magnetic field strength reference levels at frequency i as given in Table 5.2.

Following this approach, based on the assumption that all the spectral components of the field add in phase, we can overestimate exposure with a too-conservative exposure assessment for verification of compliance with the exposure limits. As an alternative approach, we can consider the 'weighted peak method'

TABLE 5.2   Reference Levels for Time-Varying Electric and Magnetic Fields for Frequencies up to 10 GHz

Frequency range	E field strength V/m	H field strength A/m	B field μT	Power Density S_{eq} W/m²
Occupational Exposure				
0 Hz Head and trunk exposure	–	0.16×10^6	2×10^6	–
0 Hz Limb exposure	–	0.64×10^6	8×10^6	–
1 Hz–8 Hz	20×10^3	$1.63 \times 10^5/f^2$	$2 \times 10^5/f^2$	–
8 Hz–25 Hz	20×10^3	$2 \times 10^4/f$	$2.5 \times 10^4/f$	–
25 Hz–300 Hz	$5 \times 10^5/f$	8×10^2	1×10^3	–
300 Hz–3 kHz	$5 \times 10^5/f$	$2.4 \times 10^5/f$	$3 \times 10^5/f$	–
3 kHz–100 kHz	170	80	100	–
100 kHz–1 MHz Nervous system effects	170	80	100	–
100 kHz–1 MHz Heating effects	610	$1.6 \times 10^6/f$	$2 \times 10^6/f$	–
1 MHz–10 MHz Nervous system effects	170	80	100	–
1 MHz–10 MHz Heating effects	$610 \times 10^6/f$	$1.6 \times 10^6/f$	$2 \times 10^6/f$	–
10 MHz–400 MHz	61	0.16	0.2	10
400 MHz–2 GHz	$3 \times 10^3 f^{1/2}$	$8f^{1/2}$	$10f^{1/2}$	$25 \times 10^3 f$
2 GHz–300 GHz	137	0.36	0.45	50
General Public Exposure				
0 Hz		3.2×10^5	4×10^5	
1 Hz–8 Hz	5×10^3	$3.2 \times 10^4/f^2$	$4 \times 10^4/f^2$	
8 Hz–25 Hz	5×10^3	$4 \times 10^3/f$	$5 \times 10^3/f$	
25 Hz–50 Hz	5×10^3	160	200	
50 Hz–400 Hz	$2.5 \times 10^5/f$	160	200	
400 Hz–3 kHz	$2.5 \times 10^5/f$	$6.4 \times 10^4/f$	$8 \times 10^4/f$	
3 kHz–100 kHz	83	21	27	
100 kHz–1 MHz Nervous system effects	83	21	27	
100 kHz–1 MHz Heating effects	87	0.73/f	0.92/f	
1 MHz–10 MHz Nervous system effects	83	21	27	
1 MHz–10 MHz Heating effects	$87/f^{1/2}$	0.73/f	0.92/f	
10 MHz–400 MHz	28	0.073	0.092	2
400 MHz–2 GHz	$1.375f^{1/2}$	$0.0037f^{1/2}$	$0.0046f^{1/2}$	f/200
2 GHz–300 GHz	61	0.16	0.20	10

Source: ICNIRP. 1998. Guidelines for limiting exposure to time-varying electric, magnetic and electromagnetic fields (up to 300 GHz). *Health Physics* 74(4): 494–522. Available from http://www.icnirp.org; ICNIRP. 2009. Guidelines on limits of exposure to static magnetic fields. *Health Physics* 96(4): 504–514. Available from http://www.icnirp.org; ICNIRP 2010 Guidelines for limiting exposure to time-varying electric and magnetic fields (1 Hz–100 kHz). *Health Physics* 99(6): 818–836. Available from http://www.icnirp.org.

that takes into account both the amplitude and the phases of the spectral components. According to this method, the waveform of the field is weighted with a filtering function related to the frequency-dependent reference levels. In this case, the exposure index is the amplitude of the weighted waveform, EI_{AWP}, which is calculated by the following expression:

$$EI_{AWP} = \left| \sum_i \frac{A_i}{RL_i} \cos(2\pi f_i t + \theta_i + \varphi_i) \right| \tag{5.7}$$

where A_i and θ_i are the amplitudes and phase angles of the field, respectively, RL_i is the reference level and φ_i are phase angles of the filter. All these quantities are referred to the ith frequency f_i.

The 'weighted peak method' can be adopted at low frequencies, below 100 kHz, when the waveform of the fields can show complex patterns with a great number of spectral components.

For electromagnetic fields with frequencies above 100 kHz, thermal effects become relevant and exposure index EI_{TE} will be given by:

$$EI_{TE} = \sum_{i=100\,\text{kHz}}^{300\,\text{GHz}} \frac{A_i^2}{RL_i^2} \tag{5.8}$$

where A_i and RL_i are the field strength and the reference level at the ith frequency, respectively.

5.4 Exposure Assessment

As a useful approach for exposure assessment aimed at the evaluation of compliance with limits, it is recommended to first use the simplest method, with an overestimation of exposure levels, even if it does not give higher accuracy. If the overestimated exposure levels are below the limits, methods that are more sophisticated are not necessary.

For some exposure conditions, especially when the body is extremely close to an RF field source and for highly localised exposure, the exposure assessment may be referenced against basic restrictions (e.g. SAR limits) instead of reference levels. In these cases, electromagnetic emission from the source is strongly affected by the presence of the exposed person, so that the source and the exposed person can be described as 'mutually coupled'.

Generally, exposure assessment can be performed either by measurement or by calculation. Calculations are necessary for exposure assessment in areas where new sources are planned.

5.4.1 Measurements

EMF sources generate fields that are not time-constant and spatially uniform. Therefore, measurement protocols should incorporate both spatial and time averaging. In the investigated area, a preliminary scan may be performed to determine the distribution of the field and the maximum field value (CENELEC 2008a). For spatial averaging, field values are measured over a grid covering the area (IEC 2017, CENELEC 2008b). To assess the whole-body human exposure to RF fields, spatial averaging is also carried out over an area equivalent to the vertical cross section of the human body: three points at the heights of 1.1 m, 1.5 m and 1.7 m above the terrain are basically recommended by International Telecommunication Union (ITU) (ITU 2008).

For the RF EMF measurements, a 6-minute averaging time is recommended (ICNIRP). This averaging time is based on the heating phenomena, not on the time variation of RF fields.

There are two methods for measuring EMF exposure levels: broadband measurements and narrow band measurements. The broadband measurements, performed by a field meter and a broadband probe, give a single exposure value for the whole frequency range. Narrow band measurements can be performed using a combination of a receiving antenna and a spectrum analyser. Narrow band measurements are selective frequency measurements that allow you to obtain an exposure value for each of the considered RF signals and to identify the RF sources that contribute more to the total RF exposure.

Broadband measurements are the simplest method for measuring exposure levels and can be adopted as a first approach in survey monitoring. Narrow band measurements should be used to get results that are more accurate.

Exposure measurements can be performed by personal meters (sometimes called 'dosimeters') worn on the body. These meters allow you to measure exposure to fields from all sources and at all places by individual encounter. Since the human body does not perturb the ELF magnetic field, personal meters have been used in many monitoring campaigns for assessment of exposure to ELF sources. There are also personal meters for RF exposure used to ensure RF safety and prevent accidental overexposure to dangerous electromagnetic fields near broadcast-transmitting antennas. However, measurements performed by RF personal meters are not accurate because RF fields are strongly affected by the presence of the human body.

For exposure to mobile phones, which are used next to the body, the metrics used to quantify the RF EMF is SAR and not electric field or power density. The compliance assessment shall be achieved by measuring SAR according to measurement procedures as defined in the IEC technical standard (IEC 2016).

5.4.2 Theoretical Calculations

Unperturbed fields can be calculated using a simplified theoretical approach or by numerical methods. In the second case, a very detailed model of the source is needed.

Regarding exposure to ELF magnetic fields, power lines are likely to be the most important source of exposure for a population.

Simplified 2-D models based on the Biot-Savart law are widely used for computation of magnetic field of an overhead power line. In these models, power-line conductors are approximated by single thin wires in straight lines parallel to the Earth's surface. The magnetic flux density, $B(r)$, generated by infinitely long straight wires can be obtained by the following equation:

$$B(r) = \frac{\mu_0 I}{2\pi r} \text{ T} \tag{5.9}$$

where I is the current flowing through the wire in amperes (A), μ_0 is the permeability of free space, equal to $4\pi 10^{-7}$ H/m (henry per meter), and r is the radial distance from the wire in meters.

Each power-line conductor can be considered a line source that contributes to the total magnetic field generated by the power line. The contribution of each power-line conductor at a given distance from the power line span is calculated by the Equation 5.9.

Considering the above-mentioned simplified model, the calculation of the magnetic field from a power line requires the following data:

- current flowing through the conductors (load current);
- relative positions of conductors on suspension towers;
- minimum clearance to ground of conductors.

A more accurate computation of the magnetic field produced by an overhead power line can be obtained using more sophisticated 3D models, which can take the catenary form of power-line conductors into account (Modric et al. 2015). The catenary form of the conductor line is reconstructed by simulating it with a sequence of a large number of straight segments. The magnetic field generated by i-th conductor, B_i, is determined as the sum of the contributions of all power line conductor segments over the entire conductor line:

$$\vec{B_i} = \sum_{j=1}^{N_S} \frac{\mu_0 I}{4\pi} \int_{\Gamma_j} \frac{\vec{dl} \times \vec{r}}{r^3} \text{ T} \tag{5.10}$$

where N_S is the number of conductor segments, dl is the infinitesimal conductor segment whose direction coincides with the direction of the current I, r is the distance between the segment axis and the calculation point and Γ_j is the integration path along the j-th conductor segment axis.

Taking into account that in a power line there are more conductors generating the magnetic field, the total magnetic field from the power line is obtained by superposition of the contributions produced by each conductor.

Transmitting antennas for radio-TV broadcasting and mobile services are among the most important sources of RF-EMF exposure for the population.

The simplest method for the theoretical exposure assessment of RF-EMF emitted by transmitting antennas consists in calculating the field level using the point-source model. According to this model, the transmitting antenna is represented by one point source, located in the antenna electric centre and having its radiation pattern. In this case, the electric field strength at a distance R from the antenna is given by:

$$E = \frac{\sqrt{30 \cdot P \cdot G \cdot F(\theta, \phi)}}{R} \quad \text{V/m} \tag{5.11}$$

where P is the average power, in watts (W), supplied to the transmitting antenna; G is the maximum gain of the transmitting antenna; and $F(\theta,\phi)$ is the normalised gain in the direction given by spherical coordinates (θ,ϕ).

Gain, G, is an antenna property dealing with its ability to radiate power in a desired direction, and it is defined as the ratio of power radiated by the antenna per unit solid angle, multiplied by 4π, to the total power supplied to the antenna.

This model can estimate exposure levels with good accuracy in the far field region and in the free space, but it is not applicable in the vicinity of the antenna, in the near field region. In this region, the dimensions of the antenna need to be taken into account for an adequate estimation of exposure levels.

Data required for calculation of exposure levels using a point source model are:

- distance to the transmitting antenna;
- power supplied to the transmitting antenna;
- gain and radiation pattern of the transmitting antenna.

Higher accuracy in calculating the exposure levels to RF radiation emitted by transmitting antennas is achieved by numerical methods that are based on solving Maxwell's equations in frequency or time domain. These methods, which include method of moments (MoM), finite difference time domain (FDTD) and many others, may be used for any field region and allow you to take into account all conductive objects causing reflection and diffraction phenomena (Taflove and Hagness 2000, Kunz and Luebbers 1993, Glassner 1989, Hansen 1990).

Numerical methods require detailed data concerning the geometry and feeding arrangement of the radiating antennas, such as the spatial position of each panel and the power of each panel feeding.

5.5 Sources of Electromagnetic Fields

In the last 50–100 years, levels of time varying EMFs in our environment have increased because of technological progress. Manmade electromagnetic field strengths are now a lot higher than natural background field strengths, so artificial sources are the dominant sources of environmental electromagnetic fields (Mild and Greenebaum 2007, SCENIHR 2015).

Natural background electromagnetic fields, at frequencies below about 30 MHz, come from thunderstorm activity, and they are characterised by random high peak transients. The generated electromagnetic pulse gives a peak electric field strength of about 10 kV/m at close range and from 5 to 20 V/m at 30 km (Willet et al. 1990). Natural background levels of ELF B field are about 50 picoTesla (pT).

At frequencies above 30 MHz, two sources contribute to the RF natural background: extraterrestrial sources and terrestrial sources. Extraterrestrial sources are the sun and the so-called cosmic background radiation remaining after the big bang. The power density of this component is a few microwatts per square meter. The terrestrial component of the RF natural background is the black-body radiation arising from the warm surface of the ground and from any other surface, including the skin surface of human bodies. The power density of RF electromagnetic fields associated with this background component is a few milliwatts per square meter, so it is much higher than the extraterrestrial component.

As we will see in the next paragraphs, the above-reported naturally occurring electromagnetic field levels are much lower than the levels of ELF fields and RF radiation emitted from artificial sources.

At ELF frequencies, in the range between 50 Hz and 300 Hz, the wavelength of the EM fields in air varies from 6000 to 1000 km. Since the reactive near field is taken to extend to a distance of $1/2\pi$ times the wavelength ($\lambda/2\pi$) from the source surface, human exposure to ELF sources always occurs in the reactive near field. Consequently, to assess human exposure to EMF generated by ELF sources, both electric and magnetic fields have to be determined.

All artificial sources of electromagnetic fields can be divided into two classes: intentional and unintentional. Intentional sources are devices that are designed to emit EMFs to perform their tasks. Important intentional sources are the telecommunication transmitting antennas that determine the RF EMF distribution near a transmitting station. Other intentional sources include electronic devices, such as cordless telephones and garage doors openers. For the unintentional sources, the emission of EMFs from the device is not the aim for which it was built and could be reduced by shielding the device itself. For example, all power-supplied devices, such as household appliances, are unintentional sources of ELF electricand magnetic fields.

In the following sections, the typical emission characteristics of some of the widespread sources and those producing the highest exposures will be described.

5.5.1 Sources of Electromagnetic Fields in the Living Environment

General public exposure to static, extremely low-frequency (ELF), intermediate-frequency (IF) and radio-frequency (RF) electromagnetic fields is due to a great variety of sources, present both indoors and outdoors. The main sources of EMF exposure for the general population are reported in Table 5.3.

5.5.1.1 Sources of Static Fields

Two conductors, positive pole and negative pole, powered by direct voltage, schematically constitute high-voltage direct current transmission lines (HVDC). Typical 350 kV ÷ 700 kV HVDC lines can be underground, in ditches ranging from 1.3 m to 2 m deep, where power conductors (two for each pole with a diameter of about 30 mm) are buried in a lean concrete mix with good heat capacity. HVDC overhead lines rely on the air gap between the conductors and the ground as electrical insulation overhead lines. They can be used up to 800 kV and are capable of transporting more than 6 GW on a single bipole route.

Such transmission lines were used in the past for undersea connections, but now are widely used due to their very low reactance, which makes them especially convenient for long-distance linking: there is no need for compensating stations along the line. Because of that, HVDCs are expected to expose large parts of the population to DC magnetic fields. The strength of such a field is, however, comparable with the natural geomagnetic field (variable between about 0.035 and 0.07 mT). Magnetic field levels up to about 10 times the Earth's magnetic field can be reached. In any case, exposure levels to HVDC are much lower than the limit value of 400 mT recommended by ICNIRP for the general public (ICNIRP 2009).

Electrostatic fields generated by HVDCs can be on the order of 20–30 kV/m. These field levels are of a magnitude similar to that of the naturally occurring static field which exists beneath thunder clouds, and they have no known adverse biological effects. The electric field from HVDC lines causes partial electrical breakdown of the air surrounding points on the surface of the line conductor and creates air ions (Corona

TABLE 5.3 Main Sources of Public Exposure to EMFs

Application	Frequency Range
Static Electric and Magnetic Fields	
DC high-voltage transmission lines	0
Traction power lines (railways, tramways)	0 and 16 Hz
ELF Electric and Magnetic Fields	
Electric power transmission and distribution lines	50–60 Hz
Electric power transmission and distribution lines, household appliances	50–2000 Hz
Radiofrequency Fields	
ISM (Industrial-, Scientific-, Medical) devices 2400 MHz band (example: microwave oven)	2400–2483.5 MHz
Telecommunication System and Radio Terminals	
FM broadcasting stations	87.5–108 MHz
TETRA systems	380–399.9 MHz
UHF TV broadcasting stations	470–860 MHz
Mobile radio base stations and handsets (GSM/UMTS/LTE systems)	880–960 MHz
	1710–1880 MHz
	1920–2170 MHz
	2300–2690 MHz
	3400–3800 MHz
DECT devices	1880–1900 MHz (Europe)
	1900–1930 MHz (Other Countries)
W-LAN /Wi-Fi devices	2400–2500 MHz
	5725–5875 MHz

effect). Air ions can charge particles in the air, including polluted particles. Air ions and charges on aerosols collectively are called 'space charge', and their presence adds to the electrostatic field generated by HVDC lines. It has been hypothesised that charged particles raise people's exposure to air pollution, but this hypothesis has not been confirmed and can be considered very unlikely. No health agencies or regulatory scientific institutions have proposed exposure limits for space charge.

5.5.1.2 Sources of ELF Electric and Magnetic Fields

ELF electric and magnetic field exposures exist in the home because of all types of electrical equipment and building wiring, as well as nearby power lines. Accordingly, in the ELF frequency range the main sources of public concern are transmission lines, power transformers installed inside residential buildings, typical household appliances and their power supplies (d'Amore et al. 2001, Azoulay et al. 2009). Depending on the source, human exposure can be very localised. In recent years, attention has been directed towards people living next to MV/LV transformer electrical substations. They consist of the set of devices dedicated to the transformation of the voltage supplied by the distribution network at medium voltage (e.g. 15 kV or 20 kV) into voltage values suitable for the power supply of the low voltage lines (e.g. 380 V or 400 V). Electrical substations include terminations of the distribution lines, switchgears and power transformers. If the power transformer is really adjacent to dwellings, long-term exposure of inhabitants to ELF magnetic fields can occur. In Figure 5.2 the spatial distribution of magnetic field levels due to devices inside a MV/LV substation, at a height of 1.5 m above the ground, is shown. At distances greater than 3 ÷ 5 m

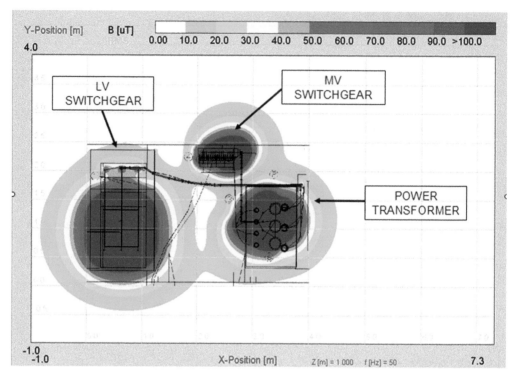

FIGURE 5.2   Spatial distribution of magnetic field levels inside a MV/LV substation at a height of 1.5 m above ground.

outside the walls of the substation, depending on the position of the primary and secondary windings of the transformer and the current flowing through the windings, magnetic field levels are 20 times lower than 200 µT, which is the reference level for general public exposure according to ICNIRP guidelines.

Electric and magnetic fields generated by power lines depend on the line voltage and on the current flowing through the conductors, respectively. Electric fields are time constant and can be easily shielded by conductive materials. As many construction materials are not electrical insulators, it is uncommon to find high electric field levels inside buildings. Magnetic and electric field levels from power lines depend not only on the load current and line voltage but also on the line geometry (tower configuration and the of the conductors height from the ground) (CENELEC 2009).

The spatial profile of an electric field near a high voltage transmission line is shown in Figure 5.3. In this figure, the E field levels are reported as a function of distance from the centre of the span of a typical 220 kV single circuit three-phase transmission line, at a height of 1.5 m above the ground.

The clearance of the lowest conductor from the ground is one of the factors affecting field levels in the vicinity of a power line. Figure 5.4 shows spatial profiles of magnetic fields produced by single circuit 220 kV transmission lines that have their lowest conductors at different heights from ground level.

For transferring very high power, the transmission lines are usually designed as double circuit three-phase overhead lines. To reduce the mutual interference between the two circuits, the overhead lines are usually transposed by changing the position of the phase conductors, as shown in Figure 5.5.

The effect of the phase arrangement on the spatial profile of a magnetic field generated by an overhead power line is illustrated in Figure 5.6. This figure shows the spatial profiles for double circuit 380 kV transmission lines with untransposed and transposed phases. The spatial profile for a single circuit 380 kV transmission line is also reported. Magnetic field levels generated by the three types of transmission lines

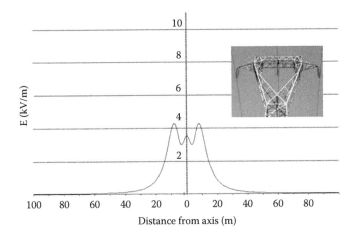

FIGURE 5.3   Typical electric field profile for a 220 kV power line with the tower design shown.

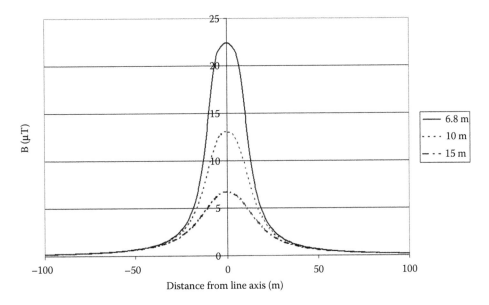

FIGURE 5.4   Typical magnetic field profiles for three 220 kV overhead power lines with clearance of the lowest conductor from the ground equal to 6.8 m, 10 m and 15 m.

are calculated at a height of 1.5 m above the ground, and the amount of current flowing through the circuits is assumed to be 2300 A.

As we can see from Figure 5.6, the optimised arrangement of the conductors, with transposed phases, can reduce magnetic field strengths generated by the power line and exposure levels near the line itself. Figure 5.6 shows also that single circuit power lines generate higher magnetic fields than double circuit power lines with the same current loads.

ELF fields in homes are primarily caused by indoor electric appliances such as alarm clocks, bedside lamps or the building's electricity supply. Moreover, most of these devices (such as drilling tools, switched power supplies to laptops, mobile phone chargers and similar devices) use electronics for power regulation. As a consequence, the frequency content of the magnetic field includes odd harmonics. In particular, the third harmonic (150 Hz) has become another ELF dominating frequency in the environment.

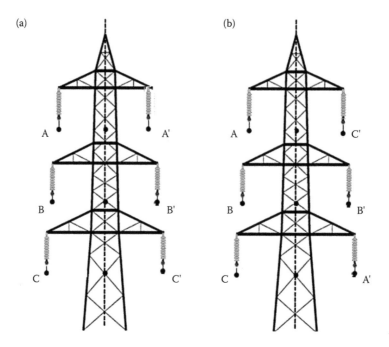

FIGURE 5.5 Double circuit power line with (a) untransposed phase arrangement and (b) transposed phase arrangement.

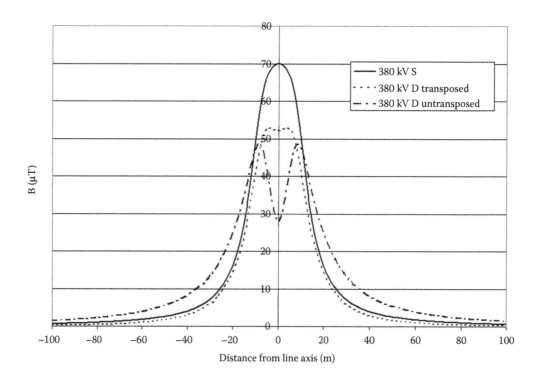

FIGURE 5.6 Typical magnetic field profiles for three types of 380 kV power lines: single circuit (S), double circuit with transposed phase arrangement (D transposed) and double circuit with untransposed phase arrangement (D untransposed).

More household appliances generating EMFs in the intermediate frequency (IF) range have recently appeared. Among them, the main source of exposure in the general public is induction hobs, which can expose their users to IF magnetic fields higher than the reference levels recommended by exposure guidelines (Christ et al. 2012).

5.5.1.3 Sources of RF Electromagnetic Fields

In the RF frequency range, from 100 kHz up to 300 GHz, human exposure to environmental sources is dominated by broadcasting antennas and mobile communication base stations (Anglesio et al. 2001). In recent years, the shift from analogous systems of signal transmission to digital ones led to a dramatic development of telecommunication facilities and services. Digital technology has improved the rational use of radio frequency resources, as digital devices need less power to provide services. Regarding mobile phone services, 2nd-generation cellular systems (2G) were the first to apply digital transmission technologies such as Time Division Multiple Access (TDMA) for voice and data communication. In TDMA the radio resources are shared in time so as to offer multiple digital channels using different time slots on a shared frequency carrier. According to this multiple access technique, a fixed sending frequency is assigned to a transmission channel between a sender and a receiver for a certain amount of time (time slot). Global System for Mobile Communications (GSM) was the first commercially operated digital cellular system. It was first developed in the 1980s through a pan-European initiative involving the European Commission, telecommunications operators and equipment manufacturers. GSM systems use an FDMA (Frequency Division Multiple Access)/TDMA technique for voice and data transmission (Lee 2008). The available spectrum band is divided into carrier frequencies whose bandwidth is 200 kHz, and each carrier frequency is also divided into eight physical channels (time slots). Each user is assigned a pair of frequencies for uplink and downlink and a time slot during a frame. In the following Figures 5.7 and 5.8, respectively, the FDMA/TDMA signals and the frame structure of GSM signals are schematically illustrated.

Technical progress led to the evolution of the GSM standard by introducing new applications such as General Packet Radio Service (GPRS) and Enhanced Data Rates for GSM Evolution (EDGE) to fulfill the need for higher data rates per data connection.

Each base station in a GSM network provides a broadcast control channel (BCCH) with information about the network and the base station itself. It is transmitted at practically constant field strength. One or more frequency channels, called traffic channels (TCH), are added to this for transmitting voice and data signals. The field strength of these channels varies with the load, and they can also be switched off completely. The worst case field strength, i.e. the maximum field strength when all TCHs are fully loaded, can be calculated from the time constant field strength of the BCCH.

Universal Mobile Communication System (UMTS) represents the third generation of mobile standards and technology. UMTS employs the radio access technique Wideband Code Division Multiple Access (WCDMA). According to this technique, several users share the same bandwidth, equal to 5 MHz for each UMTS signal, at the same time. User information bits are spread over the bandwidth by multiplying the user data with the spreading code. Consequently, each channel is coded and mixed with other

FIGURE 5.7 Scheme of channel allocation for GSM systems with FDMA/TDMA multiple access technique.

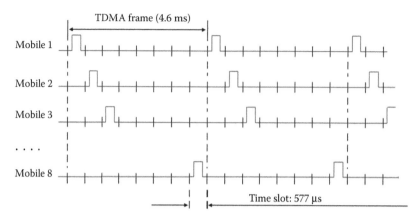

FIGURE 5.8   GSM frame structure. A single frame is composed of 8 time slots.

channels. Original information can be extracted by again multiplying the signal with the same code sequence (de-spreading process). A comparison between GSM and UMTS carrier signals is shown in Figure 5.9.

Like BCCH in GSM networks, Primary Common Pilot Channel (P-CPICH) has a constant, known power level in UMTS networks. P-CPICH is an unmodulated channel that is scrambled with the primary scrambling code of the base station. The power of P-CPICH is defined by the factor β between the power of the P-CPICH and the maximal power of the base station (%). Since the provider knows the factor β for each base station, it is possible to extrapolate the exposure level at maximum output power from a measurement of the reference channel P-CPICH.

An example of frequency domain measurement of GSM and UMTS signals performed by a spectrum analyser is shown in Figure 5.10. The frequency spectrum shows UMTS and GSM carriers with bandwidths of 5 MHz and 200 kHz, respectively.

The evolution from 3rd- to 4th-generation (4G) mobile technologies has recently led to Long-Term Evolution (LTE) systems. Furthermore, new wireless technologies, referred to as 5th-generation (5G), are rapidly being developed. The new 5th generation for cellular systems will probably start to come to fruition around 2020, with deployment following afterward. These newest developments in wireless systems will be discussed in Chapter 8.

Field strength levels near a radiating antenna, such as broadcasting antennas or base station antennas for mobile communications, depend on the radiation pattern, supplied power, gain and beam tilt of the antenna. In the vicinity of these antennas the incident electric field is not uniform due to the high directivity on the vertical plane of the radiation pattern. Consequently, in the area close to the antenna, general population exposure levels can be higher at greater distances from the antenna. Figure 5.11 shows the

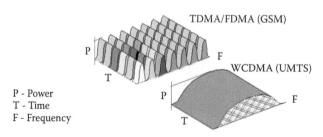

FIGURE 5.9   Scheme of channel allocation for GSM and UMTS systems using FDMA/TDMA and WCDMA multiple access techniques, respectively.

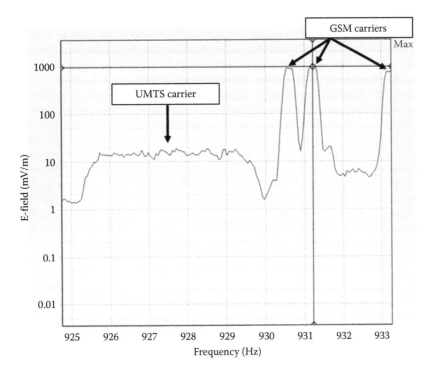

FIGURE 5.10   Frequency spectrum of GSM and UMTS signals.

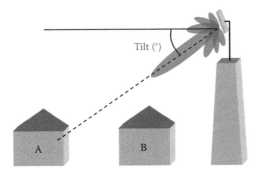

FIGURE 5.11   Effect of a typical vertical radiation pattern of a base station antenna for mobile communication on exposure levels in two neighbouring houses located at different distances from the base station. Electromagnetic field exposure level is higher in house (A), which is in the main lobe of the radiation pattern.

vertical radiation pattern of an antenna installed near two houses. The EMF exposure level at house (A), which is farther from the antenna location, is higher than that at closer house (B), as house (A) is in the main lobe of radiation emitted from the antenna, while house (B) is outside of the main beam of the antenna.

On the ground level, the actual variation of EMF levels with distance from the antenna could be rather complicated, due to the existence of side lobes of radiation patterns. Figure 5.12 shows the typical trend of the electric field levels emitted by a base station for mobile phones at a height of 1.5 m from the ground. As we can see from this figure, there are several relative maximum and minimum points from the antenna position to the distance where the main lobe reaches the ground.

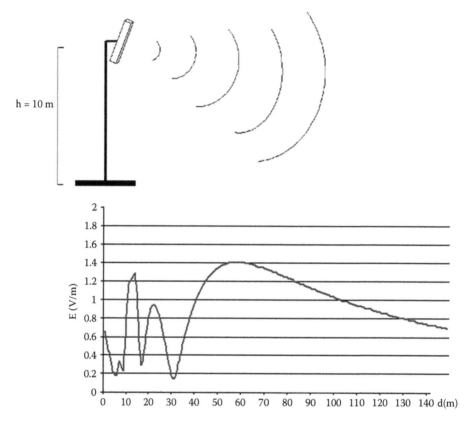

FIGURE 5.12 Typical RF electric field strength as a function of distance from a transmitting antenna at the base station for mobile phones, at a height of 1.5 m above ground. The antenna's electrical centre is 10 m above ground.

In the last few years, the number of RF sources used in residential indoor environments has increased. The installation of access points and short-range base stations, such as 3G femtocells, Wi-Fi hotspots and DECT devices, has given rise to new sources of exposure to RF EMFs. Generally, at distances greater than 1 m from these appliances, the emitted EMF levels do not exceed common background levels, so they can be considered near field sources, which are responsible for highly localised exposure. Consequently, the EMF emissions from these devices, even when added together, do not significantly affect the average exposure of the general public in indoor environments.

Among the devices that are carried on the person and can transmit in the immediate proximity to the body, mobile phones are the most important for human exposure to EMFs. They produce highly non-uniform EMFs over the body which diminish rapidly in strength with increasing distance. Human exposure to mobile phones is also influenced by the technical features of mobile phone networks, such as the adaptive power control (APC) and discontinuous transmission (DTX). APC and DTX reduce the time-averaged output power of a mobile phone during a call by about 50% and 30%, respectively (Wiart et al. 2000, Vrijheid et al. 2009). Furthermore, average output power varies depending on the environmental reception level. Measurements of the output power of mobile phones performed by the Regional Environmental Protection Agency of Piemonte (Arpa Piemonte) showed that levels of exposure to GSM mobile phones in outdoor areas with higher environmental electric field strengths and a good reception quality are 60 times lower than they are in indoor situation with bad signal reception. For UMTS systems, this variability is much higher, reaching a factor of about 2000 (d'Amore et al. 2014).

5.5.2 Sources of Electromagnetic Fields in Workplaces

There exists a great variety of sources of electromagnetic fields in workplaces, depending on the type of workplace, the purpose of the source itself, the characteristics of the electromagnetic fields emitted, the duty cycles, etc.

In general, in a workplace the geometry and location of the EMF source, as well as the frequency and level of EMF produced by it, can be unstable. Consequently, the characteristics of EMFs in the workplace are often more complex than those of fields experienced in a general-public environment, and there is a need for a specific exposure assessment protocol.

In addition to that, EMF exposure assessments in workplaces often have to take into account the coexistence of devices emitting fields in different frequency ranges and with various time-domain evolution of the signals.

In the following, we will focus on some sources of particular interest in terms of human exposure to EMFs, choosing them from among the sources that can be found in some categories of workplaces.

5.5.2.1 Sources in Offices and Shops

The EMF sources that can be found in this type of workplace are, for example, computers and office devices, power supply and electric networks, WLAN networks, mobile phones, radiofrequency identification systems (RFID), anti-theft devices and RF activated lighting devices.

In general, these sources produce EMFs that are not critical for human exposure, because there exist specific production standards that limit the emissions: EN 50360 for mobile phones, EN 50364 for RFID and electronic article surveillance (EAS) devices, IEEE802.11 for WLAN in freely accessible frequency bands, etc.

Following is some information about two types of sources that can require, under particular conditions, a specific exposure assessment.

In recent years there has been a peak in the production and sale of compact fluorescent lamps (CFLs). They are nothing but compact versions of classic neon lamps. The main difference is that CFLs incorporate the ballast power supply, while neon lamps have a separate starter and power supply. The power supply transforms the alternating current to the network frequency (50 Hz) into medium-frequency alternating current (25 kHz–70 kHz). Medium-frequency current goes through the tube containing a mixture of several gases, including mercury, which, when excited, emits ultraviolet radiation (UV). The internal lining of the tube, made up of phosphors, converts the UV into visible light. CFLs can emit significant levels of optical radiation (UV and blue light), and, in a limited space, electromagnetic fields as well.

Results of a survey of EMFs emitted by 17 types of CFLs, carried out by Arpa Piemonte, indicate that EMF emissions from 17 types of lamps are in the frequency range from 30 kHz to 60 kHz, with electric field levels varying from 20 V/m to 86 V/m at a distance of 5 cm from the lamp (Adda et al. 2012). The highest electric field level measured at 5 cm from the lamp is comparable to the limits set for population exposure in ICNIRP guidelines, but a very rapid decrease of the field levels in the first 30–40 cm was observed.

Radio frequency identification (RFID) systems and electronic article surveillance (EAS) systems are used for contactless identification and tracking of objects with electromagnetic fields. They basically consist of a tag on an object and a reader to read and write data on the tag. There are several applications for these devices that include, for example, ID cards, electric locks, ski passes, toll systems and animal identification systems. Depending on the application, the manufacturer and the technology, different frequencies are used: from 20 Hz to 18 kHz at the lower frequencies, radiofrequency devices range from 1.8 MHz to 10 MHz and microwave devices range from 902 MHz to 2.45 GHz (Wout et al. 2012).

RFID devices operate at low power levels, with a very short duty cycle (generally about 10%), but the electromagnetic fields they emit are usually scattered across the area surrounding the device, and at distances of less than 1 to 2 meters from the source they can, in some cases, exceed the limits set for population exposure. In this case, supplementary information comes from numerical modeling to evaluate dosimetric parameters (Fiocchi et al. 2013).

In Fiocchi et al. 2013, the second class of RFID systems reviewed (UHF RFID) presents different issues. Most of them, and in particular the ones that are fixed in one place, such as a cash register at a store check-out, are designed to be continuously turned on and to generate power allowing communication at a distance of more than 1 m. While negligible exposure levels have been found for children and adults, pregnant women and their fetuses could be overexposed. Indeed, the local SAR in the fetus can exceed the maximum allowable levels recommended by ICNIRP for public RF exposure (ICNIRP 1998).

5.5.2.2 Sources in Industry

Industry applications often use devices producing high-level electromagnetic fields with very specific spectral and time-domain characteristics. EMF sources which provide high-level occupational exposure include, for instance, welding devices (arc and plasma welding above all), heating devices (induction heaters, dielectric heaters and microwave ovens), demagnetisers and electrolysis devices.

Arc welding is a process that is used to join metal to metal by employing electricity to create enough heat to melt metal. It uses a power supply to create an electric arc between an electrode and the base material to melt the metals at the welding point. It can use either direct (DC) or alternating (AC) current. The pulsed current levels can reach thousands of ampères, and the supply cable is usually very close to the worker's body, so high-level exposure to fast transient magnetic fields can occur.

Capacitive (or dielectrical) heating installations produce heat in non-conductive materials. Therefore, electromagnetic radiation with frequencies within the ISM bands (typical frequencies 13.56 MHz, 27.12 MHz) is used to penetrate the non-conductive material. For this type of source, the level of EMF exposure strongly depends on the duty cycle, the working procedures and the grounding of metal parts of the device and the objects surrounding the installation area (because of re-irradiated fields).

Induction heating devices exploit the currents induced in materials exposed to a magnetic field generated by a coil to obtain heating. In industry, different frequencies of the currents powering the coil can be used: lower-frequency generators work at frequencies below 5 kHz, intermediate-frequency generators work from 5 to 30 kHz and higher-frequency generators work at frequencies up to 1–2 MHz. These heating devices are used in different production divisions for various applications, such as precious metal casting (for dental or jewellery purposes), casting or forming of brass, bronze, iron alloys, steel, copper, aluminium (foundry, metallurgical industry), wire drawing and temper.

They can generate strong magnetic fields, due to the high power often necessary for these applications. Furthermore, the working procedures used often lead to extreme proximity of the worker to the coils in

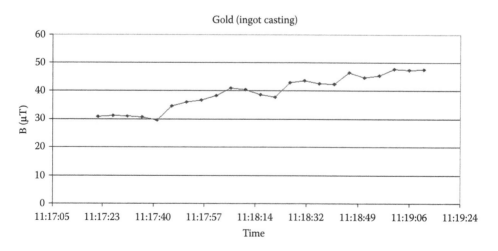

FIGURE 5.13 Time-evolution of magnetic flux density during a casting application.

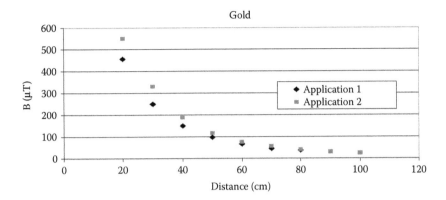

order to control the process. Therefore, using this kind of device can have a non-negligible potential impact on worker exposure, with respect to the number of exposed workers and the exposure levels.

For example, the magnetic field levels measured by Arpa Piemonte on precious metal casting furnaces, at a distance of about 20 cm from the crucible, ranged from a few µT to almost 400 µT (depending on the furnace) in the low- and intermediate-frequency bands and from 0.1 A/m to 0.4 A/m for higher-frequency generators (Adda et al. 2009).

In general, there is great variation in the EMF harmonic content from device to device. For some devices the harmonic components are significant, for others they are low and the fundamental component is predominant.

Figures 5.13 and 5.14 show the time-evolution of B field during a gold-casting application (Figure 5.13) and the decrease of B field with increasing distance from the gold-casting device (Figure 5.14).

A review of the results of measurement campaigns on workers' exposure to induction heating devices used in industry is presented in Chapter 9.

5.5.2.3 Sources in Medical and Beauty Centres

The working principle of many medical devices is based on the generation of EMFs: electrosurgery devices, trans-cranial magnetic stimulators, MRI scanners and diathermy devices are some examples of sources with significant emissions.

High-frequency electrosurgery uses high-frequency alternating current, which is converted to heat by resistance as it passes through the tissue. This device can expose the surgeon to substantial levels of EMFs.

Diathermy devices are used in medicine to heat specific regions and tissues of the body with electromagnetic radiation or currents. In contrast to infrared (IR) devices, diathermy can also heat tissues beneath the skin. The frequencies used are part of the ISM bands and are applied with electrodes or antennas.

There is some variety in the design of diathermy devices, and probably more diversified designs will be introduced in the future. In any case, certain common features can be distinguished. A diathermy device consists of an applicator, an RF generator and a control console. The applicator, also called an electrode, applies RF energy to a certain portion of a patient's body. The RF generator is usually housed in the control console, although there are designs where it is incorporated into the applicator. RF power is delivered from the generator to the applicator by two cables forming an unshielded two-wire transmission line.

Two basic types of electrodes (applicators) are in use, the capacitor type and the inductor type such as a 'pancake' coil or diplode. The heating mechanism and profile are somewhat different for the two types. In the case of capacitive electrodes, tissue heating is due to the RF electric field, while inductive electrodes (coils) produce heat via eddy currents induced in tissue by the magnetic field.

The stray fields close to the applicator but outside the treatment area are relatively strong and highly spatially inhomogeneous. High-intensity fields are also produced near the cables. There is no simple formula to predict the intensities of stray fields, except that the higher the power setting the stronger they are, if all other parameters (e.g. electrode placement) remain unchanged.

Another example of a treatment that exploits EMF is coblation (typically used for facial rejuvenation), which uses an electrical current to ionise a conduction medium such as isotonic saline. The ionised medium is then used to transmit heat to tissue.

Coblation, or 'cold ablation', technology has been applied in a variety of surgical disciplines. Originally created for the vaporization of cartilage during arthroscopic procedures, it relies on the creation of a plasma field within an electrically conductive solution, or gel, between closely approximated electrodes. The resulting plasma field causes a predictable de-epithelialization at temperatures significantly lower ($<90°C$) than those seen with conventional electrosurgical devices. However, the most potent effect with regard to treatment of the aging face or deep acne scarring results from controlled injury to the superficial dermis, leading to new collagen formation, re-epithelialization and subsequent facial rejuvenation.

References

Adda S., Anglesio L., d'Amore G., Fontana M., Tura P., 2009 (in italian). *Valutazione dell'esposizione a campi elettromagnetici di lavoratori in comparti che utilizzano riscaldatori ad induzione*, Atti del Quarto Convegno Nazionale 'Controllo Ambientale degli Agenti Fisici: nuove prospettive e problematiche emergenti', Vercelli, 24–27 marzo 2009.

Adda S., Alviano M., Anglesio L., Bonino A., d'Amore G., Facta S., Saudino Fusette S. (in italian) 2012. *Le nuove lampade a risparmio energetico: valutazione dell'esposizione ai campi elettromagnetici e alla radiazione ultravioletta.*, Atti del V Convegno Nazionale Il Controllo degli Agenti Fisici: Ambiente, Salute e Qualità della Vita, Novara 6-8 giugno 2012 (ISBN 978-88-7479-118-7).

Anglesio L., Benedetto A., Bonino A., Colla D., Martire F., Saudino Fusette S., d'Amore G. 2001. *Population Exposure to Electromagnetic Fields Generated by Radio Base Stations: Evaluation of the Urban Background by Using Previsional Model and Instrumental Measurements. Radiat. Prot. Dosimetry,* Vol. 97 (4), pp. 355–358.

Azoulay A., Merckel O., Letertre T. 2009. *ELF Environmental Exposure of the Population: A Case Study in a Village South of Paris. EMC'09/Kyoto*, pp. 619–622. Available from: https://www.researchgate.net/publication/278774452

CENELEC EN 50383. 2010. *Basic Standard for the Calculation and Measurement of Electromagnetic Field Strength and SAR Related to Human Exposure from Radio base Stations and Fixed Terminal Stations for Wireless Telecommunication Systems.*

CENELEC EN 50413. 2008a. *Basic Standard on Measurement and Calculation Procedures for Human Exposure to Electric, Magnetic and Electromagnetic Fields (0 Hz–300 GHz).*

CENELEC EN 50492. 2008b. +A1:2014 *Basic Standard for the in-situ Measurement of Electromagnetic Field Strength Related to Human Exposure in the Vicinity of Base Stations.*

CENELEC EN 62110. 2009. *Electric and Magnetic Field Levels Generated by AC Power Systems. Measurement Procedures with Regard to Public Exposure.*

Christ A., Guldimann R., Buhlmann B., Zefferer M., Bakker J.F., van Rhoon G.C., Kuster N. 2012. Exposure of the human body to professional and domestic induction Cooktops Compared to the Basic Restrictions. *Bioelectromagnetics* 33(8): 695–705.

d'Amore G., Anglesio L., Benedetto A., Mantovan M., Polesel M. 2014. *Human exposure evaluation to mobile phone electromagnetic emissions* Atti del III convegno nazionale 'Interazioni tra campi elettromagnetici e biosistemi', Napoli 2–4 luglio 2014 http://www.icemb.org/napoli/Atti .del.terzo.Convegno.Nazionale.pdf d'Amore G., Anglesio L., Tasso M., Benedetto A., Roletti S. 2001. Outdoor background ELF magnetic fields in an urban environment, *Radiat. Prot. Dosimetry*, Vol. 94 (4), pp. 375–380.

EFHRAN Report D2. 2012. *Risk Analysis of Human Exposure to Electromagnetic Fields*, October 2012. Available from http://efhran.polimi.it

EFHRAN Report D4. 2010. *Report on the Level of Exposure (Frequency, Patterns and Modulation) in the European Union Part 1: Radiofrequency (RF) Radiation*, august 2010. Available from http://efhran.polimi.it

Fiocchi S., Markakis I.A., Ravazzani P. 2013. SAR exposure from UHF RFID reader in adult, child, pregnant woman, and fetus anatomical models. *Bioelectromagnetics* 34(6): 443–452.

Gajsek P., Ravazzani P., Wiart J., Grellier J., Samaras T., Bacos J., Thuroczy G. 2016. Review of studies concerning *Electromagnetic Field (EMF) exposure assessment in Europe: Low frequency fields (50 Hz–100 kHz)*. International Journal of Environmental Research and Public Health 13: 875.

Gajsek P., Ravazzani P., Wiart J., Grellier J., Samaras T., Thuroczy G. 2015. Electromagnetic field exposure assessment in Europe – radiofrequency fields (10 MHz–6 GHz). *Journal of Exposure Science and Environmental Epidemiology* 25: 37–44.

Glassner A.S. 1989. *An Introduction to Ray Tracing*, San Francisco: Morgan Kaufmann Publishers.

Hansen R.C. 1990. *Moment Methods in Antennas and Scattering*, Artech House: London.

ICNIRP. 1998. Guidelines for limiting exposure to time-varying electric, magnetic and electromagnetic fields (up to 300 GHz). *Health Physics* 74(4): 494–522. Available from http://www.icnirp.org

ICNIRP. 2009. Guidelines on limits of exposure to static magnetic fields. *Health Physics* 96(4): 504–514. Available from http://www.icnirp.org

ICNIRP. 2010. Guidelines for limiting exposure to time-varying electric and magnetic fields (1 Hz–100 kHz). *Health Physics* 99(6): 818–836. Available from http://www.icnirp.org

IEC 62209-1. 2016. *Measurement Procedure for the Assessment of Specific Absorption Rate of Human Exposure to Radio Frequency Fields from Hand-Held and Body-Mounted Wireless Communication Devices – Part 1: Devices used Next to the Ear (Frequency Range of 300 MHz to 6 GHz)*.

IEC 62232. 2017. *Determination of RF Field Strength, Power Density and SAR in the Vicinity of Radiocommunication Base Stations for the Purpose of Evaluating Human Exposure*.

IEEE C95.1. 2005. IEEE Standard for Safety Levels with Respect to Human Exposure to Radio Frequency Electromagnetic Fields, 3 kHz to 300 GHz.

ITU Recommendation T K.61. 2008. *Guidance on Measurement and Numerical Prediction of Electromagnetic Fields for Compliance with Human Exposure Limits for Telecommunication Installations*. Available from http://www.itu.int/rec/T-REC-K.61/en

Kunz K.S., Luebbers R.J. 1993. *The Finite Difference Time Domain Method for Electromagnetism*, Boca Raton: CRC Press. Available from http://b-ok.org/book/450548/bae755

Lee W.C. 2008. *Mobile Communications Engineering*. 2nd ed. TataMcGraw-Hill: New Delhi.

Mild H.K., Greenebaum B. 2007. Environmental and occupationally encountered electromagnetic fields. In: Barnes F.S., Greenebaum B., editors. *Handbook of Biological Effects of Electromagnetic Fields*. 3rd ed. (Vol. 1. Bioengineering and biophysical aspects of electromagnetic fields). Boca Raton, FL, USA: CRC Press; pp. 1–34.

Modric T., Vujevic S., Lovri D. 2015. 3D computation of the power lines magnetic field. *Progress In Electromagnetics Research M*, Vol. 41, 1–9.

SCENIHR. 2009. *Health Effects of Exposure to EMF. Scientific Committee on Emerging and Newly Identified Health Risks*. Brussels: European Commission, Health & Consumer Protection DG. Available from https://ec.europa.eu/health/electromagnetic_fields/key_documents_en

SCENIHR. 2015. *Potential Health Effects of Exposure to Electromagnetic Fields (EMF)*. ISBN 978-92-79-30134-6. Brussels: European Commission, Health & Consumer Protection DG. Available from: https://ec.europa.eu/health/electromagnetic_fields/key_documents_en

Stratton J.A. 2007. *Electromagnetic Theory*. McGraw-Hill, New York, 1941. Reprinted, IEEE Press: Piscataway, NJ.

Taflove A., Hagness S.C. 2000. *Computational Electrodynamics: the Finite-Difference Time-Domain*. Method Artech House: Boston, MA.

Vrijheid M., Mann S., Vecchia P., Wiart J. 2009. Determinants of mobile phone output power in a multinational study: Implications for exposure assessment. *Occupational and Environmental Medicine* Vol. 66, n. 10, pp. 664–671.

Wiart J., Dale C., Bosisio A.D. et al. 2000. Analysis of the influence of the power control and discontinuation tranmission on RF exposure with GSM mobile phone. *IEEE Transactions on Electromagnetic Compatibility* 42: pp. 376–385.

Willet J.C., Bailey J.C., Leteinturier C., Krider E.P. 1990. Lightning of electromagnetic radiation field spectra in the interval from 0.2 to 20 MHz. *J Geophys. Res.* 95(D12): 20367–20387.

Wout J., Verloock L., Vermeeren G., Martens L. 2012. In situ magnetic field exposure and ICNIRP-based safety distances for electronic article surveillance systems. *Radiation Protection Dosimetry* 148(4): 420–427.

Zhu G., Gong X., Luo R. 2017. Characterizing and mapping of exposure to radiofrequency electromagnetic fields (20–3,000 Mhz) in Chengdu. *China Health Physics* 112(3): 266–275.

6

Biological Effects from Electromagnetic Field Exposure: The Experience of the European EFHRAN Project*

Paolo Ravazzani
Emma Chiaramello
Serena Fiocchi
Istituto di Elettronica, Ingegneria dell'Informazione e delle Telecomunicazioni CNR IEIIT

James Grellier
University of Exeter Medical School

Marta Parazzini
Istituto di Elettronica, Ingegneria dell'Informazione e delle Telecomunicazioni CNR IEIIT

Aslak Harbo Poulsen
Danish Cancer Society Research Center

Joachim Schüz
International Agency for Research on Cancer (IARC)

Zenon Sienkiewicz
Public Health England, Centre for Radiation, Chemical and Environmental Hazards

---

* The views and opinions expressed in this chapter are those of the authors and do not necessarily reflect the official policy or position of the authors' institutions.

6.1 Introduction

European populations face increasing exposure to novel physical and chemical agents, some of which may be detrimental to human health. Among these agents, electromagnetic fields (EMFs) are one of the most widespread and ubiquitous, especially as many new technologies and applications based on high-frequency fields are being developed and commercialised. Research on the possible health and biological effects of EMFs is being carried out by many centres in Europe, North America, Japan and other countries. The extent and diversity of these research activities, encompassing many areas of medical and biological research, as well as the latest developments in physics and engineering, make it particularly difficult to provide relevant, authoritative and timely input for the development of scientifically sound policies that have an impact on public health. Furthermore, complexities inherent to the hazards, exposures and potential risks may result in misinterpretation of evidence resulting from research or inappropriate application of risk assessment information to conditions or situations outside of those specifically tested.

In 2009, in order to help meet the needs of public health policy-makers in these areas, the European Commission funded the European Health Risk Assessment Network on Electromagnetic Fields Exposure (EFHRAN) project[*]. The EFHRAN project established a network of scientists from 17 European countries to bring together relevant published information about exposure to and effects of electromagnetic fields.

This chapter presents the main outcomes of the EFHRAN project in terms of health risk analysis, which are based on in-depth review of the most important published research exploring the possible effects of EMF on humans, in order to identify potential public health concerns[†]. These outcomes are explained in the context of more recent statements issued by various scientific bodies and state of the art results of European Commission–funded projects.

In order to evaluate the strength of evidence for adverse effects arising as a consequence of exposure to EMFs, EFHRAN made use of the four-point classification system originally proposed by the FP6 Coordination Action EMF (Figure 6.1).

Classification	Necessary Inclusion Criteria
Sufficient evidence	• When a positive relationship is observed between the exposure and the effect investigated • When the effect is replicated in several studies by independent investigators or under different protocols, and when there is a consistent exposure-response relationship • When confounding factors could be ruled out with reasonable confidence
Limited evidence	• When the evidence of the effect is restricted to a few studies, or when there are unsolved questions regarding the adequacy of the design, conduct or interpretation of the study • When confounding factors could not be ruled out in the studies with reasonable confidence
Inadequate evidence	• When the studies are of insufficient quality, consistency or statistical power to permit a conclusion
Evidence suggesting a lack of effects	• When no effects are reported in several studies by independent investigators under different protocols involving at least two species or two cell types and a sufficient range of field intensities

FIGURE 6.1   The four-point system used in EFHRAN to classify the strength of evidence for any particular effect. *Note:* Clearly, a classification of sufficient evidence requires there to have been a large amount of high-quality research producing a consistent outcome; independent replication of results is also considered a key element. Similarly, evidence suggesting a lack of effects indicates that several studies have reported the absence of field-related effects using a range of appropriate models and relevant exposure conditions.

---

[*] The European Health Risk Assessment Network on Electromagnetic Fields Exposure (EFHRAN) project was funded by the European Commission – Executive Agency for Health and Consumers (EAHC) – AGREEMENT NUMBER 2008 11 06 – (February 1, 2009–July 31, 2012)

[†] EFHRAN Report on Risk analysis of human exposure to electromagnetic fields – revised version, October 2012. Available at: http://efhran.polimi.it/docs/D2_Finalversion_oct2012.pdf (last accessed June 23, 2017)

6.2 Low Frequencies (Up to 300 Hz)

For more than a century, exposure to extremely low-frequency (ELF) electric and magnetic fields has been ubiquitous, related to the production, transmission, distribution and use of electric currents. Research into the possible adverse health effects of such exposure intensified in the late 1970s, with epidemiological and experimental studies focusing mainly on cancers, neurodegenerative diseases, cardiovascular diseases, reproductive effects and non-specific symptoms affecting health and well-being. In terms of exposure, research has been focused on residential exposure, for instance people living close to power lines, on occupational exposure, such as for electricians, and on the use of electrical household appliances. While some studies have estimated exposure in a crude way, measuring or estimating distances between residences and the nearest power line, using broad job titles to categorise occupational exposures or asking study participants about their use of electrical appliances, assessment methods have been refined over the years and comprehensive stationary or personal measurements, as well as detailed job-exposure-matrices based on work activities, have been developed. In addition to studies on health effects (WHO, 2007), many measurement surveys have been conducted to better understand the distribution of exposure in time and space and the relative contribution of various exposure sources to an individual's total exposure. For all European countries where measurement data are available (Gajšek et al., 2016), outdoor average flux densities of extremely low-frequency magnetic fields (ELF MF) in public areas in urban environments range between 0.05 and 0.2 μT, but stronger values (on the order of a few μT) may occur directly beneath high-voltage power lines, at the walls of transformer buildings, and at the boundary fences of substations. In the indoor environment, high values have been measured close to several domestic appliances (up to the mT range), some of which are held close to the body, e.g. hair dryers and electric shavers.

Although numerous studies have been conducted to assess whether any health effects are associated with exposure to electric or magnetic fields, the evidence in support of such hypotheses remains ambiguous. Results of individual studies are inconsistent, in part due to methodological shortcomings. It is therefore important to continuously review the body of evidence. In the last decade this has been done by three international organisations: namely, the World Health Organization (WHO, 2007), the EMF-NET project of the European Union (EMF-NET, 2009) and the Scientific Committee on Emerging and Newly Identified Health Risks (SCENIHR) of the European Commission (SCENIHR, 2007, 2009, 2015). The outcomes of the most recent one, i.e. the SCENIHR report of 2015, are substantially in line with the previous outcomes of EFHRAN: existing studies do not provide convincing evidence for a causal relationship between ELF MF exposure and behavioural outcomes, cortical excitability and self-reported symptoms. There is also no convincing evidence of an increased risk of neurodegenerative diseases, including dementia, or adverse pregnancy outcomes in relation to ELF MF. The studies on childhood health outcomes in relation to maternal residential ELF MF exposure during pregnancy suggest implausible effects and need to be replicated. Recent results do not show an effect of the ELF fields on the reproductive function in humans.

The epidemiological studies are consistent with earlier findings of an increased risk of childhood leukaemia with estimated daily average exposures above 0.3 to 0.4 μT; this resulted in the classification of ELF MF as 'possibly carcinogenic' within the International Agency for Research on Cancer (IARC) Monograph program on the evaluation of carcinogenic risks to humans (IARC, 2002). An assessment conducted as part of EFHRAN used available ELF MF exposure data to estimate the attributable burden of childhood leukaemia in Europe at around 1%–2% of all cases, but with wide confidence intervals (Grellier et al., 2014). However, no mechanisms have been identified that explain the potential effects observed in epidemiological studies and no convincing evidence has arisen from experimental studies in support of these findings. The European Commission FP7 Framework Programme funded the 'Advanced Research on Interaction Mechanisms of electroMagnetic exposures with Organisms for Risk Assessment' (ARIMMORA) project (2011–2015), which was specifically aimed at finding possible mechanisms that might explain the observed association between ELF MF and childhood leukaemia. No such mechanisms

Outcome	Strength of Evidence
Cancer Outcomes	
Leukaemia in children	Limited
Brain tumours in children	Inadequate
Brain tumours in adults	Inadequate
Breast cancer in adults	Lack of effect
Other cancer (children or adults)	Inadequate
Neurodegenerative Diseases	
Alzheimer's disease	Inadequate
Amyotrophic lateral sclerosis (ALS)	Inadequate
Other neurodegenerative diseases	Inadequate
Reproductive Outcomes	
All outcomes	Inadequate
Cardiovascular Diseases	
All diseases	Lack of effect
Well-Being	
Electrical hypersensitivity (EHS)	Lack of effect
Symptoms	Inadequate

FIGURE 6.2   The EFHRAN project strength of evidence for any health outcome being associated with exposure to low frequency magnetic fields.

were identified. Hence, the risk assessment of ARIMMORA confirmed the previous IARC classification of ELF MF as 'possibly carcinogenic' (Schüz et al., 2016). The EFHRAN strength of evidence for each health outcome is summarised in Figure 6.2.

6.3 Intermediate Frequencies (300 Hz–100 kHz)

Exposure to intermediate-frequency (IF) fields has in the past largely been restricted to long-range radio, welding devices, cathode ray tube–based monitors and magnetic resonance imaging (MRI). However, sources of and exposure to these fields are now increasing due to the development of a number of new and emerging technologies, such as those incorporated into anti-theft devices, badge readers and induction hobs and hotplates; compact fluorescent lighting also produces fields in the IF range. IF fields can induce electric fields and currents in the human body, much as is seen with ELF MF exposure, but they can also induce heating effects in the body similar to those observed in radiofrequency exposure. Very little research has been carried out on the possible health effects of IF fields (SCENIHR, 2015) and, accordingly, the EFHRAN classification of the related strength of evidence as 'Inadequate' remains valid.

New evidence relating to the possible effects of exposure to IF will be available after the conclusion of the FP7 European Commission–funded 'Generalised EMF Research Using Novel Methods' (GERoNiMO) project (2014–2018), which includes studies looking at exposures and potential health and biological effects of IF*.

---

* http://radiation.isglobal.org/index.php/geronimo-home, Last accessed February 1, 2018

6.4 High Frequencies (100 kHz–300 GHz)

Research into the possible effects of exposure to low-level radiofrequency (RF) fields has increased over the last two decades following the widespread increase in mobile phone usage and the rollout of base station networks. Concerns have also been raised about DECT cordless phones, and interest in the potential health effects of wireless local area networks (LANs) and Wi-Fi has followed the introduction of these applications into schools, homes and workplaces. However, the effects of RF fields associated with commonly occurring sources in the environment, such as broadcasting, radar and microwave communication links, have been subject to scientific scrutiny for some decades, and an extensive literature on this subject is available (Ahlbom et al., 2004, 2009; van Rongen et al., 2009; Swerdlow et al., 2011).

The EFHRAN risk analysis is summarised below. Early epidemiological investigations centred on a variety of occupational groups with the potential for high exposure to RF fields, such as radar technicians and radio and telegraph operators, with interest in potential associations with brain tumours and leukaemia. In general, results from these studies were inconsistent and no conclusions could be drawn, due to the generally small size and/or methodological limitations of many of these studies as well as very limited exposure assessment. Other studies investigated risks to people living near radio or TV transmitters. These studies did not demonstrate the existence of any risk; again, results have been inconsistent, but were dependent on very crude measures of exposure, such as using distance from broadcasting masts. Other studies concentrated on cancer risks from the use of mobile phones, but other endpoints and sources have been considered. The evidence from the various national studies and pooled analyses from parts of the European Commission FP5 Interphone project on exposure to mobile phone fields and some type of brain cancer have been reviewed by SCENIHR 2009. It was concluded that this evidence, combined with the results of animal and cellular studies, indicated that exposure to RF fields was unlikely to lead to an increase in brain cancer or parotid gland tumours in humans. However, in 2011, IARC completed the procedure to assess the carcinogenicity of RF electromagnetic fields with frequencies between 30 kHz and 300 GHz. These fields were classified as 'possibly carcinogenic to humans – Group 2B (IARC, 2013).

On non-cancer outcomes, the available scientific evidence failed to provide support for an effect of RF fields on self-reported symptoms. Regarding effects of RF fields on the brain and nervous system, several studies using volunteers have reported no consistent effects on various behaviours or cognitive functions, although sporadic effects were noted in some studies. A large number of studies have reported that exposure has no detectable effect on either the auditory or visual system. Some, but not all, studies have reported effects on sleep and sleep encephalogram (EEG) patterns, and others have reported an effect on specific EEG components during exposure.

The conclusions of the SCENIHR report of 2015 are still substantially in line with those of EFHRAN. As to epidemiological studies, SCENIHR stated that overall, the epidemiological studies on mobile phone RF EMF exposure do not show an increased risk of brain tumours. Furthermore, they do not indicate an increased risk for other cancers of the head and neck region. They also observed that some studies raised questions regarding an increased risk of glioma and acoustic neuroma in heavy users of mobile phones. But they concluded that the results of cohort and incidence time trend studies do not support an increased risk for glioma, while the possibility of an association with acoustic neuroma remains open. Furthermore, they stated that epidemiological studies on RF exposure do not indicate increased risk for other malignant diseases, including childhood cancer. They stressed similar considerations also about RF exposure and self-reported symptoms, the possible effects of RF fields on the brain and nervous system, in which no consistent effects on various behaviours or cognitive functions have been found. SCENIHR highlighted a few possible effects of RF on the frequency content of the EEG during sleep. However, one should still consider the previous comment of SCENIHR (2009), in which they questioned the relevance of these subtle changes to health and noted that no interaction mechanism could be identified, although no firm conclusion could be arrived at.

In addition to studies on health effects, many measurement surveys have been conducted. Within the framework of the EFHRAN project, Gajsek et al. (2015) conducted a survey on RF levels which concluded

Outcome	Strength of Evidence
Cancer Outcomes	
Leukaemia in children	Inadequate
Brain tumours in children	Inadequate
Brain tumours in adults	Limited
Breast cancer in adults	Inadequate
Other cancer (children or adults)	Inadequate
Neurodegenerative Diseases	
Alzheimer's disease	Inadequate
Amyotrophic lateral sclerosis (ALS)	Inadequate
Other neurodegenerative diseases	Inadequate
Reproductive Outcomes	
All outcomes	Inadequate
Cardiovascular Diseases	
All diseases	Inadequate
Well-Being	
Electrical hypersensitivity (EHS)	Lack of effect
Symptoms	Inadequate

FIGURE 6.3   The EFHRAN strength of evidence for any health outcome being associated with exposure to RF fields.

that all EMF exposure assessments, surveys of sources and epidemiological studies so far conducted in EU countries indicated that everyday exposure levels of the general public are well below the current guidelines.

The EFHRAN strength of evidence for each health outcome is summarised in Figure 6.3.

6.5 Overall Summary and Conclusions

EFHRAN aimed to monitor and search for evidence of health risks associated with exposure to EMFs at low, intermediate and high frequencies: low frequencies are defined as time-varying EMFs with frequencies of up to 300 Hz; intermediate frequencies as EMFs of 300 Hz to 100 kHz; and high frequencies as EMFs with frequencies between 100 kHz and 300 GHz.

In 2012, the EHFRAN project concluded that there was not sufficient evidence of an adverse health outcome with exposure at any frequency, but there was limited evidence of an association between childhood leukaemia and low-frequency magnetic fields and between brain tumours in adults and high-frequency fields. More recent scientific evidence has not necessitated any change to these findings, and for many endpoints the available evidence remains inadequate to permit a conclusion. The EFHRAN health risk analysis can be summarised in Figure 6.4.

Acknowledgements

This study was funded by the Project EFHRAN European Health Risk Assessment Network on EMF Exposure, European Commission, Executive Agency for Health and Consumers (EAHC), Agreement Number 20081106 (2009–2012).

Adverse Health Outcome		Low-Frequency	IF	High-Frequency
Cancer	Leukaemia in children	▮		
	Brain tumour in children			
	Brain tumour in adults			▮
	Breast cancer in adults	▮		
	All other cancers			
Neurodegenerative diseases	Alzheimer's			
	ALS			
	Other diseases			
Reproductive outcomes	All			
Cardiovascular diseases	All	▮		
Well-being	EHS	▮		▮
	Symptoms			

FIGURE 6.4   Summary of health risk assessments: the strength of evidence for any adverse outcome being associated with exposure to low-, intermediate- (IF) or high-frequency electromagnetic fields. *Note:* For no outcome at any frequency is there sufficient evidence of an effect, but there is limited evidence of an association between childhood leukaemia and low-frequency magnetic fields, and between brain tumours in adults and high-frequency fields (shown in orange). There is evidence suggesting a lack of effects for four outcomes (shown in green), and for all other outcomes the available evidence is inadequate to permit a conclusion (shown in yellow).

References

Ahlbom A., Feychting M., Green A., Kheifets L., Savitz D., Swerdlow A. 2009. ICNIRP (International Commission for Non-Ionizing Radiation Protection) Standing Committee on Epidemiology. Epidemiologic evidence on mobile phones and tumor risk: a review. *Epidemiology*, 20(5), 639–652.

Ahlbom A., Green A., Khiefets L., Savitz D., Swerdlow A. 2004. ICNIRP (International Commission on Non-Ionizing Radiation Protection) Standing Committee on Epidemiology. Review of the Epidemiologic Literature on RF and Health. *Environ Health Perspect*, 112(17), 1741–1754.

EMF-NET. 2009. Report on new epidemiological studies on static fields, ELF, intermediate frequencies, and RF. Deliverable D15c. EMF-NET FP6 European Commission Project.

Gajšek P., Ravazzani P., Grellier J., Samaras T., Bakos J., Thuroczy G. 2016. Review of studies concerning electromagnetic field (EMF) exposure assessment in Europe: Low frequency fields (50 Hz–100 kHz), *International Journal of Environmental Research and Public Health*, 13(9), Article number 875. doi: 10.3390/ijerph13090875.

Gajšek P., Ravazzani P., Wiart J., Grellier J., Samaras T., Thuróczy G. 2015. Electromagnetic field (EMF) exposure assessment in Europe. radio frequency fields (10 MHz–6 GHz). *Journal of Exposure Science and Environmental Epidemiology*, 25, pp. 37–44.

Grellier J., Ravazzani P., Cardis E. 2014. Potential health impacts of residential exposures to extremely low frequency magnetic fields in Europe. *Env Int.* 62, 55–63.

IARC. 2002. *Non-Ionizing Radiation, Part 1: Static and Extremely Low-Frequency (ELF) Electric and Magnetic Fields. IARC Monographs on the Evaluation of Carcinogenic Risks to Humans.* Volume 80. Lyon. International Agency for Research on Cancer.

IARC. 2013. *Non-Ionizing Radiation, Part II: Radiofrequency Electromagnetic Fields [IARC Working Group on the Evaluation of Carcinogenic Risks to Humans, 2011: Lyon, France, IARC Monographs on the Evaluation of Carcinogenic Risks to Humans.* Volume 102, Lyon, International Agency for Research On Cancer.

SCENIHR. 2007. The Possible Effects of Electromagnetic Fields (EMF) on Human Health. Scientific Committee on Emerging and Newly Identified Health Risks. *European Commission, Health & Consumer Protection DG*. Available from: http://ec.europa.eu/health

SCENIHR. 2009. Health Effects of Exposure to EMF. Scientific Committee on Emerging and Newly Identified Health Risks. *European Commission, Health & Consumer Protection DG*. Available from: http://ec.europa.eu/health

SCENIHR. 2015. Opinion on Potential Health Effects of Exposure to Electromagnetic Fields (EMF). *Scientific Committee on Emerging and Newly Identified Health Risks*. European Commission, Health & Consumer Protection DG. Available from: http://ec.europa.eu/health

Schüz J., Dasenbrock C., Ravazzani P., Röösli M., Schär P., Bounds P.L., Erdmann F. et al. 2016. Extremely low-lrequency magnetic fields and risk of childhood Leukemia: A risk assessment by the ARIMMORA consortium, *Bioelectromagnetics*, 37, 183–189, 2016.

Swerdlow A.J., Feychting M., Green A.C., Kheifets L., Savitz D.A. (International Commission for Non-Ionizing Radiation Protection Standing Committee on Epidemiology) 2011. Mobile phones, brain tumours and the Interphone study: Where are we now? *Environ Health Perspect*, 119(11), 1534–1538. http://dx.doi.org/10.1289/ehp.1103693.

van Rongen E., Croft R., Juutilainen J., Lagroye I., Miyakoshi J., Saunders R., de Seze R. et al. 2009. Effects of radiofrequency electromagnetic fields on the human nervous system. *J Toxicol Environ Health B Crit Rev.* 12(8), 572–597.

WHO. 2007. Extremely Low Frequency Fields. *Environmental Health Criteria No 238*. Geneva, World Health Organization.

7

LF Sources – Electromagnetic Field Exposure Assessment: Source Modelling and Theoretical Estimation of Exposure Levels

Aldo Canova
Politecnico di Torino

Luca Giaccone
Politecnico di Torino

7.1 Introduction

In this chapter, the interaction between electromagnetic fields and the human body is analysed. Generally speaking, time-varying magnetic fields induce eddy currents in the human body but the interaction is different at low or high frequency [1]. In the first case, eddy currents flow through the whole body volume. Consequently, the main interaction is with the central and peripheral nervous system [2,3]. At high frequency, due to the skin effect, eddy currents are confined close to the surface of the human body ($\approx$skin tissues). The effective cross-section where eddy currents flow is small and, therefore, the main effect is a thermal energy transfer [4]. This chapter focuses on low-frequency fields ($f < 100$ kHz). In this frequency range, the exposure is always in the near-field region, therefore, electric and magnetic fields have to be handled separately and usually the interaction with magnetic fields is stronger than the one with electric fields.

The protection against the exposure to electric and magnetic fields is regulated by safety standards and guidelines. Standards are often related to specific products whereas safety guidelines are more general. In the low-frequency range two important references are provided by the IEEE and the International

Commission on Non Ionising Radiation Protection (ICNIRP). Both institutions makes a distinction between *measurable quantities* and *induced quantities* (i.e. quantities directly related to the physiological interaction) [2,3]. At low frequency, measurable quantities are the electric and the magnetic field whereas induced quantities are the current density or the induced electric field. Limits have been provided for both cases using the following rationale:

1. Perception levels related to the induced quantities are defined [5,6].
2. Safety factors are introduced to build conservative limits for the induced quantity.
3. Dosimetry analyses are performed to find the external field that induces the quantities at point 2. Conservative conditions are used as homogeneous field and maximum coupling.
4. Additional safety factors are introduced to build conservative limits for external quanties.

The intrinsic conservatism used to define these limits ensures that, if the magnetic field is below reference levels, basic restrictions are automatically respected [2,3]. For this reason, both IEEE and ICNIRP promote a two-step approach that entails to verify the magnetic field first and then, only if it exceeds reference levels, to carry out a dosimetric assessment. It is important to stress that, in the case reference levels are exceeded, it is possible to work on the field source in order to reduce the generated magnetic field. The literature covers many solutions to mitigate the magnetic fields, some of them act directly on the source requiring a reconfiguration [7–10]. When this is not possible of not sufficient, many shielding systems are available and they are divided into passive [7,11–15] and active shields [7,16–18].

This chapter deals with the exposure assessment of low-frequency (LF) sources. First, the modelling of the most common sources that can be found in power systems and in industrial processes is analysed. Afterward, a summary of the possible mitigation techniques is provided. Finally, the methodology to perform a human exposure assessment is presented.

7.2 Source Modelling

7.2.1 Power System Components

The main power sources that can be found in civil and industrial environment are the electrical components that provide transmission, distribution, conversion and utilization of electricity. The evaluation of the impact generated by the above sources is provided by numerical models which has to consider the following aspects:

- Three dimensionality of the field sources.
- Open domain.
- Superposition of the source effects.
- Time dependence of the source currents or voltages.

During the environmental impact assessment, it is generally accepted that the conservative hypothesis of neglecting the field mitigation effects is due to factual shields such as metal barriers, electrowelded grids, pylons, casings and metal boxes of various types and also forth. Thanks to this assumption, all the components can be modeled by means of integral models [19]. In the section paragraph the different sources are examined from the point of view of their numerical modelling.

7.2.2 Power Lines: Cables and Busbars

The simplest component is, of course, the current carrying cable. The source models are mainly based on two types of geometry:

- Indefinite straight-line wire conductors (two-dimensional model).
- Segments of straight length filament conductors (three-dimensional model).

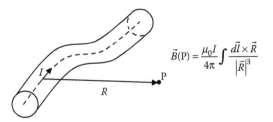

$$\vec{B}(P) = \frac{\mu_0 I}{4\pi} \int \frac{\vec{dl} \times \vec{R}}{|\vec{R}|^3}$$

FIGURE 7.1 Biot–Savart law.

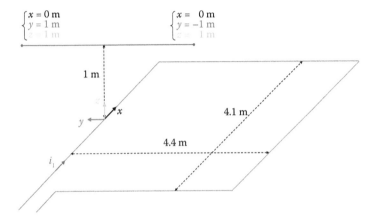

FIGURE 7.2 Example of a 3D source: square coil.

The first models are usually adopted for evaluating magnetic fields generated by buried cables or busbar systems. In fact, in these cases, the parallelism of the different conductors is guaranteed for big path lengths, compared with the distance between the source and the field point. In the case of a 3D path, it is split in several contiguous straight conductors. Their magnetic field is computed by means of the Biot–Savart law (see Figure 7.1) with the analytical expression that can be found in Reference 20. An example of the application of Biot–Savart law is reported here in the case of a square loop employed for the identification of big metallic objects. The loop is reported in Figure 7.2 and the comparison between measurements and simulation can be observed in Figure 7.3. The 3D model is often used in the case of overhead power lines for considering the catenary of the conductors [21] and in the case of all the electrical connections in primary and secondary substations [20].

7.2.3 Power Transformers

The transformer is a complex field source. The presence of a laminated magnetic core that links primary and secondary windings is often the main reason for choosing a finite element approach to analyze it. On the other hand, in this work the objective is not the accurate modelling of the internal field of the transformer. On the contrary, it is of high importance to model the outer magnetic field in order to find the distance where the limit is reached. That is why in this study an approximated integral model is used. Basically, the leakage flux is modelled taking into account the three columns of the transformer composed of primary and secondary windings as shown in Figure 7.4. Each column is represented in a local cylindrical coordinate system and it is discretised in several filamentary loops. The magnetic flux density produced by the current carrying filamentary loop is computed as explained in Reference 19. Therefore, considering a constant current distribution within the cross-section, the magnetic field generated by a single column is

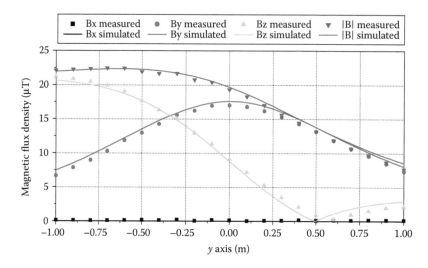

FIGURE 7.3   Comparison between experimental and numerical results of a square coil.

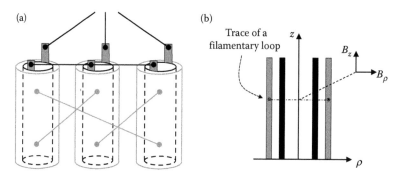

FIGURE 7.4   (a) represents the geometry of a cast resin transformer. The MV windings and their relative connections are represented in grey while the LV windings, the star point and the LV cables are represented in black. (b) represents a single column in the cylindrical coordinate system. In the same figure a filamentary loop that will be used for the magnetic field computation is shown.

computed integrating the contribution of all filamentary loops that constitute the column. Finally, the number and position of the filamentary loops is chosen according to the adaptative quadrature rule based on Kronrod coefficients. The validation of the model is performed by a comparison between simulation and measurement of the magnetic flux density along two inspection lines as shown in Figure 7.5. In Figure 7.6a and b the comparison is reported. The contribution of the windings has to be integrated with the contribution of the low-voltage (LV) and medium-voltage (MV) connections. The first contribution is the dominant one and the path of the conductors exit from the LV terminals is very important for an accurate evaluation of the global magnetic impact. In the case of power cast resin transformers, the modelisation of the windings and of the LV connections is sufficient but in the case of oil transformers the contribution of the oil tank has to be considered [22].

7.2.4 Power Switchgears

The approach used for switchgears (usually MV and LV components) consists in building the geometry of the internal conductors, splitting them in several contiguous straight conductors, and applying the

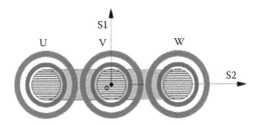

FIGURE 7.5 Scheme of the inspection lines for the validation of the transformer model.

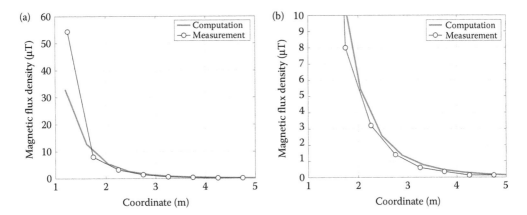

FIGURE 7.6 Comparison between simulation and experimental results along the inspection line S1 (a) and line S2 (b).

Biot–Savart law in order to calculate the magnetic flux density due to each segment. An example of MV and LV switchgear discretisation is shown in Figures 7.7 and 7.8. The proposed model neglects the shielding effect of the metallic enclosure. Therefore, the use of this model to calibrate the simplified method assures some degrees of safety. However, it must be stressed that, the distance of compliance is often referred to a very low magnetic field level (some microT). Hence, it usually falls in the range of 2–4 m and. The effect of the metallic enclosure is quite negligible at these distances from the switchgears. A comparison between simulation and measurements for the MV switchgear (shown in Figure 7.7) is reported in Figure 7.9.

7.2.5 Industrial Devices

The industrial sector is full of devices that generates magnetic fields. Unlike the magnetic field sources previously described, the industrial devices generate magnetic fields not only at 50 Hz. A good and up-to-date review of the most important industrial devices is the second volume of the Non-binding guide to good practice for implementing Directive 2013735/EU [23]. In this section, we present some details of an important device used in the automotive sector: resistance spot welding (RSW) guns. This device will be later considered in the section about dosimetry assessment.

RSW devices are mainly subdivided into two categories: *alternating current* (AC) and *medium frequency direct current* (MFDC). Regarding the AC guns, there are other two sub-families: with or without on-board transformers. The AC guns without on-board transformers are supplied by two long cables that carry the welding current (several kiloamperes). These cables often create a large coil that generates a significant magnetic field [24]. On the contrary, in AC guns with on-board transformers the welding current flows only in the electrodes and the magnetic field produced by the supply system becomes negligible.

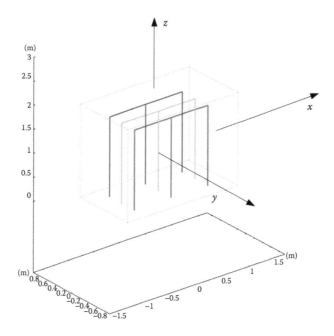

FIGURE 7.7 3D model of a three-unit MV switchgear: Unit 1 and 2 are arranged as ring-main, unit 3 supplies a transformer.

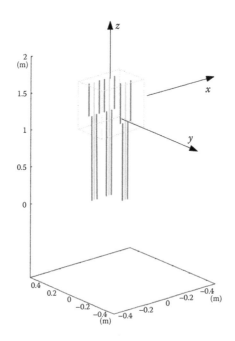

FIGURE 7.8 3D model of a two-unit LV switchgear supplied by one power line.

MFDC guns are often preferred to the AC ones because they require a shorter weld time resulting in a significant energy saving. Moreover, MFDC systems are very stable in working conditions far from the rated power (common range: 20%–95%). Conversely, AC systems are unstable and inefficient when used outside the 70%–90% range of the rated power. In this work we describe both technologies to include the few existing plants in which the AC technology is still used.

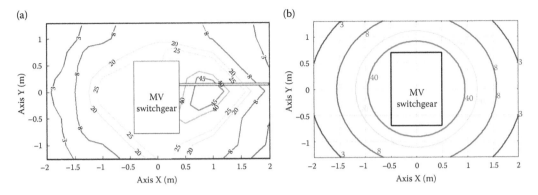

FIGURE 7.9   Comparison between measured ad simulated magnetic flux levels. (a) measurements, (b) simulations.

FIGURE 7.10   Representation of a RSW gun with on board transformer. (a) side view and (b) top view.

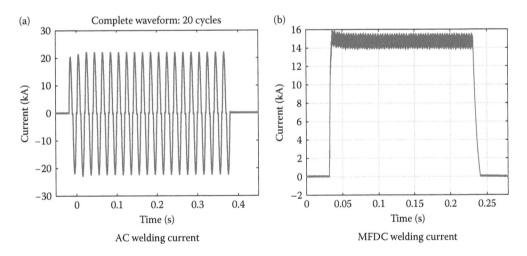

FIGURE 7.11   The typical welding current used in AC guns (a) and MFDC guns (b).

A typical gun equipped with on-board transformer looks like the representation of Figure 7.10. The typical welding current used in AC guns is represented in Figure 7.11a, it includes several cycles at 50 Hz and the single cycle is a pure sine-wave only at rated power. In the case represented in Figure 7.11a the gun is at 70% of the rated power, hence the single cycle is not a pure sine-wave. The typical welding current used in MFDC guns is represented in Figure 7.11b, it is a not an ideal rectangular waveform because it includes a ripple ad 2000 Hz and higher harmonics introduced by the power electronics of the supply system.

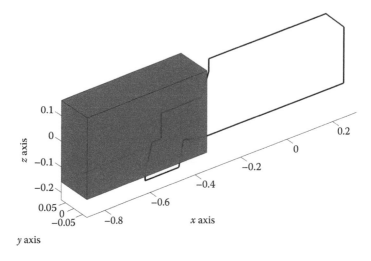

FIGURE 7.12   Example of 3D model of a MFDC welding gun.

The modelling of these devices can be again based on the integration of the Biot-Savart law. This approach has been proven to be effective comparing simulations with measurements [25] and it is also suggested by the reference standard for RSW devices [26]. An example of a 3D model is shown in Figure 7.12.

7.3 Shielding Systems

7.3.1 Source Management

One of the method to reduce the magnetic flux density impact is based on the optimisation of the position of the conductors or circuits belonging to the different electrical phases. In the case of power cables this technique is known as conductor transposition [7–10]. An example of this technique is presented in Figure 7.13 where a power line composed of 21 conductors is shown: six conductors for each phase and three neutral conductors. Figure 7.14 shows results for the base configuration and two optimised configurations for the 21 cables in the duct. Opt2 is the best solution that minimizes the magnetic flux density, Opt1 is not the best solution, but it is a good compromise between ease of installation and magnetic field emission.

7.3.2 Passive Techniques

7.3.2.1 Flat Metallic Shields

The mitigation of the magnetic field can be achieved using planar shields made of two different materials:

- Material with high magnetic permeability,
- Material with high electrical conductivity.

FIGURE 7.13   Base and optimised configurations of power lines. (a) Base, (b) Opt.1, (c) Opt.2.

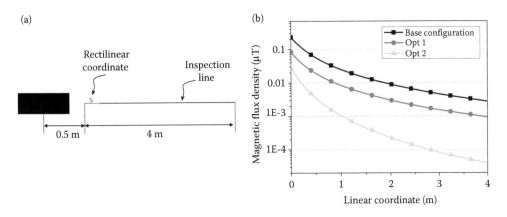

FIGURE 7.14 Inspection line considered to compare the three configurations and position (black rectangle) of the duct (a). Magnetic flux density along the inspection line (b).

In the first case, the high magnetic permeability allows the absorption of the magnetic field inside the shield. Its behaviour is similar to a shielding umbrella but the protection decays very quickly far from the shield. The materials used for ferromagnetic shields are mainly based on iron or nickel alloys [7,27]. In the second case, a layer of material with high electrical conductivity in the presence of a variable magnetic field (induction field) is a site of induced currents that generate a magnetic field of reaction (induced field). Copper or aluminium are commonly employed. Talking about open shields, the shielding factor close to shields is higher in the case of a ferromagnetic shield but far from the shield the field mitigation is guaranteed only with conductive material [28]. The combination of the two materials, ferromagnetic and conductive, allows good shielding performance both close and far from the shield [14]. In Figure 7.15a is reported an example of the application of a ferromagnetic shield: the scope is the protection of a box office placed above a buried power line. The 3D simulation allows to estimate the efficiency of the shield and in Figure 7.15b is reported the model adopted for the simulation of the shield and the source. The results are reported in Figure 7.16 where it is possible to see that the limit considered of 3 µT is out of the shielded area. An example of implementation of a conductive shield is reported in Figure 7.17. This is a very large shield, higher than 600 m², which has been installed for protecting an area of an industrial building, below a high-voltage overhead power line, dedicated to host administrative offices (with permanence higher than 4 hours). The magnetic flux density inside the building with the power line carrying about 3000 A is around 10 µT. The goal was to reach 3 µT and so a shielding factor at least equal to 3 is required. The designed shield was made of only conductive material, aluminium of 3 mm thick, installed on the ceiling and on one wall of the area to be protected. After the installation the test has been done and the

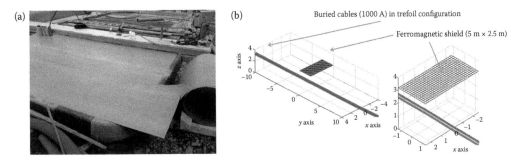

FIGURE 7.15 Example of ferromagnetic shield (a). Model of the ferromagnetic shield and source (b).

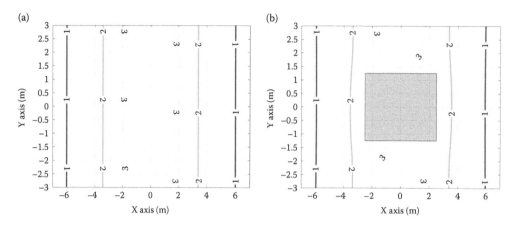

FIGURE 7.16 Magnetic flux density levels without and with the shield at ground level. (a) Without shield, (b) With shield.

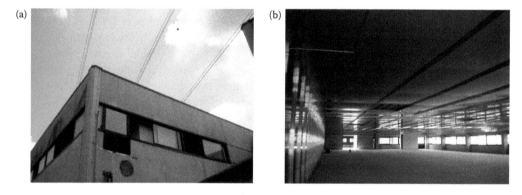

FIGURE 7.17 Example of application of large conductive shield. Power lines above the building (a). Indoor view of the shield installed (b).

chromatic map of the measured values is reported in Figure 7.18. From this figure is possible to compare the different levels obtained in the shielded and unshielded areas of the building (not all the building needs to be shield). Is important to underline that during the measurements the currents in the conductors were very low compared with the ampacity (about 450 A).

7.3.2.2 Passive Loops

The use of passive loops for shielding purposes is a well-established technique, which is often use for buried cables or overhead power lines [29,30]. The working principle is very easy: a primary magnetic field generated by one or more current sources induces, in a set of passive loops close to the sources, electric currents which generate a secondary magnetic field which tends to reduce the primary one. The main advantage is represented by the very simple working principle and installation. However, to obtain significant shielding factors, it is necessary to optimise the passive loop positions and to use a great number of passive loops [12]. The exploitation of the shielding conductors is limited by the low magnetic coupling between sources and passive loops. In order to increase the magnetic coupling a modified version of passive loops has been proposed and adopted in particular for shielding joint bays of high-voltage power

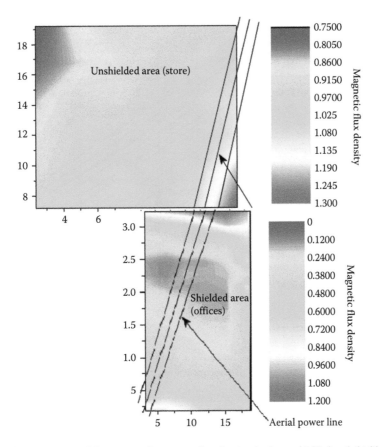

FIGURE 7.18 Chromatic map of the measured magnetic flux density: in the unshielded and shielded areas.

cables [11,13]. This solution allows a very strong mitigation but particular care has to be paid in the thermal analysis of the sources and shields cables [31].

7.3.3 Active Techniques

The compensation of magnetic fields through additional ones generated by a set of loops properly supplied is another mitigation technique which is called 'active'. This methodology requires one or more magnetic probes measuring the magnetic field to be mitigated and provide this information to a control unit which supply the set of active loops. The compensation can work at every frequency from DC (0 Hz) to AC (usually some kHz or tens of kHz). The main drawbacks are the complexity of the system, the need of supply energy, the costs and the need to maintain along the time the calibration of the system. The main advantage is the possibility to reach extremely high shielding factors. This technique is often used in the case of electronic devices (e.g. electronic microscope) where magnetic flux density lower than some nT (at quasi-DC field) is required [32]. The possibility of using active loops for shielding MV/LV substation has been proved and implemented as reported in References 16–18. It is also possibile to use active systems when other techniques cannot be used as in the case presented here. A high-voltage overhead power line generates a magnetic field higher than 3 µT and this is the actual limit for the living area in Italy. Close to the power line a private house has been built and a shielding system is required. The power-line is characterised by: two three-phase lines, rated voltage 380 kV, rated current 2950 A (each line), minimum height of the conductors from the ground 17.4 m. The geometry of the overhead power-line and the location of the pylons and of the building with respect to the power-line are reported in Figure 7.19. An optimisation

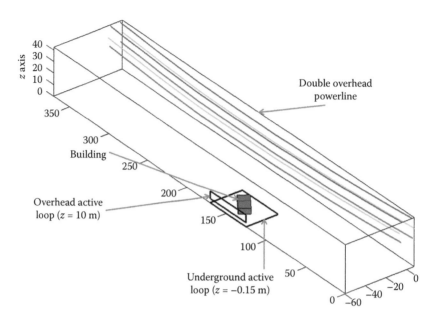

FIGURE 7.19 Example of an active shield.

procedure is performed in order to define the configuration of the active shield. For this application, two active loops are used. The constraints of the optimisation procedure are:

- 1 loop on XY plane, realised like an ordinary underground LV powerline, at $z = -0.15$ m,
- 1 loop on YZ plane, realised like an ordinary LV overhead span powerline attached to two poles with height max $= 10$ m,
- Max shielding current: 600 A.

The resulting geometry of the shielding loops is reported in Figure 7.19, and Figure 7.20 shows the resulting magnetic flux density in the volume interested by the building. The green surface is a contour

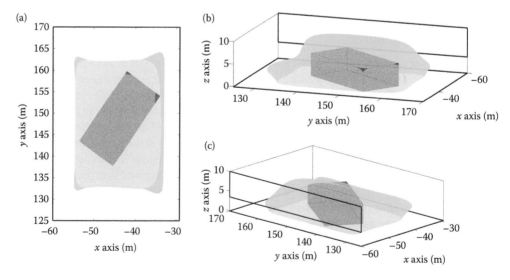

FIGURE 7.20 Volume of compliance obtained turning on the active shield. Top view (a) and perspective views (b) and (c).

plot at the level of 3 μT, inside the green surface the magnetic flux density is lower than 3 μT. It is clear that the building (represented as a red volume) is inside the green surface, therefore the magnetic flux density in the whole volume inside the building is lower than 3 μT, except for two corners in the upper part of the building, whose dimension is less than 1 meter and which is not significant for the human exposure.

7.4 Human Exposure Assessment

The exposure assessment methodology depends on the nature of the source field. The most simple source is the one that generates a steady-state single frequency field. In this case the assessment can be done referring to that single-frequency by means of simulations or measurements. In both cases the result can be directly compared with reference levels provided by safety guidelines and/or standards.

Of more interest are magnetic field sources that generates non-sinusoidal fields or pulsed fields. To give some examples, in Figure 7.21 three complex field waveforms are shown. Figure 7.21a is classified as non-coherent, that means its spectral content changes during time. Figure 7.21b is a coherent waveform because it is a periodic field. Finally, Figure 7.21c is special case of a pulsed field called sinusoidal burst. For these cases, several methods have been proposed [2,3,33] to estimate the exposure and they will be briefly summarised in the following by means of examples.

7.4.1 Non-Sinusoidal Fields

Dealing with non-sinusoidal waveforms it is not straightforward to make use of reference levels because they are defined for single-frequency fields. Therefore, both the IEEE and the ICNIRP provide suitable

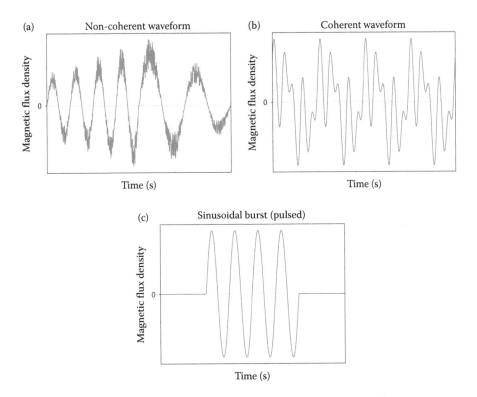

FIGURE 7.21 Examples of complex waveforms: non-coherent field (a), coherent (and periodic) field (b) and pulsed field (sinusoidal burst) (c).

methods [2,3]. The two institutions make use of a different notation but, in the end, the rationale behind the proposed methods is similar. For the sake of shortness we will focus on the two methods mostly used: the *frequency summation rule* and the *weighted peak method*. For more details about all methods we suggest Reference 34 which provides several comparisons.

7.4.1.1 Frequency Summation Rule

The frequency summation rule makes use of the spectrum of the field waveform. The exposure is considered compliant if the exposure index (*EI*) satisfies

$$EI = \sum_{j=1\,\text{Hz}}^{10\,\text{MHz}} \frac{A_j}{A_{\text{R}j}} < 1 \tag{7.1}$$

A_j is the quantity to be assessed at frequency j and $A_{\text{lim},j}$ the limit for this quantity at frequency j.

It is obvious that this method provides a conservative result because it assumes that the spectral components add in phase, that is, all maxima coincide at the same time. This is a good assumption for non-coherent waveform but, on the contrary, it leads to an unnecessary overestimation of the exposure to coherent fields. For this reason, the ICNIRP developed another method called weighted peak method (WMP).

7.4.1.2 Weighted Peak Method

The WPM takes into account not only the magnitude but also the phase of the spectrum. The exposure is considered compliant if the exposure index (*EI*) satisfies:

$$EI = \left| \sum_j \frac{A_j}{A_{\text{lim},j}} \cos(2\pi f_j t + \theta_j + \varphi_j) \right| < 1 \tag{7.2}$$

A_j and $A_{\text{lim},j}$ are the same values described for the frequency summation rule. θ_j is the phase of A_j and φ_j is the phase of $A_{\text{lim},j}$.

φ_j is defined according to the dependance of $A_{\text{lim},j}$ on the frequency. Where the limit varies directly proportional to $1/f^2$, $1/f$, f^0 (constant), and f the related phases are 180, 90, 0 and 90°, respectively. Defining a frequency-dependent weight function whose magnitude and phase are $1/A_{\text{lim},j}$ and φ_j, it is possible to interpret the weighted peak method as the application of a filter. Dealing with the magnetic flux density and considering reference levels for public exposure, the filter becomes the one shown in Figure 7.22. It is worth noting that the frequency-dependent weight function can also be approximated in time domain by an analog filter which approximates the frequency-dependent filter as shown in Figure 7.22 [2].

7.4.1.3 Comparison of the Methods for Pulsed Fields

Let us consider the magnetic flux density represented in Figure 7.23a. The application of the frequency summation rule requires the magnitude of the spectrum shown in Figure 7.23b. On the same figure, the red curve represents the reference levels for the magnetic flux density. The application of Equation 7.1 corresponds to the sum of the point-by-point division of the spectral lines and the red curve in Figure 7.23b. Therefore, it is clear that the results depends on the number of spectral lines considered in the summation. In this sense, Figure 7.23b shows the complete spectrum of the waveforms up to 1000 Hz (in blue) and also the same spectrum depurated by the spectral lines below a given threshold (orange curve, threshold taken as 1% of the maximum spectral line). It is worth noting that the suppression of negligible spectral lines is adopted in some standards [26]. The application of the frequency summation rule to the complete spectrum provides an exposure index equal to $EI = 6.1$ whereas the use of the depurated spectrum provides $EI = 1.25$.

The application of the weighted peak method is less critic because it can be applied in frequency domain (orange weight function in Figure 7.22) or in time domain (blue weight function in Figure 7.22). However, in both cases, the time-domain waveform at the left hand side of Equation 7.2 has to be evaluated. Consequently, the result is less affected by the frequency resolution of the spectrum. For example, Figure 7.24 represents the left-hand side of Equation 7.2 when the input is the magnetic flux density in Figure 7.23a. The blue curve corresponds to the time-domain application of the WPM whereas the red curve is the frequency-domain application of the WPM. The deviation between the two waveforms is due to the deviation of the weight functions in frequency and time domains (see Figure 7.22), however, the exposure index defined as the maximum value for each waveform is very similar, $EI = 0.84$ for the blue curve and $EI = 0.82$ for the red curve. In both cases, the exposure is considered compliant ($EI < 1$), and furthermore, it is confirmed that the WPM is less conservative than the multiple frequency method, which gives $EI = 6.1$ using the complete spectrum and $EI = 1.25$ using the depurated spectrum.

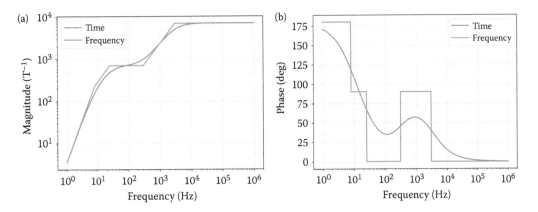

FIGURE 7.22 Example of weight function in time domain and frequency domain. (a) Magnitude and (b) phase.

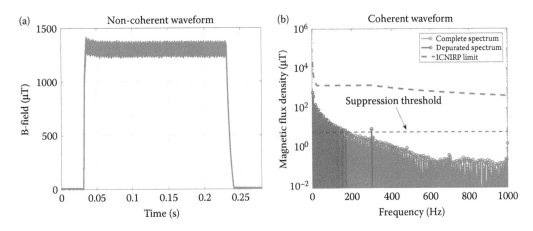

FIGURE 7.23 Pulsed magnetic field generated by a medium frequency direct current welding gun (a). Spectrum of the magnetic fields with and without depuration of negligible components (b).

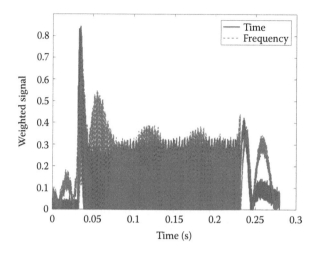

FIGURE 7.24 Comparison of the weighted waveform obtained through the weighted peak method in time and frequency domains.

7.4.2 Dosimetry Assessment

When the magnetic fields exceed reference levels there are two possible solutions: (1) design a mitigation system the lower the magnetic field values or (2) perform a dosimetry assessment to check the compliance with basic restrictions. In this section we briefly present a dosimetric assessment for a medium frequency direct current welding gun. This device is chosen because it likely exceeds reference levels and it is impossible to install a mitigation system, hence, the only solution is to perform a dosimetry assessment.

The device under test is the one introduced earlier in Section 7.2.5. It generates a pulsed magnetic field that can be assessed with the weighted peak method as explained in previous sections. Figure 7.25a represents the device and the colormap of the exposure index. Very high values are observed close to the gun (higher than 15). The white contour represents the boundary over which the exposure index falls below 1 (i.e. a compliance region). A 3D representation of this boundary is given in Figure 7.25b.

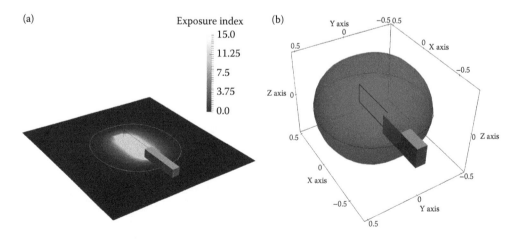

FIGURE 7.25 Exposure index computed by the weighted peak method in the region surrounding a welding gun. The white line represents the boundary outside which the exposure index is lower than 1 (a). Furthermore, (b) shows a 3D representation of the contour outside which the exposure index is lower than 1.

It is apparent that the worker who has to manually operate the gun, could not respect this safety boundary. Hence, a dosimetry assessment is carried out. The human exposure is assessed using the 2 mm × 2 mm × 2 mm Duke (34-year-old male) voxel model from the Virtual Family [35].

We consider two configurations: horizontal and vertical. The horizontal configuration is shown in Figure 7.26a. The colormap represents the exposure index related to the magnetic flux density computed on the human body. It reaches a value of 2.39. The source field induces a current density in the human body and its path is shown in Figure 7.26b. The vertical configuration is shown in Figure 7.26c. The maximum value of the exposure index referred to the magnetic flux density is 4.39 and the current density path is shown in Figure 7.26d. The current density are computed using the method proposed in Reference 36. This method makes it also possible to compute the induced electric field in each tissue in time domain for

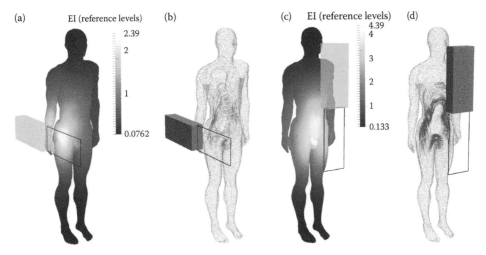

FIGURE 7.26 Configuration analysed by means of dosimetry. (a) and (b) represent the exposure index over the human body and the current density path for the horizontal configuration. (c) and (d) represent the exposure index over the human body and the current density path for the vertical configuration.

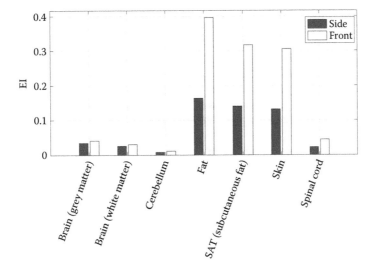

FIGURE 7.27 Exposure index related to basic restrictions at some representative tissues of the peripheral and the central nervous system.

the application of the weighted peak method. Considering some representative tissues of the peripheral and the central nervous system, it is observed that the exposure index referred to the induced electric field is lower than 1 in each tissue for both configurations (see Figure 7.27).

In conclusion, the exposure is compliant with basic restrictions even if the magnetic flux density exceeds significantly reference levels at some localised points.

References

1. ICNIRP. Guidelines for limiting exposure to time varying electric, magnetic and electromagnetic fields (up to 300 GHz). *Health Physics*, 74(4):494–522, 1998.
2. ICNIRP. Guidelines for limiting exposure to time-varying electric and magnetic fields (1 Hz to 100 kHz). *Health Physics*, 99(6):818–836, 2010.
3. IEEE Standards Coordinating Committee 28. C95.6 – IEEE Standard for Safety Levels with Respect to Human Exposure to Electromagnetic Fields, 0–3 kHz.
4. IEEE International Committee on Electromagnetic Safety (SCC39). C95.1 – IEEE Standard for Safety Levels with Respect to Human Exposure to Radio Frequency Electromagnetic Fields, 3 kHz to 300 GHz.
5. J. Patrick Reilly, V. T. Freeman, W. D. Larkin. Effects of transient electrical stimulation evaluation with a neuroelectric model. *IEEE Transactions on Biomedical Engineering*, BME-32(12):1001–1011, 1985.
6. Jp. P. Reilly. Peripheral nerve stimulation by induced electric currents: exposure to time-varying magnetic fields. *Medical and Biological Engineering and Computing*, 27:101–110, Mar 1989.
7. Mitigation techniques of power frequency magnetic fields originated from electric power systems. Technical Report Worging group C4.204, International Council on Large Electric Systems (CIGRE'), 2009, ISBN: 978-2-85873-060-5.
8. A. Canova, F. Freschi, and M. Tartaglia. Multiobjective optimization of parallel cable layout. *IEEE Transactions on Magnetics*, 43(10):3914–3920, 2007.
9. F. Freschi and M. Tartaglia. Power lines made of many parallel single-core cables: A case study. *IEEE Transactions on Industry Applications*, 49(4):1744–1750, 2013.
10. L. Giaccone. Optimal layout of parallel power cables to minimize the stray magnetic field. *Electric Power Systems Research*, 134:152–157, May 2016.
11. A. Canova and L. Giaccone. A novel technology for magnetic-field mitigation: High magnetic coupling passive loop. *IEEE Transactions on Power Delivery*, 26(3):1625–1633, 2011.
12. A. Canova, F. Fabio, L. Giaccone, A. Guerrisi, and M. Repetto. Magnetic field mitigation by means of passive loop: Technical optimization. *COMPEL – The International Journal for Computation and Mathematics in Electrical and Electronic Engineering*, 31(3):870–880, May 2012.
13. A. Canova, D. Bavastro, F. Freschi, L. Giaccone, and M. Repetto. Magnetic shielding solutions for the junction zone of high voltage underground power lines. *Electric Power Systems Research*, 89:109–115, Aug 2012.
14. D. Bavastro, A. Canova, L. Giaccone, and M. Manca. Numerical and experimental development of multilayer magnetic shields. *Electric Power Systems Research*, 116:374–380, Nov 2014.
15. D. Bavastro, A. Canova, F. Freschi, L. Giaccone, and M. Manca. Magnetic field mitigation at power frequency: Design principles and case studies. *IEEE Transactions on Industry Applications*, 51(3):2009–2016, May 2014.
16. A. Canova, J. C. Del-Pino-López, L. Giaccone, and M. Manca. Active shielding system for ELF magnetic fields. *IEEE Transactions on Magnetics*, 51(3):2015, Article no. 8001004.
17. J. C. Del-Pino-López, L. Giaccone, A. Canova, and P. Cruz-Romero. Design of active loops for magnetic field mitigation in MV/LV substation surroundings. *Electric Power Systems Research*, 119:337–344, Feb 2015.

18. A. Canova and L. Giaccone. Real-time optimization of active loops for the magnetic field minimization. *International Journal of Numerical Modelling: Electronic Networks, Devices and Fields*, In press.

19. D. Bavastro, A. Canova, L. Giaccone, and M. Manca. Integral and analytical models for evaluating the distance of compliance. *International Journal of Numerical Modelling: Electronic Networks, Devices and Fields*, 27(3):590–599, May 2014.

20. A. Canova, F. Freschi, M. Repetto, and M. Tartaglia. Description of power lines by equivalent source system. *COMPEL – The International Journal for Computation and Mathematics in Electrical and Electronic Engineering*, 24(3):893–905, 2005.

21. G. Lucca. Magnetic field produced by power lines with complex geometry. *European Transactions on Electrical Power*, 21(1):52–58, 2011. cited By 9.

22. L. Li, W. N. Fu, S. L. Ho, S. Niu, and Y. Li. Numerical analysis and optimization of lobe-type magnetic shielding in a 334 mva single-phase auto-transformer. *IEEE Transactions on Magnetics*, 50(11), 2014. cited By 3.

23. Non-binding guide to good practice for implementing Directive 2013/35/EU – Electromagnetic Fields – Volume 2 – case studies.

24. P. Mair. Effects on the human body and assessment methods of exposure to electro-magnetic-fields caused by spot welding. In Proceedings of the 4th International Seminar on Advances in Resistance Welding, page 17, 15 November, 2006.

25. A. Canova, F. Freschi, and L. Giaccone. Occupational exposure to the magnetic field produced by spot welding guns. *IEEE Industry Applications Magazine*, in press.

26. EN 50505 – Basic standard for the evaluation of human exposure to electromagnetic fields from equipment for resistance welding and allied processes.

27. O. Bottauscio, M. Chiampi, D. Chiarabaglio, F. Fiorillo, L. Rocchino, and M. Zucca. Role of magnetic materials in power frequency shielding: Numerical analysis and experiments. *IEE Proceedings: Generation, Transmission and Distribution*, 148(2):104–110, 2001. cited By 22.

28. E. Cardelli, A. Faba, and A. Pirani. Nonferromagnetic open shields at industrial frequency rate. *IEEE Transactions on Magnetics*, 46(3 PART 2):889–898, 2010. cited By 5.

29. B. -Y. Lee, S. -H. Myung, Y. -G. Cho, D. -I. Lee, Y. -S. Lim, and S. -Y. Lee. Power frequency magnetic field reduction method for residents in the vicinity of overhead transmission lines using passive loop. *Journal of Electrical Engineering and Technology*, 6(6):829–835, 2011. cited By 1.

30. J. C. Del Pino López and P. C. Romero. The effectiveness of compensated passive loops for mitigating underground power cable magnetic fields. *IEEE Transactions on Power Delivery*, 26(2):674–683, 2011. cited By 3.

31. A. Canova, F. Freschi, L. Giaccone, and A. Guerrisi. The high magnetic coupling passive loop: A steady-state and transient analysis of the thermal behavior. *Applied Thermal Engineering*, 37:154–164, 2012. cited By 7.

32. H. Nowak, B. Hilgenfeld, and J. Haueisen. Active shielding to reduce low frequency disturbances in biomagnetic recordings. *Physica Medica*, 20(Suppl. 1):16–18, 2004. cited By 0.

33. Guidance on determining compliance of exposure to pulsed and complex non-sinusoidal waveform below 100 kHz with icnirp guidelines. *Health Physics*, 84(3):383–387, 2003.

34. V. De Santis, X. L. Chen, I. Laakso, and A. Hirata. On the issues related to compliance of LF pulsed exposures with safety standards and guidelines. *Physics in Medicine and Biology*, 58(24):8597–607, 2013.

35. A. Christ, W. Kainz, E. G. Hahn, K. Honegger, M. Zefferer, E. Neufeld, W. Rascher et al. The virtual family–development of surface-based anatomical models of two adults and two children for dosimetric simulations. *Physics in Medicine and Biology*, 55(2):N23–38, Jan 2010.

36. A. Canova, F. Freschi, L. Giaccone, and M. Manca. A Simplified procedure for the exposure to the magnetic field produced by resistance spot welding guns. *IEEE Transaction on Magnetics*, 52(3), 2015.

8

RF Sources – Exposure Assessment of Mobile Phone Base Stations: 5th-Generation Wireless Communication Systems

Giorgio Bertin
Telecom Italia

Enrico Buracchini
Telecom Italia

Paolo Gianola
Telecom Italia

8.1 Introduction

5G Wireless Technology is rapidly developing and is one of the biggest areas of research from industry and academia, also involving standardisation organisations and regulatory authorities. 5G technology offers many advantages, such as increased capacity, flexibility coverage and service ubiquity [1–3]. It is expected that the peak rate service will rise by an order of magnitude and, at the same time, the energy required per bit transmitted will drop significantly, allowing the simultaneous development of applications conceived for rapidly transferring large amounts of data and applications running on small devices, such as sensors and meters, that are powered by conventional batteries. Another important feature of 5G, the reduction of data latency, will push the development of new real-time applications, including robotised systems and autonomous vehicles.

It is expected that to address this challenge, there will be a deployment of a large network of human-to-human, human-to-machine, and machine-to-machine connections, and as a consequence a dramatic rise in the number of user devices and radio base stations radiating RF power. Moreover, to support the increase in cellular capacity, there will be growing interest in a new frequency band in the so-called millimeter wave, between 30 and 300 GHz [4].

Large new portions of the spectrum being assigned to mobile communication and the consequent influx of new radio technologies, such as new modulation schemes, Massive MIMO and Beamforming, will create a need to manage the approach for the evaluation of human exposure to electromagnetic (EM) fields.

8.2 Reference Standards and Existing Methodologies for Exposure Assessment in Current Wireless Systems

8.2.1 Reference Standards Summary and Overview

In the past, concerns about the potential harmful effects of radiofrequency radiation have been deeply investigated by institutions and research centers, and safety guidelines [5,6] and standards [7–11] addressing both occupational and public exposure have been published.

But now, considering all the factors mentioned above, new questions are arising about the possible health risks from electromagnetic radiation [12].

The following sections of this chapter will address these questions in the following order: first the existing standards regarding radiofrequencies emission will be briefly reviewed, with a focus on the assessment of fields generated by 2G–4G radio systems. Next an overview of the 5G system will be provided, and finally a discussion of potential risk related to exposure to 5G signals and possible methods of mitigation.

The International Electrotechnical Commission (IEC) has recently published Standard IEC 62232, 'Determination of RF field strength and SAR in the vicinity of Radiocommunication Base Stations for the purpose of evaluating human exposure' [13].

This standard addresses the evaluation of radiofrequency (RF) field strength, power density or specific absorption rate (SAR) levels in the vicinity of radiocommunication base stations (RBS) radiating in the frequency range 110 MHz to 100 GHz. For our purposes, reference is made to Appendix F, 'Technology-specific guidance', where guidance on how to apply the evaluation methods for 2G–4G technologies is provided.

The basic principle of the assessment is to measure the power received from a constant radio frequency source, typically a pilot signal, at a given location, and apply an extrapolation factor as described in [13] Annex B.5. This method ensures that the resulting field is the maximum obtainable at that location for the considered radiofrequency source. When multiple sources are present, the summation of all significant RF fields is required, as described in [13] Annex B.6.

8.2.2 2G Field Measurement Techniques

GSM/GPRS and EDGE signals are measured with a conventional spectrum analyser following the guidance reported in [13] Annex F.3. Special care must be taken when selecting the RBW filter in order to maximise the contribution of the signal to be measured and prevent the adjacent channel from being included in the measurement. If P_{max} represents the required assessment configuration and E_{max} the corresponding RF field strength, the following relationship applies:

$$P_{max} = P_{BCCH} \cdot N_c \qquad (8.1)$$

and

$$E_{max} = E_{BCCH} \cdot \sqrt{N_c} \tag{8.2}$$

where P_{BCCH} is the power associated with the *BCCH* pilot channel and E_{BCCH} the corresponding field. The extrapolation factor N_c represents the total number of radio channels (control and traffic) that feed into the antenna.

8.2.3 3G Field Measurement Techniques

WCDMA/UMTS mobile phone systems use spread spectrum technology employing a constant power control/pilot channel, the Common Pilot Channel (*CPICH*), which has a fixed power relationship to the maximum allocated power [13] Annex F.4.2. The ratio of the maximum allocated power to the power of the pilot channel is a parameter set by a telecommunication operator whose typical value is 10 (i.e. 10% of total power allocated to *CPICH*). Field assessment requires a code-domain analyser for detecting the scrambling codes and decoding the *CPICH* signals. The extrapolated field is then easily obtained with the following relationship:

$$E_{max} = E_{CPICH} \cdot \sqrt{\beta} \tag{8.3}$$

where E_{CPICH} is the electric field associated with the *CPICH* channel and β the extrapolation factor.

If multiple *CPICH* channels are detected, the total extrapolated field, E_{ext} for one carrier frequency can then be expressed as the quadratic sum of all M detected and extrapolated *CPICH* channels:

$$E_{ext} = \sqrt{\sum_{i=1}^{M} (E_{max}^2)_i} \tag{8.4}$$

8.2.4 4G Field Measurement Techniques

LTE mobile phone systems use Orthogonal Frequency Division Multiplexing (OFDM) technology, defined in 3GPP TS 36.104, to enhance the capacity related to the data throughput [13] Annex F.7. In time domain the LTE standard defines for FDD system a frame of 10 ms, as shown in Figure 8.1.

A frame is subdivided into 10 subframes of 1 ms, and each subframe consists of two slots. In the frequency domain, the OFDM system divides the operating frequency band in subcarriers of 15 KHz bandwidth, grouped into a physical resource block (PRB) of 12 subcarriers. Figure 8.2 shows the resource grid representation for a 1.4 MHz bandwidth LTE system.

Each unitary element (66.6 μs × 15 KHz) is a resource element (RE). All REs are modulated following one of these schemes: QPSK, 16QAM or 64QAM. The LTE downlink spectrum is totally flexible and the

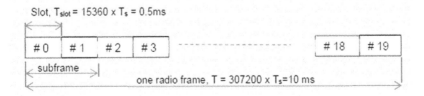

FIGURE 8.1   LTE generic frame structure.

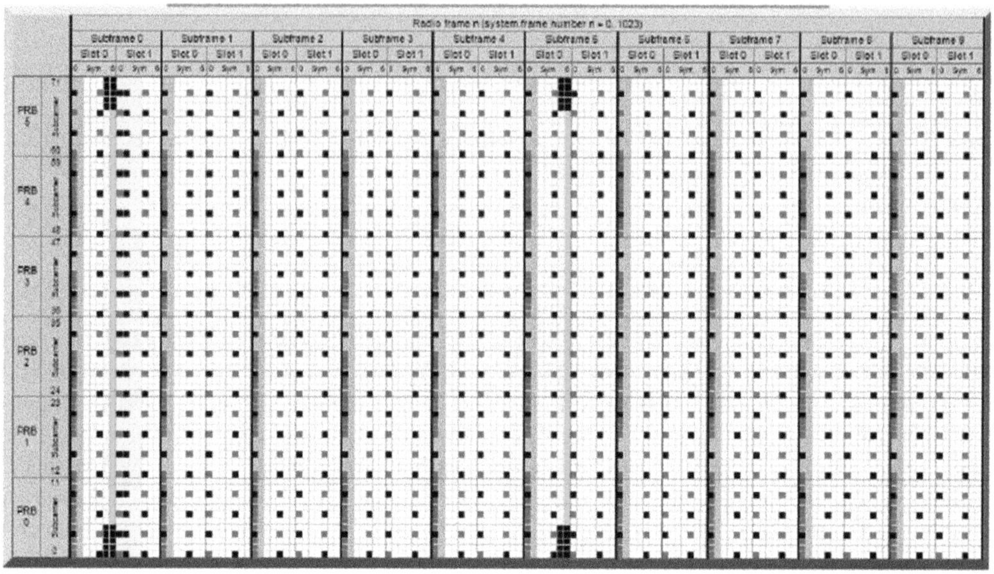

FIGURE 8.2 Resource grid of a LTE FDD signal for a 1.4 MHz bandwidth with the main signals reported: red squares are the reference signal (RS), green squares are the primary synchronisation signals (PSS), the yellow squares are the secondary synchronisation signals (SSS) and the blue squares are the physical broadcast channel (*PBCH*) (Adapted from LTE resource grid, http://niviuk.free.fr/lte_resource_grid.html.) [14].

bandwidth may vary from 1.4 to 20 MHz, as shown in Table 8.1. This table reports the corresponding number of PRBs and subcarriers (n_{RS}). This last number, also expressed in dB, corresponds to the ratio of the maximum output power emitted by the base station to the power of a single reference signal.

To evaluate the exposure level for maximum traffic conditions by extrapolation, it is important that the transmitted power of the received signal or channel is not dependent on the amount of traffic. The reference signal (RS), the primary synchronisation signal (P-SS), the secondary synchronisation signal (S-SS) and the physical broadcast channel (*PBCH*) can be used as the received signal in the extrapolation-based evaluation method, because the power levels are constant. The *RS* is well suited for this because the locations of the LTE reference signals are uniformly distributed over the occupied radio bandwidth to reduce the effects of frequency selective fading.

To assess the maximum exposure level from an LTE base station, two evaluation methods can be used: the first makes use of a dedicated decoder, the second requires a basic spectrum analyser.

TABLE 8.1   Theoretical Extrapolation Factor n_{RS} as a Function of the Bandwidth, Assuming That All Subcarriers Are at the Same Power Level

Bandwidth	PRBs	n_{RS}	n_{RS}[dB]
1.4	6	72	18.75
3	15	180	22.55
5	25	300	24.77
10	50	600	27.78
15	75	900	29.54
20	100	1200	30.79

8.2.4.1 Method Using a Dedicated Decoder

In this method, the field strength corresponding to the *RS* of an LTE cell is measured. If multiple antennas are used for transmission by the same cell (MIMO), the *RS* should be determined for each antenna (or antenna port) [13] Annex F.7.2.2.

The maximum electric field strength (V/m), E_{max}, is:

$$E_{max} = E_{RS} \cdot \sqrt{N_{RS}} \tag{8.5}$$

where E_{RS} is the field level (V/m) of the *RS* and N_{RS} is the extrapolation factor.

In the case of a MIMO antenna system,

$$E_{ext} = \sqrt{\frac{N_{RS}}{BF}} \sqrt{\sum_{i=1}^{M} \left(E_{RS}^2\right)_i} \tag{8.6}$$

where *BF* denotes the power-boosting factor for the *RS*. This value may be obtained from the operator.

8.2.4.2 Method Using a Basic Spectrum Analyser

A basic spectrum analyser is less expensive and more commonly available than a dedicated LTE decoder [13] Annex F.7.2.3. However, when using a basic spectrum analyser, the powers of the *RSs* cannot be accurately detected because they are transmitted on single resource elements spread in frequency and time. To overcome this issue, this method focuses on the physical broadcast channel which is transmitted, regardless of the configuration or service bandwidth, and spans a bandwidth of 6 RBs (approximately 1 MHz) over the centre frequency of the LTE signal (see the blue squares in Figure 8.2). The PBCH power can be measured in scope mode (with the frequency span set to zero) with the resolution bandwidth set to 1 MHz to integrate the signal over its bandwidth. The extrapolation to the maximum level is then obtained using the equation:

$$E_{max} = N_{PBCH} \cdot \sqrt{E_{PBCH}} \tag{8.7}$$

where N_{PBCH} is the extrapolation factor which is the ratio of the maximum transmission power to the transmission power corresponding to the *PBCH* over 6 RBs.

N_{PBCH} can be provided by the network operator or can be calculated theoretically according to

$$N_{PBCH} = \frac{N_{RS}}{72} \tag{8.8}$$

where N_{RS} denotes the number of subcarriers in the used transmission bandwidth.

8.3 5G Overview

8.3.1 Categorisation of 5G Use Cases

Recently, several research and development activities have been performed all over the world, such as EU project METIS [15], NGMN and 5G Americas white papers [16], and Korean, Japanese, United States and European trials [17,18], in order to study, define and test the generation of mobile wireless systems overcoming LTE and its evolution, often referred to as 4G.

The first rationalisation of the main concepts, use cases and main capabilities of this new generation, often referred to as 5G, is introduced by the ITU-R Recommendation 'IMT Vision – Framework and Overall Objectives of the Future Development of IMT for 2020 and Beyond' [19], published in 2015, on the vision of the 5G mobile broadband–connected society and future international mobile telecommunications (IMT). This Recommendation defines the framework and overall objectives of the future development of IMT for 2020, which includes further enhancement of existing IMT and the development of IMT-2020 and beyond, in light of the roles that IMT could play to better serve the future needs of networked societies in both developed and developing countries. It includes a broad variety of capabilities associated with envisaged usage scenarios.

IMT Vision introduced three main categories of usage scenario, broadly adopted for the 5G ecosystem by the time of its publication, and depicted them in the well-known 'triangle' diagram: enhanced mobile broadband (eMBB), ultra-reliable and low-latency communications (URLLC) and massive machine type communications (MmTC). These are all described in IMT Vision [19].

8.3.1.1 Enhanced Mobile Broadband

Mobile broadband addresses human-centric use for access to multimedia content, services and data. The demand for mobile broadband will continue to increase, leading to enhanced mobile broadband. The enhanced mobile broadband usage scenario will come with new application areas and requirements in addition to existing mobile broadband applications for improved performance and an increasingly seamless user experience. This usage scenario covers a range of cases, including wide-area coverage and hotspots, which have different requirements. For the hotspot case, i.e. for an area with high user density, very high traffic capacity is needed, while the requirement for mobility is low and the user data rate is higher than that of wide-area coverage. In the case of wide-area coverage, seamless coverage and medium to high mobility are desired, along with a much-improved user data rate compared to existing data rates. However, the data rate requirement may be relaxed compared to hotspot.

8.3.1.2 Ultra-Reliable and Low-Latency Communications

This use case has stringent requirements for capabilities such as throughput, latency and availability. Some examples include wireless control of industrial manufacturing or production processes, remote medical surgery, distribution automation in a smart grid, transportation safety, etc.

8.3.1.3 Massive Machine Type Communications

This use case is characterised by a very large number of connected devices typically transmitting a relatively low volume of non-delay-sensitive data. Devices are required to be low cost and have a very long battery life (Figure 8.3).

To serve this plethora of different use cases and scenarios, the key design principles are flexibility and diversity, for which related capabilities are described in the recommendation, with different level of relevance and applicability, as depicted hereafter in the widely adopted spider diagram [19] (Figure 8.4).

8.3.2 5G Technical Solutions and New Radio

As described in the previous section, the 5G ecosystem includes several business areas and related enablers that can be grouped in 3 macro-categories: enhanced mobile broadband (eMBB,), massive machine type communication (mMTC) and ultra-reliable and low-latency (URLLC) communication, each of them with different requirements, as described in ITU-R IMT-VISION [19].

Besides that, 5G is natively based on radio access and core network virtualisation in order to rapidly deploy a plethora of new services with different requirements and technical characteristics.

Furthermore, it is foreseen that the 5G system will have to work in frequency ranges both under and above 6 GHz, depending on the availability and regulation of the spectrum, as described; for example, in the Action Plan for 5G published by the European Commission [20], which also indicates the need for 'pioneer bands' for 5G deployment in Europe, such as 700 MHz and 3.4–3.8 GHz under 6 GHz, and 26 and 31 GHz above 6 GHz.

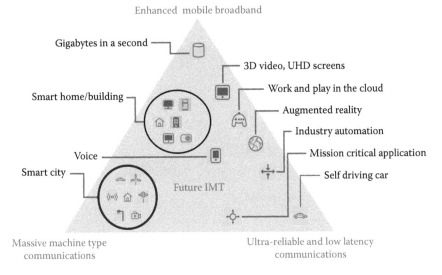

FIGURE 8.3   Usage scenarios of IMT for 2020 and beyond (Adapted from ITU-R IMT Vision, https://www.itu.int/dms_pubrec/itu-r/rec/m/R-REC-M.2083-0-201509-I!!PDF-E.pdf.) [19].

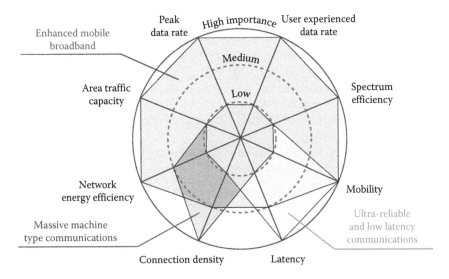

FIGURE 8.4   The importance of key capabilities in different usage scenarios (Adapted from ITU-R IMT Vision, https://www.itu.int/dms_pubrec/itu-r/rec/m/R-REC-M.2083-0-201509-I!!PDF-E.pdf.) [19].

First of all, it is very important to note that the 3rd Generation Partnership Project (3GPP) has decided that, since Release 15, LTE and its evolution will be considered an integral part of 5G radio technologies in order to respond to the different scenarios addressed by 5G. One key point is that connectivity performance should no longer be considered the evolutionary 'main driver'; the availability of a new, flexible and agile framework that can meet different service requirements from the three macro-categories mentioned above should become the most important goal.

In this optic, the system, and in particular the 5G radio access, which encompasses new radio components (NR) and LTE evolutions, must not only guarantee improved performance (or 'better performance

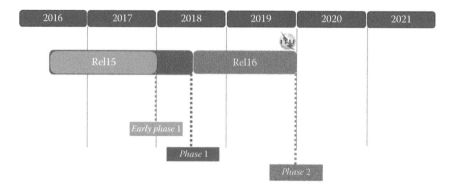

FIGURE 8.5   3GPP RAN roadmap for 5G releases.

than today') in terms of bit rates, a target that is also reachable via a continuous and constant evolution of LTE, but must enable, not necessarily simultaneously, different use cases related to the 3 macro-categories: bit-rate considerably higher (e.g. 20 Gbps peak rate in an indoor environment in downlink) for eMBB services, radio latencies on the order of 1 ms for URLLC services and a low-cost sensor density of 10^6 per square km in case of mMTC services.

It should be noted that in 3GPP, Release 15 is focused mainly on the eMBB use case for New Radio (NR), with native support and 'forward compatibility' for later releases of URLLC use cases. As far as the detail of the mMTC case use radio support is concerned, it will be standardised starting from Release 16, when the feasibility study for New Radio solutions supporting such use cases will begin. As already stated, it is important to assume that, as LTE evolution is part of 5G radio access, the underlying technology solutions of URLLC use cases (e.g. VTX, industry 4.0, robotics) and mMTCs will be based, at least until the availability of the standard and subsequent products of New Radio components, on LTE components such as NB-IoT for mMTC use cases.

In the following, the roadmap for 3GPP phases and releases is reported in Figure 8.5, including the preliminary release known as 'Early Drop'.

As already mentioned, the 5G features will be released in different phases, since it is not possible to standardise everything in time for the completion of Release 15 and early deployments (2018–2020):

- Release 15 'Early drop' (Dec. 17): born to intercept early market scenarios in Korea and the US
- Release 15 (Stage 1, June 18) targets deployment 2020
- Release 16 (Phase 2, Dec. 19) also targets submitting to ITU-R for IMT-2020

8.3.2.1 New Radio Features

The main features of the 'New Radio' under study and standardisation in 3GPP are:

- adoption of centimeter/millimeter waves/technologies that in turn imply:
  - densification of radiating points (UDN: Ultra Dense Network);
  - New waveform design and 'ultra-lean signaling';
- Massive/full dimensional MIMO and beamforming.

The potential benefits of using cm/mm waves may be summarised in:

- Availability of large portions of the spectrum, on the order of hundreds of MHz;
- Extremely high data rates (for example, the 20 Gbps peak downlink indoor environment);
- Very high spatial reuse thanks to beamforming techniques;
- 'Flexible deployment': it is possible to use the radio interface for both user terminals and back-hauling/front-hauling access.

The main challenges, however, can be summarised as:

- High link attenuation (partly attenuated by beamforming gains) and high sensitivity to 'blocking' and absorption phenomena, in addition to the difficulties associated with indoor penetration;
- Need for robust and efficient algorithms for tracking and searching the beams and complex system management with numerous 'directional' connections.

Under these assumptions, the main features of the new waveform design currently being discussed in 3GPP are summarised in:

- Use of OFDM as in LTE, but increasing efficiency in the use of available bandwidth (90% LTE at 95–98%);
- Different pilot symbols (RS) with respect to LTE, in order to manage the effects of the radio channel above 6GHz, also trying to reduce overhead (OH) and interference generated;
- Adoption of different carrier spacing values (30, 60, 120, 240, 480 KHz and not just 15 kHz as in LTE) to handle different bandwidths and different use cases, even dynamically and possibly simultaneously;
- Adoption of several Cyclix Prefix values to manage different coverage ranges as the frequency range varies;
- 'Ultra-Lean Signaling': attempts to reduce overhead control channels, both common and dedicated, adoption of 'grant free operation' for low-latency use and 'self-contained signaling'.

The fundamental principles of the Full Dimensional (FD)/Massive MIMO adopted in 5G are summarised in Figure 8.6:

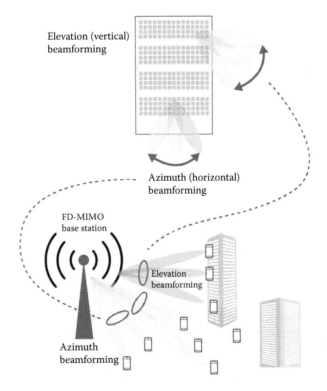

FIGURE 8.6 Main principles of FD/Massive MIMO (http://www.sharetechnote.com/html/5G/5G_ MassiveMIMO_FD_MIMO.html) [33].

- High number of antenna elements at the base station;
- Tens of users simultaneously served on the same radio resources, thanks to the Multi-User MIMO (MU-MIMO) realised using beamforming techniques, both horizontally and vertically.

8.4 5G Exposure Assessment

8.4.1 SotA (State of the Art) of Studies on Exposure to mmWaves

As described, the deployment of 5G networks adopting mmWaves will feature a high proliferation of radio base stations (RBSs), e.g. in order to meet the increasing demand for bandwidth in the eMBB use case: this may raise further interest on the level of electromagnetic radiation exposure to the public. In this section, an overview of the existing literature on this item and related topics is provided.

The World Health Organisation (WHO) established 'The International EMF Project' [21] to assess the health and environmental effects of exposure to static and time-varying electric and magnetic fields in the frequency range 0–300 GHz.

Some recent projects financed by the European Commission were involved in the investigation of items such as possible health hazards, reduction of EM radiation and the transmitting power of mobile communication systems, such as LEXNET [22], EARTH [23] and MiWaveS [4].

The paper [24] is a good tutorial survey of electromagnetic radiation exposure and reduction in mobile communication systems: it provides a comprehensive survey of the most recent literature on dosimetry metrics, international projects and guidelines and limits on exposure to EM radiation in advanced mobile communication systems adopting MIMO and beamforming techniques. It also proposes possible techniques for reducing EM radiation exposure by exploring known concepts related to SAR and transmission power reduction of mobile systems.

Investigations of interactions between the human body and millimeter-wave devices are analysed in [25], giving hints on today's regulatory requirements and mainly focusing on a 60 GHz transceiver. Also, the propagation characteristics of millimeter waves in the presence of the human body are studied, and four models representing different body parts are considered to evaluate the thermal effects of millimeter-wave radiation on the body.

The article [26] includes a review of the literature and a detailed analysis of the current understanding of the potential biological effects of nonionising mmWave radiation on the human body (mainly on skin and eyes), focusing on what is required to ensure the safety of emerging mmWave technologies for next-generation (5G) mobile communications networks.

The use of large array RBS antennas and the application of millimetric wave band for 5G networks poses new challenge in RF EMF assessments. One of the challenges is the mutual coupling between the antenna arrays and determination of the field distributions. In RF EMF assessments for large array multiport antennas, the combined exposure from all ports needs to be considered. Different ways of assessing antenna arrays intended for beamforming applications have been investigated in [27–31]. In [27] it is shown that accurate and efficient EMF compliance assessments can be conducted using the embedded element approach, in which each array element is excited sequentially while other elements are terminated in matched loads. This approach is suitable for small array antennas, e.g. with a maximum of eight antennas. In [29], numerical EMF exposure assessments are carried out using a conservative field procedure combined with the embedded element approach. While this particular procedure is suitable for MIMO arrays, it is rather costly in terms of assessment time and resources for very large array antennas. Degirmenci et al. [31] proposes approximate methods for EMF compliance assessments of large array antennas. These important aspects are very well summarised in [32].

8.4.2 Preliminary Evaluation Methodologies for 5G

In the previous section, the state of the art of the literature on exposure to mmWaves was summarised. In this section, some preliminary considerations of possible methodologies related to 5G radio peculiarities are briefly highlighted.

As described above, New Radio is adopting OFDMA and FD/MU-MIMO advanced techniques with high processing beamforming capabilities. In this optic, EMF exposure assessment strongly depends on transmission frequency, number of antennas, type of beamforming algorithm, gain(s) of the array and related directivity of the various beams. As a first approximation, an upper bound for a certain direction can be obtained in a single point by measuring the power associated with the reference symbols, which, as in LTE, are supposed to be transmitted at constant value, supposing that the point is reached with the maximum gain of the beamformer; to obtain the maximum received power, as nowadays, it is necessary to multiply the measured *RS* value by the number of resource blocks in the channel bandwidth being considered. In order to keep into account the angular and time variation of the power associated with the beamforming, the value obtained has to be further multiplied by an attenuating factor, depending on the exact characteristics of the beamforming algorithm that influences the variations.

Finally, the ability of beamforming to aim signals towards the desired users has the advantage of minimising or eliminating EM radiation towards unintended users. Incorporating EM radiation exposure metrics into the optimisation process of joint power allocation and beamforming schemes could significantly reduce EM radiation exposure levels in all users in the system in both the uplink and downlink directions [24].

References

1. ZTE COMMUNICATIONS. 2015. *SPECIAL TOPIC: 5G Wireless: Technology, Standard and Practice.* New York: An International ICT R&D Journal Sponsored by ZTE Corporation.
2. PREPARED REMARKS OF FCC CHAIRMAN TOM WHEELER *The Future of Wireless: A Vision for U.S. Leadership in a 5G Word.* Washington, DC: National Press Club June 20, 2016.
3. GSMA ANALYSIS Understanding 5G: *Perspectives on future technological advancements in mobile.* December 2014.
4. MiWaves-Beyond. 2020. Heterogeneous Wireless Network with Millimeter-Wave Small-Cell Access and Backhauling: http://www.miwaves.eu/
5. ICNIRP. Guidelines for limiting exposure to time-varying electric, magnetic, and electromagnetic fields (up to 300 GHz), *Health Physics*, vol. 74, no. 4, pp. 494–522, 1998.
6. ICNIRP. Exposure to high frequency electromagnetic fields, biological effects and health consequences (100 kHz–300 GHz), ISBN 978-3-934994-10-2, 2009.
7. 1999/519/EC, Council Recommendation of 12 July 1999 on the Limitation of Exposure of the General Public to Electromagnetic Fields (0 Hz to 300 GHz).
8. 2004/40/EC, Directive of the European Parliament and of the Council of 29 April 2004 on the Minimum Health and Safety Requirements Regarding the Exposure of Workers to the Risk Arising from Physical Agents (Electromagnetic Fields).
9. EN 50413 – 2008, Basic Standard on Measurement and Calculation Procedures for Human Exposure to Electric, Magnetic and Electromagnetic Fields (0 Hz–300 GHz).
10. EN 532 – 2008, Report from the Commission on the Application of Council Recommendation of 12 July 1999 (1999/519/EC) on the Limitation of the Exposure of the General Public to Electromagnetic Field (0 Hz to 300 GHz).
11. IEEE Standard for safety levels with respect to human exposure to radio frequency electromagnetic fields, 3 kHz to 300 GHz, ISBN 0-7381-4835-0 SS95389, April 2006.

12. Colombi, D., Thors, B., Törnevik, C. Implication of EMF exposure limits on output power lever for 5G devices above 6 GHz. *IEEE Antenna and Wireless Propagation Letters*, vol. 14, 2015.
13. IEC62232 Ed. 2.0. 2016. *Determination of RF Field Strength and SAR in the Vicinity of Radiocommunication Base Stations for the Purpose of Evaluating Human Exposure*. International Standard.
14. LTE resource grid. http://niviuk.free.fr/lte_resource_grid.html
15. https://www.metis2020.com/
16. NGMN – 5G WHITE PAPER. https://www.ngmn.org/5g-white-paper.html
17. http://www.rcrwireless.com/20170110/fundamentals/five-5g-trials-tag23-tag99
18. https://www.sdxcentral.com/articles/news/5g-trials-and-tribulations-a-guide-to-global-5g-operator-tests/2016/12/
19. ITU-R IMT Vision. https://www.itu.int/dms_pubrec/itu-r/rec/m/R-REC-M.2083-0-201509-I!!PDF-E.pdf
20. 5G for Europe: An Action Plan. http://ec.europa.eu/newsroom/dae/document.cfm?doc_id=17131
21. WHO-The International EMF Project. http://www.who.int/peh-emf/project/en/
22. LEXNET, Low EMF Exposure Future Networks, http://www.lexnet.fr/
23. EARTH, Driving the Energy Efficiency of Wireless Infrastructure to its Limits, https://www.ict-earth.eu/
24. Yusuf A. Sambo, Student Member IEEE, Fabien Heliot, Member IEEE and Muhammad Ali Imran, Senior Member IEEE. *A Survey and Tutorial of Electromagnetic Radiation and Reduction in Mobile Communication Systems.*
25. Wu, T., Rappaport, T. S., Collins, C. M. *The Human Body and Millimeter-Wave Wireless Communication Systems: Interactions and Implications*, accepted in 2015 IEEE International Conference on Communications (ICC), June 2015.
26. Wu, T., Rappaport, T. S., Collins, C. M. *Safe for Generations to Come*, IEEE Microwave Mag. 2015 March.
27. Kelley, D. F. Embedded element patterns and mutual impedance matrices in the terminated phased array environment, *IEEE Antennas Propag. Soc. AP-S Int. Symp.* vol. 3 A, pp. 659–662, 2005.
28. Chen, Q. et al., SAR investigation of array antennas for mobile handsets, *IEICE Trans. Commun.*, vol. E90-B, no. 6, pp. 1354–1356, 2007.
29. Thors, B. et al., Radio frequency electromagnetic field compliance assessment of multi-band and MIMO equipped radio base stations, *Bioelectromagnetics*, vol. 35, no. 4, pp. 296–308, 2014.
30. Degirmenci, E. Assessment of compliance with RF EMF exposure limits: approximate methods for radio base station products utilizing array antennas with beam-gorming capabilities, *IEEE Trans on Electromagnetic Compatibility*, vol. 58, no. 4, pp. 1110–1117, 2016.
31. Degirmenci, E. et al., Approximate methods for EMF compliance assessments of large array antennas, Proc.2015 Int. Conf. Electromagn. Adv. Appl. ICEAA 2015, pp. 1141–1144, 2015.
32. Nor Adibah Ibrahim, Tharek Abd. Rahman, Olakunle Elijah, *Recent Trend in Electromagnetic Radiation and Compliance Assessments for 5G Communication*. Int. Jour. of Elec. and Comp. Eng. (IJECE), April 2017.
33. http://www.sharetechnote.com/html/5G/5G_MassiveMIMO_FD_MIMO.html

9

Electromagnetic Field Exposure Assessment in Workers and the General Public: Measurement Techniques and Experimental Dosimetry

Peter Gajšek

Institute of Nonionizing Radiation (INIS)

9.1 Introduction

In recent decades, man-made sources of electromagnetic fields (EMFs) have steadily increased due to electricity demand, domestic appliances, telecommunications, wireless technologies, broadcasting, various applications in industry and medical equipment such as magnetic resonance imaging (MRI). This leads to increased EMF exposure in our living and working environments, and the intensity varies substantially with the situation. Since humans in most cases do not perceive EMFs (except visible light and warmth due to high-level fields), the presence of EMFs must be measured by instruments or approximated by theoretical calculations. Measurements may provide the most realistic EMF exposure assessment. For example, measurements must frequently be made, even after computations have been performed, due to the uncertainties inherent in the particular exposure environment. In a multiple-source environment, or in the case of leakage sources (such as RF heat sealers, induction heaters and other unintentional radiators), the calculations may become so cumbersome that measurements may be the most expedient method for assessing exposure.

9.2 EMF Measurement Techniques

In general, measurements determine whether the limit values for EMFs specified in laws, national regulations and recommendations are being adhered to. This is essential for occupational safety and general public protection. EMF measurements are performed using particular sensors or probes, such as EMF meters. These sensors can be generally considered as antennas, although they have different characteristics. In fact, the sensors should not perturb the EMFs and must prevent coupling and reflection as much as possible in order to obtain precise results.

In general, there are two main types of EMF measurements:

- *Broadband measurements* are performed using a broadband probe, that is, a device which senses any signal across a wide range of frequencies. An EMF probe that is usually made with diode or thermocouple detectors may respond to fields only on one axis or may be tri-axial, showing components of the field in three directions at once. Amplified, active probes can improve measurement precision and sensitivity, but their active components may limit their speed of response.
- In *Narrowband or frequency/code selective measurements* the measurement system consists of a field antenna and a frequency selective receiver or spectrum analyser allowing monitoring of the frequency range of interest.

Several commercially available instruments permit direct broadband field strength measurement. Special care must be taken to avoid measurement errors in low-frequency, multiple-frequency, amplitude-modulated and intermittent-field environments when using broadband instruments.

Frequency selective measurements can overcome the issue of the unknown spectral composition of the field. Specialized instrumentation is also commercially available that provides for spectrum analysis of measured EMFs and can display both the frequency-specific and wide-band values of the field as a percentage of the applicable exposure limit. Code selective measurements are necessary in LTE (FDD/TDD) and UMTS operating modes for evaluating pilot signal information and extrapolation to maximum exposure levels (Joseph et al. 2012a).

9.2.1 Low-Frequency Measurements (1 Hz–10 MHz)

For low-frequency (LF) fields, the electric (E) and magnetic (B) fields must be measured separately because the two components are independent of one another, and we are almost always in the near-field region. Isotropic (non-directional) sensors designed to measure all three spatial directions simultaneously are needed to measure the electric and magnetic field exposures properly. Broadband measuring devices show the overall exposure for all of the field strengths within a specified frequency range. Individual signals can be assessed selectively using filters or calculations or analysed in terms of their frequency components using computational techniques (e.g. fast Fourier transformation). Some measuring devices provide convenient band pass and band stop filters along with suitable spectrum analysis methods.

Multi-frequency signals in the LF range can be assessed quickly with spectrum analysis. The time-domain version of the signal captured using a probe is automatically transformed into the frequency domain using a fast Fourier transformation (FFT). The spectral components are analysed at the same time. A look at the spectrum quickly reveals the distribution of the field strengths, fundamental frequency and harmonics. To assess exposure to low-frequency fields properly, it is often necessary to have in-depth knowledge of the field and the measuring devices. Measurement methods that automatically evaluate the frequency response (weighted peak, or shaped time domain = STD) can greatly simplify everyday work. The way the limit values depend on frequency is automatically taken into account. Suitable detectors are needed to measure the RMS and peak values. When making isotropic measurements, the phases of the individual components are taken into account. The B or E field is measured in real time and displayed as a percentage of the limit over the entire frequency range. Signals occupying one or more frequencies are evaluated correctly, as are pulsed signals. Whereas ambient conditions have little effect on the magnetic field, the presence of people, condensation or humidity can affect the electric field. To eliminate any possible influence by the body (particularly by the test personnel), measurements must always be made at a proper distance, using remote E field measuring devices or probes with a tripod if necessary.

9.2.2 High-Frequency Measurements (10 MHz–100 GHz)

Since it is impossible to precisely determine the propagation direction of waves in a free field, isotropic probes must be used. This is even more important when making high-frequency (HF) measurements in environments involving multiple field sources. To suppress brief, irrelevant limit violations, results are averaged – over an interval of 6 minutes as required by many standards – using the averaging function provided in the measuring device. Hot spots (e.g. under antennas) and blank spots due to standing waves and reflections can result in local field maxima and minima. This sort of problem can be handled using a higher density of test points and by taking measurements near objects that cause reflections. The field distribution is rarely homogeneous. Reasons for this include reflections due to neighboring antennas, buildings with metallic panels, screens, fences and cranes. To assess full-body exposure, measurements must be made at multiple locations. The quadratic mean (spatial average) of these values is then determined.

Since exposure limits vary according to frequency, field strengths at the different frequencies must be evaluated. Some measuring devices use 'shaped probes' to automatically provide this kind of frequency response evaluation. The device then displays the exposure level as a percentage of the relevant exposure limit. This type of equipment is particularly useful in multi-frequency environments. For pulsed signals used in radar facilities, thermocouple probes are very useful for making precise measurements of EMF radiation with extreme pulse/pause ratios. These probes are much better at determining RMS values than diode probes.

9.3 EMF Exposure Assessment

Different methods of exposure assessment in the low- or high-frequency spectrum have been used, including (i) spot or long-term measurements, (ii) personal exposimetry (PEM) and (iii) exposure related to specific devices.

Spot measurements represent the simplest form of measurement that rely on readings made at a point in time at one location. The major drawback of spot measurements is their inability to capture temporal variations. Long-term measurements were mainly introduced for monitoring the fields at one of more locations for longer periods. The long-term EMF monitoring system is made up of a series of EMF monitors installed wherever the EMF presence needs to be assessed continuously or by long-term observation. Data collected from spot or long-term measurements do not provide sufficient insight into the exposure levels of individuals, who are often exposed to multiple sources at the same time and who are frequently moving through a variety of living and working environments. Capturing the true EMF exposure experienced by an individual requires that explicit measurements be made across time and space, using, for example, personal exposimeters (PEM). Monitoring the individual exposure of a subject using personal exposimeters (also called 'dosimeters') that are worn on the body is an attractive proposition because it allows the recording of exposure to fields from all sources and in any environment in which the individual is located.

The uncertainty analysis when using a PEM is crucial and should follow the next issues: a) body shielding; b) residual uncertainties due to calibration; c) measurement errors due to true RMS response; d) how well the shaped frequency response of probes follows sensitivity variation with frequency; e) measurement artefacts (out-of-band pick-up/especially for the H-field probe; multi-signal error; static pick-up). In addition, these measurements with PEM tend to underestimate or overestimate the actual RF EMF exposure

(Bolte et al. 2011, Knafl et al. 2008). A potential solution would be to determine correction factors based on the calibrations in order to modify the measurement results (Lauer et al. 2012).

9.3.1 Public Exposure to Low-Frequency Fields

The general public is exposed to an ever-increasing number of sources of LF EMFs from various electric devices, appliances and technologies (i.e. transmission lines, transformers, typical household appliances and their power supplies). The related exposures are localised and their magnitude strongly depends on the distance of the user from the appliance itself (Leitgeb et al. 2008). The available exposure assessment studies done on the exposure of the general public to low-frequency electric and magnetic fields show that outdoor average extremely low-frequency magnetic fields in public areas in urban environments range between 0.05 and 0.2 µT in terms of magnetic fields, but stronger values (on the order of a few µT) may occur directly beneath high-voltage power lines, at the walls of transformer buildings and at the boundary fences of substations. A large electricity pylon carrying a 380/400 kV conductor produces a magnetic field of around 10–20 µT directly under the pylon itself, and an electric field of around 3–5 kV/m. These levels fall with increased distance (inverse square law) from the sides of the line (WHO 2007). In the indoor environment, high magnetic fields have been measured close to several domestic appliances (up to the mT range), some of which are held close to the body, e.g. hair dryers and electric shavers (Gajšek et al. 2016).

9.3.1.1 Spot and Long-Term Measurements in Different Environments

In summary, outdoor average ELF-MFs in public areas in urban environments range from 0.05 to 0.2 µT, but stronger values may occur directly beneath high-voltage power lines, at the wall of transformer buildings, and at the boundary fences of substations. In the case of the latter, the maximum field intensity can be as high as 20–80 µT. Underground cables can produce magnetic fields directly above them (i.e. along the line of the route itself) that are stronger than those associated with aerial lines, due to the smaller distance between the external environment and the cable itself (typically a few meters). For instance, a 400 kV underground cable can produce a magnetic field intensity of over 30 µT at ground level, falling to 10 µT at 2 m above the ground, since the field intensity falls very rapidly with increasing distance from either side. On the other hand, measurement studies of indoor exposure assessment showed that only in a small percentage of homes were the median magnetic fields above 0.2 µT (Gajšek et al. 2016).

Some studies were focused on magnetic field exposure due to transformer stations (from 0.4 to 10 kV) installed in multi-level residential buildings (Transexpo 2010). The source of these magnetic fields is mainly the distribution bars (so-called 'bus bars') that are typically mounted on the wall and/or ceiling of the room containing the transformer. These magnetic fields have been observed to reach some tens of µT on the floor above the transformer room, but decreased to a few µT at a distance of 1 m above the floor. These results are consistent with a study performed by Thuroczy et al. (2008) in which mean magnetic field exposure to 50 Hz magnetic fields was 0.98 µT, whereas in the apartments on other upper floors it was only 0.1 µT. Similar studies have been conducted by Röösli et al. (2011), Ilonen et al. (2008) and Yitzhak et al. (2012), where the magnetic field was up to 0.59 µT on average in apartments that were either directly above the transformer or in contact with the transformer room, but magnetic fields in apartments which did not touch any wall of the transformer building were from 0.06 to 0.14 µT (Table 9.1).

9.3.1.2 Personal Exposimetry

Since for all practical purposes the human body does not perturb low-frequency magnetic fields, the recorded magnetic field may represent a reliable estimation of the real exposure levels. The values of magnetic field exposure recorded by personal (isotropic) meters tend to be stronger than those derived from spot or long-term measurements in homes, because the device will measure the field from all sources.

TABLE 9.1   Measurement Data from Several Personal Exposure Monitoring Studies Including Geometric Mean (nT) Magnetic Field Measurements

Author	Study Type	Measurement	Geometric Mean (nT)
Struchen et al. (2016)	Children ($n = 172$) living close to power lines	Personal 24 h exposure	40
		Bedroom	50
Liorni et al. (2016)	Highly exposed children ($n = 86$)	Personal 24 h exposure	20 to 80
Brix et al. (2001)	1952 volunteers	Personal 24 h exposure	64
Van Tongeren et al. (2004)	81 volunteers	Personal 24 h exposure	80

WHO summarised measurement studies in the ELF range conducted until 2007 (WHO 2007) where the majority of studied persons were exposed to magnetic field levels below 0.1 μT (73.6%–89.9%), but a few (0.5%–4.5%) had exposure levels above 0.3 μT (based on arithmetic means). Geometric means were only available for a small number of studies, but more than 90% had exposures of more than 0.1 μT and 0.4%–1.2% had exposures of more than 0.4 μT. The electric field component was more difficult to assess because it is more susceptible to shielding and perturbation by conduction bodies.

The exposure to LF magnetic fields of the general population of Europe was estimated as a log normal distribution with a median of 0.02 μT and a standard deviation of 0.06 μT; 0.54% of the population was considered as having an exposure of ≥ 0.3 μT (Grelier et al. 2014). Importantly, this study highlighted that many existing exposure assessments and epidemiological studies have concentrated on estimating or measuring exposure among those specific subgroups of the population with stronger than average exposures, and the proportion of the general population represented by such groups is unclear.

9.3.1.3 Exposure Related to Specific Devices

For the general public, the highest LF magnetic fields are found in close proximity to household and other consumer electric appliances, and these fields may reach up to a few mT. In a survey of 50 homes, magnetic fields generated by a few appliances were in excess of 0.2 μT at a distance of 1 m: microwave cookers produced 0.37 μT; washing machines 0.27 μT; dishwashers 0.23 μT; some electric showers 0.11 μT; and can openers produced 0.2 μT. Of continuously operating devices, only three devices produced fields of some tens of μT at 0.5 m: central heating pumps produced 0.51 μT, central heating boilers 0.27 μT and fish-tank air pumps produced 0.32 μT. In any case, one should note that persons spend on average about 4.5 h per day in the kitchen, where the strongest sources of magnetic fields were located (Preece et al. 1997). Moreover, these data suggest that one-third of the total exposure to magnetic fields can be attributed to personal appliance use.

However, these high fields are very localized and are limited to very short distances (less than some centimetres) from the surface of the equipment. The maximum possible exposure next to a specific source often differs by some orders of magnitude from the average individual exposure (SCENIHR 2015). The highest magnetic field exposures up to 2 mT occur during the use of electric appliances that are held in close proximity to the body, such as electric razors or hair dryers (Gajšek et al. 2016).

Figure 9.1 illustrates two main points: First, the magnetic field strength around all appliances rapidly decreases the further you get away from them. Secondly, most household appliances are not operated very close to the body. At a distance of 30 cm the electric and magnetic fields surrounding most household appliances are more than 10 times lower than the given guideline limit of 5 kV/m and 200 μT at 50 Hz for the general public.

9.3.2 Public Exposure to Intermediate Frequencies (IF)

Only a few studies have dealt with exposure of the general public to IF fields (i.e. electronic article surveillance, energy saving bulbs, induction cookers). Recently, Aerts et al. (2017) performed an extensive

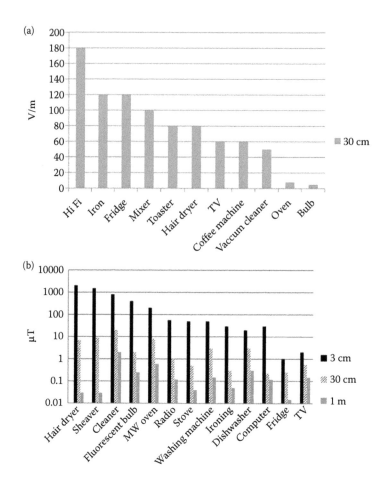

FIGURE 9.1 Typical electric (a) and magnetic (b) field strengths measured near household appliances at various distances.

measurement survey on the levels of electric and magnetic fields in the IF range typically present in residences and emitted by a wide range of household appliances under real-life circumstances. Using spot measurements, residential IF field levels were found to be generally low, while the use of certain appliances at close distance (20 cm) may result in a relatively high exposure. Overall, appliance emissions contained either harmonic signals, with fundamental frequencies between 6 kHz and 300 kHz, which were sometimes accompanied by regions in the IF spectrum of rather noisy, elevated field strengths, or much more capricious spectra, dominated by 50 Hz harmonics emanating far in the IF domain. The maximum peak field strengths recorded at 20 cm were 41.5 V/m and 2.7 A/m, both from induction cookers.

9.3.3 Public Exposure to High-Frequency Fields

Public exposure to HF EMFs in the living environment is continuously increasing. Although older studies showed that the principal sources of HF EMFs were signals in the broadcast bands (Tell and Mantiply 1980), some recent studies showed that almost 65% of HF exposure was due to wireless telecommunication devices (Boursianis et al. 2012). This trend seems to be continuously increasing. The major part of HF EMF public exposure comes from mobile and wireless portable devices, and not from fixed transmitters.

9.3.3.1 Spot or Long-Term High-Frequency EMF Measurements

Narrowband and broadband measurement methods of assessing exposure levels to HF EMFs have been applied in the range from several MHz to 10 GHz. Most of them were focused on exposure in the frequency range of mobile telecommunications (base stations). On-site (spot and long-term) HF EMF measurement campaigns were performed either as part of the planning permission process or upon request by public or local authorities (see Table 9.2).

Many cross-sectional studies in urban and rural areas have shown that total HF EMF exposure, including mobile telecommunication signals, was below 1 V/m. The mean value of electric field strength was slightly higher in a rural area (0.13 V/m) than in an urban area (0.08 V/m) (Gajšek et al. 2015, Hutter et al. 2006) (Table 9.3).

9.3.3.2 Personal Exposimetry

Many personal exposimetry measurements were performed in different microenvironments, such as offices or outdoor urban areas, to characterise typical exposure levels in these places (microenvironmental studies) (see Table 9.4). The studies were population surveys in which personal exposure distribution in the population of interest was determined. The strategies for the recruitment of the study participants, as well as the data analysis methods, differed between studies; therefore, a direct comparison of their results is difficult. Exposure in all countries was found to be on the same order of magnitude. All studies concluded that mobile telecommunication is the largest source of exposure. In all studies the mean exposure levels were found to be well below the ICNIRP exposure guidelines, with the highest exposure levels measured in transport vehicles (trains, cars and buses).

According to the PEM data, the mean total electric field strength is generally between 0.10 V/m and 0.27 V/m (Table 9.4). The overall results of the personal exposure assessment indicate that the average exposure to HF EMFs is less than levels measured using spot or long-term measurement methods. This

TABLE 9.2 Measurement Data on HF EMF Exposure Using On-Site (Spot or Long-Term) Campaigns in Mean Total Electric Field Strength in V/m or Average Exposure in % of the ICNIRP Level

Author	Study Type	Results
Paniagua et al. (2007)	Outdoor magnetic fields in five Spanish cities	Average 0.2 µT, ranging from 0.1 to 7 µT.
Lindgren et al. (2001)	Outdoor magnetic fields in centre of Goteborg, Sweden	Arithmetic mean (AM) of 0.17 µT, median magnetic field 0.2 µT.
D'Amore et al. (2001)	Outdoor measurements while walking at normal speed along an established path – 100,000 samples	Arithmetic mean (AM) of 0.19 µT, median of 0.08 µT and a geometric mean (GM) of 0.06 µT.
Maslanyi et al. (2005)	Survey conducted on 27 substations	Typical values at the perimeter fence were 10 µT for 275 and 400 kV substations, and 1.6 µT for an 11 kV substation. Mean field at the substation boundary was 1.1 µT, with a field of 0.2 µT up to 1.5 m from the boundary.
Schüz et al. (2000)	Indoor 24 h magnetic field measurements in 1835 residences	Median magnetic fields above 0.2 µT were found only in 1.4% of homes; 0.3 µT was found in 0.4%, and 0.4 µT was found in 0.2% of residences.
Tomitsch et al. (2010)	Spot measurements at the bedside in 226 households	Average night time arithmetic mean above 0.1 µT in 2.3% of households
Khan and Silva (2010)	Indoor measurements in 100 houses	90% of the measured houses had magnetic fields below 0.2 µT Mean value of 0.11 µT Median value of 0.05 µT

TABLE 9.3   Measurement Data on HF EMF Exposure Using On-Site (Spot or Long-Term) Campaigns in Mean Total Electric Field Strength in V/m or Average Exposure in % of the ICNIRP Level

Author	Country	Mean Total Electric Field Strength (V/m)	Description
Bornkessel et al. (2007)	Germany	1.31	Locations nearby the base stations and publicly accessible places (i.e. hospitals or schools)
Tomitsch and Dechant (2011)	Austria	0.39	Spot measurements in 226 households
Hutter et al. (2006)	Austria	0.13 – rural 0.08 – urban	Spot measurements of 10 randomly selected base stations
Altpeter et al. (2006)	Switzerland	0.56	AM radio transmitter
Gotsis et al. (2008)	Greece	1.64	46 monitoring stations 4 million measurements
Cooper et al. (2006)	UK	0.1–1.8	20 randomly selected GSM microcell and picocell
Rufo et al. (2011)	Spain	0.6	18 outdoor locations
Joseph et al. (2010)	Belgium, Netherlands, Sweden	0.74 – urban 0.46 – suburban 0.40 – residential 0.09 – rural	311 locations, 68 indoor and 243 outdoor, in three European countries
Rowley and Joyner (2012)	20 countries	0.52 global average	173,000 measurement surveys of mobile phone base stations in more than 20 countries
Troisi et al. (2008)	Italy	68.8% below 1 22.6% from 1–3 6.3% from 3–6 2.2% from 6–20 0.1% >20	Nationwide EMF long-term monitoring network

TABLE 9.4   Measurement Data on the HF EMF Exposure of Individuals Using Personal Exposimeters (PEM) in Mean Total Electric Field Strength in V/m or Average Exposure in % of the ICNIRP Level

Authors	Country	Mean Total Electric Field Strength (V/m)
Frei et al. (2009)	Switzerland	0.24
Trček et al. (2007)	Slovenia	0.27
Thuróczy et al. (2008)	Hungary	0.10
Bolte et al. (2008)	Netherlands	0.15
Viel et al. (2009)	France	0.21
Mann (2010)	UK	0.11
Joseph et al. (2012b)	Belgium	0.12
Breckenkamp et al. (2012)	Germany	0.09
Bolte and Eikelboom (2012)	Netherlands	0.26

could be because most of the time subjects are indoors, where mobile signals are weaker, whereas most in situ measurements are performed outdoors and close to emitters. Another cause may be that, in the case of personal exposure meters, body shielding explains part of the underestimation. The majority of studies concluded that mobile telecommunications (particularly the use of mobile and DECT/cordless phones) can be considered the main contributor to personal HF exposure.

9.3.3.3 Exposure Related to Specific HF EMF Sources

The exposure assessment of HF sources close to the human body needs dosimetric evaluation due to the complex HF electromagnetic field pattern near HF devices. The most important devices, those producing the highest exposure to HF in the public, are mobile phones. Other portable HF wireless devices which also can be close to the body expose the individual at levels many times lower than mobile phones. The averaged local exposure in the head induced by the mobile phone is considerably higher than that of far-field sources such as base stations and broadcasting. Although the EU limit of permissible SAR in the head is 2 W/kg, real-life exposure to HF fields from mobile devices is less than the results of compliance tests. According to Kuhn et al. (2007), compliance tests of more than 600 mobile phones frequently used in the European market and at maximum radiated HF power provided a Gaussian distribution with around 1 W/kg peak SAR, measured according to EN standards in a liquid head phantom.

In general terms, the brain exposure induced by a phone is about 0.5 W/kg for GSM and about 0.01 W/kg for UMTS. For a base station with 0.1 V/m it is about 10^{-6} W/kg (Wiart et al. 2000). It is important to emphasize that at a distance of 10 cm from the mobile phone, the absorbed power in the head was 10 times lower than when assessed close to the ear. At 40 cm in front of the head the maximum SAR over 10 g is close to 1% of the SAR obtained when touching the phone to the head.

Usually, exposure from DECT handsets and Bluetooth, WLAN and Wi-Fi devices is generally smaller than it is from cellular phones, but present for longer exposure periods. The maximum electric field values recorded at 0.5 meters around laptops and access points were 2.89 and 5.72 V/m, respectively (Peyman et al. 2011). The electric field reduced rapidly with distance for all tested Wi-Fi devices (maximum electric field strength at 0.5 m of 2.87 and 5.72 V/m, for laptops and access points, respectively; at 1 m, 1.22 and 2.60 V/m, respectively).

9.3.4 Occupational Exposure to EMF

Low-frequency field occupational exposure is associated with the generation and distribution of alternating current electricity and the use of office appliances and industrial/commercial equipment. HF EMF occupational exposure is associated with object-detection systems, telecommunications and some heating-based manufacturing and medical equipment. Some newer technologies, such as object identification and induction heating, emit intermediate frequency (IF) fields (Vila et al. 2016). The highest EMFs, where overexposure could occur, is reported for workers involved in electricity generation, production and distribution, welding, induction heating, transcranial magnetic stimulation, electrolysis, broadcasting and telecomunication and magnetic resonance imaging (MRI) (Stam 2014, Bolte and Pruppers 2006, Capstick et al. 2007).

9.3.4.1 Electricity Generation, Production and Distribution

The production and distribution of electricity involves a wide variety of equipment that operates at DC (e.g. high voltage direct current lines) or 50/60 Hz AC. Among them, the most significant from a human EMF exposure perspective are: power stations, overhead power lines, switchyards, transformers and substations (Bottura et al. 2009, Cooper 2002).

At *power stations*, electricity is generated from other forms of energy. Inside the power plant, the voltage is relatively low (between 5 and 20 kV) and is typically transformed to 110 kV or higher voltage for transport via overhead power lines. Because the voltage in the power station is low, there are strong currents producing strong magnetic fields. Near generators, they are below the limit values, but near bus bars the limit values can be exceeded by a factor of 10 or higher. Electric field strength in power stations is usually very small because the current-carrying conductors are shielded cables, but in switchgear yards the electric fields can be large (Heitanen et al. 2002).

Since it is necessary to distribute large amounts of power, the currents and voltages are high. Typical values are between 110 kV and 1000 kV for electric voltage and 300 A to 2 kA for electric current. The

TABLE 9.5   Exposure Assessment of Workers in Electricity Generation, Production and Distribution Sector

Type of Exposure	Source	Exposure
Power plant	10 kA 10 kV	5000 μT near bus bar or conductors for extremities 1000 μT for head and torso Up to 20 kV m^{-1}
High voltage overhead power lines	2 kA 380 kV and more	8 kV/m under the line 100 μT under the line 20 kV/m on the mast 500 μT on the mast
High voltage overhead power lines	240 kV	5 kV m^{-1} under the line 50 μT under the line 10 kV m^{-1} on the mast 200 μT on the mast
High voltage overhead power lines	110 kV	2 kV m^{-1} under the line 20 μT under the line 10 kV m^{-1} on the mast 100 μT on the mast
HVDC transmission lines	800 kV	20 kV m^{-1} under the line
HVDC transmission lines	3 kA	10 μT 10 cm away from the line
Switchyard	2 kA 400 kV	Up to 20 kV m^{-1} 400 μT

fields from overhead power lines at ground level can reach 8 kV m^{-1} for electric field strength and 100 μT for magnetic flux density. During 'live-line work', when maintenance work is carried out on energised conductors, electric field strength at the position of the worker could be up to 20 kV m^{-1} (although protective suits are worn to protect against high electric field levels) and magnetic flux density up to 2 mT (Bolte and Pruppers 2006). Computational modelling has been used to show that in many cases the exposure limit values are not exceeded during live-line work (Korpinen et al. 2009).

High voltage direct current (HVDC) transmission lines are used to transmit large amounts of power over long distances, since it can be more economical to transmit the energy using direct current instead of alternating current. Besides the benefit of smaller losses, DC lines reduce the wiring profile for a given power transmission capacity.

HVDC transmission lines operate at voltages up to 800 kV (typically 500 kV). The resulting electric fields may reach 20 kV m^{-1} under the lines (Leitgeb 2014). The currents in HVDC transmission lines can be up to 3 kA. The action value for static magnetic fields is 200 mT, and this not exceeded close to HVDC power lines. (Table 9.5)

9.3.4.2 Industrial Sector

Induction heaters and furnaces are used in industrial applications for metal processing, for example deep and surface hardening, tempering, etc. During the industrial process the products are placed inside or close to the induction coil, which generates a strong magnetic field. Consequently, the operators of induction furnaces and heaters are exposed to some of the highest magnetic fields existing in any industrial workplace. Heaters operating at a frequency of 50 Hz may produce 5000 μT fields at 20 cm and over 100 μT at a distance of several meters. Thus, the action value can potentially be exceeded when workers are close to the heaters (Kos et al. 2011). Values for electric field strength lie between 3 and 1000 V m^{-1}. Similarly, an induction furnace operating at 250 Hz was reported to produce a magnetic flux density over 700 μT at the operator's normal position. In addition, 50 μT was produced at the operator's position by an induction furnace operating at 150 kHz. Therefore, the relevant action values (100 μT for 250 Hz and 13.3 μT for 150 kHz) were exceeded at the operator's location when performing normal work tasks near furnaces (Bolte and Pruppers 2006). For 1 kHz and 10 kHz equipment, the magnetic flux density at a distance of 1 m can be up to 500 μT and may reach 5000 μT at 10 cm. The action value (30.7 μT for 10 kHz and 1 kHz) can therefore be exceeded close to the heaters. Measured values for electric field strength of up to 4000 V m^{-1} at 0.1 m and 1400 V m^{-1} at 0.5 m have been reported, exceeding the action value by a significant margin (Heitanen et al. 2002).

Many different welding processes can be used in industry. Regarding electromagnetic field exposure, the most significant welding processes are *arc welding, resistance welding and induction welding*. At the surface of the welding cable, magnetic flux densities reach 1–2 mT, exposing the hands of the welder to strong fields (Hamnerius and Persson 2005). Moreover, the body of the welder is exposed to a magnetic flux density of 100–200 μT. Electric field strength, however, is low, typically 10–100 V m^{-1} (Mair 2006). Resistance welding uses a power supply to pass an AC or pulsed DC current across the gap between two firmly butted components for a short period, often less than a second (Skotte and Hjollund 1997). Usually resistance welding equipment operates at 50 Hz with current up to 100 kA. Measurements have shown that the action values can be exceeded during resistance welding at 15 kA up to 30 cm away from the electrodes. For a resistance welding system operating at 11 kA and 50 Hz the measured magnetic flux density was 3.7 mT at 26 cm and 1.1 mT at 42 cm (Nadeem et al. 2004).

Radiofrequency dielectric heating (also known as capacitive or radio frequency heating) is a process used in industry to heat non-metallic objects. Dielectric heating is used for drying materials such as ceramics, leather, tobacco and paper and for gluing and drying wood. The same process is used for plastic sealing, where RF heating is used in the plastics manufacturing and processing industry. These devices generally operate within the frequency range of 10 to 100 MHz with output power between 0.2 and 1000 kW. Typical operating frequencies are 13.56 MHz, 27.12 MHz and 40.68 MHz. Electric field strengths at the operator's position can typically range from 10 to 1000 V m^{-1}, and magnetic field strengths from 0.1 to 20 A m^{-1} (Chen et al. 1991, ILO 1998, Wilen et al. 2004) (Table 9.6).

9.3.4.3 Broadcasting/Telecommunications

Radio and TV broadcasting antennas are amongst the most powerful sources of radiated RF energy. The typical maximum transmitting power for an amplitude-modulation (AM) radio broadcasting antenna is about 500 kW. For frequency-modulation (FM) radio broadcasting, systems and antennas work in the frequency range 87.5–108 MHz, with typical maximum transmitting power about 100 kW. In the last few years digital transmitters have been introduced for both radio and television. Digital transmitters – including terrestrial digital audio broadcasting (T-DAB), a successor to analogue, digital radio mondial

TABLE 9.6   Exposure Assessment of Workers in Industrial Sector

Type of Exposure	Source	Exposure
Induction heaters and furnaces	Frequency 50 Hz, Up to 5 MW	5000 μT at 20 cm 500 μT at 1 m 80 μT at 2.5 m 4000 Vm^{-1} at 10 cm 1400 Vm^{-1} at 50 cm
Arc welding	Up to 1 kA, DC, pulsing DC, AC	2 mT at the surface of the cable, 200 μT at the body of the welder
Resistance welding	Up to 100 kA, AC	Typically up to 10 mT at 10 cm, 1 mT at 1 m
Induction welding	Up to 20 kW, frequency below 1 MHz	300 μT at 1 m 2.5 mT on power supply cable
RF heaters	10–100 MHz, 1000 kW	1000 Vm^{-1} at operator location 20 Am^{-1} at operator location

(DRM), the successor to analogue AM, and digital video terrestrial broadcasting (DVB-T) – use less power than their analogue counterparts.

Even though radio and TV broadcasting antennas are mounted on towers, and the exposure at the ground is low, occupational exposure can be high for workers who have to climb these towers to perform maintenance and other tasks. Sometimes the towers must be climbed, usually for maintenance purposes, whilst the antennas are energized (Valič et al. 2012a).

At the bases of FM towers and within distances of typically 30 m, field strength may reach values higher than the action value. For example, electric field strengths range from 2 to 200 V m^{-1} on the ground beneath the towers. High exposure can occur for personnel working near or climbing on FM antenna towers. Thus, for FM frequencies, the action values are exceeded in some worker-accessible locations near towers (Valič et al. 2017) (Table 9.7).

High exposures occur for workers climbing TV antenna towers, but are low on the ground beneath towers. Under the towers, typical exposures are below 30 V/m, whereas on the towers near DVB-T antennas, electric field strength is up to 600 V m^{-1} and 2.5 µT. Thus, for workers climbing antenna towers, action values could be exceeded (Valič et al. 2012b).

Base station antennas are usually mounted on towers in rural areas and on buildings in cities. Since the radiation patterns of the antennas are narrow in the vertical plane, and the antennas are positioned vertically, EMF exposure at ground level below the antennas is small. Near the antennas exposure levels can be higher, but it is necessary to be only half a meter below the antenna for the exposure level to be below the action value. The exposure limit values of 23 W m^{-2} (900 MHz) and 45 W m^{-2} (1800 MHz) will be exceeded at about 60 cm in both cases in the direction of the main beam, based on worst-case near-field calculations. Occupational exposure is possible during maintenance of the base station, as well as during construction and similar tasks that are performed on the roof in close proximity to the transmitting antenna (Martinez et al. 2005).

Radar systems produce strong EM fields in the axis of the antenna. In a worst-case situation (for stationary antennas), peak power densities of 10 MW m^{-2} may occur up to few metres from the source and more than 100 W m^{-2} at distances over 1 km. The rotation of the antenna reduces the average power

TABLE 9.7   Exposure Assessment of Workers in Broadcasting/Telecommunication Sector

Type of Exposure	Source	Exposure
AM radio tower	153 kHz–26.1 GHz, 500 kW	1000 V m^{-1} at the surface of the tower 16 A m^{-1} at the surface of the tower 700 V m^{-1} at 1 m 1.6 A m^{-1} at 1 m
FM radio tower	87.5–108 MHz, 100 kW	1000 V m^{-1} on the tower 0.4 A m^{-1} on the tower 5 A m^{-1} in front of the antenna at 30 cm 200 V m^{-1} on the ground
DVB-T TV tower	VFH 174–230 MHz, 300 kW UHF 470–862 MHz, 5000 kW	400 V m^{-1} on the tower 2 A m^{-1} on the tower 30 V m^{-1} on the ground
Multiband GSM/UMTS base station	1770–1880 MHz, 80 W	300 W m^{-2} at 10 cm 45 W m^{-2} at 40 cm
TETRA base station	380–465 MHz, 4–40 W	100 W m^{-2} at 10 cm 10 W m^{-2} at 1 m
Radar	500 MHz–50 GHz, 2200 kW peak, 3 kV mean	10 MW m^{-2} peak in the axis of the antenna at a few metres 80 W m^{-2} mean in the axis of antenna at few metres 1 W m^{-2} at typical working locations

density (compared to stationary conditions) by a factor of about 100–1000. At a distance of 20 m, power density for large air traffic radar still could be more than 80 W m^{-2} (Puranen and Jokela 1996). However, for rotating radars, people are generally outside the main beam and the average power density in areas accessible to workers is rarely higher than 1 W m^{-2}. In areas surrounding military radars, workers can be exposed to power densities up to tens of kW m^{-2}, but normally those encountered are below 0.4 W m^{-2}.

9.3.4.4 Medical Sector

Magnetic resonance imaging (MRI) is a medical diagnostic and research method based on absorption and re-emission of radiofrequency (RF) fields by protons in a strong static magnetic field. The static magnetic fields inside MRI scanners are very high; the typical magnetic flux density is between 1 and 3 T. The tendency for the future is to use even stronger fields (up to 5 or 9 T) to improve the resolution of the images.

In addition to the static magnetic field, during the scanning procedure there are pulsed RF fields in the range of 10–400 MHz and also rapidly changing gradient magnetic fields. The time derivatives of gradient fields (dB/dt) typically vary from 1 to 3 mT/ms (1–3 T/s). Because of the large static magnetic fields used in MRI, exposure levels of MRI workers working close to the scanner can be relatively high. Close to the MRI scanner, magnetic flux density is up to 1 T, whereas at typical working locations (such as a nurse attending a patient) it is up to 200 mT. Outside the MRI room the magnetic field is low; normally magnetic flux density is less than 1 mT (Stam 2014, Bolte and Pruppers 2006).

Because the static magnetic field around the MRI scanner rapidly decreases with increasing distance, there is strong gradient of magnetic field presented. When moving in a gradient magnetic field, a current is induced inside the human body. For a 4 T MRI scanner, estimated maximum induced current density is approximately 300 mA m^{-2} when a body is moving at a constant speed of 0.5 m s^{-1} close to the MRI scanner. At distances greater than 1 m, induced current densities are below 40 mA m^{-2}. Pulsed gradient magnetic fields could induce high currents inside the human body. Normally, gradient magnetic fields are pulsed at low frequencies–about 1 kHz, where the value of the exposure limit for current density is 10 mA m^{-2}. Close to the gradient coils, 1 cm^2 averaged current density could be as high as 900 mA m^{-2} (Capstick et al. 2007).

A special case of occupational exposure is interventional MRI, where in some cases the worker can be exposed to levels similar to those experienced by the patient. The field strengths are lower than they are in normal MRI procedures (static magnetic field up to 1.5 T, gradient fields 20 mT m^{-1} and a slew rate of 40 T m^{-1} s^{-1}), but even so, the surgeon can be exposed to field strengths exceeding the action values for static magnetic fields. Sometimes the exposure limit for current density can be exceeded by exposure to the pulsed gradient magnetic field (Table 9.8).

Electrosurgery uses radiofrequency (300 to 600 kHz) electromagnetic currents to cut and coagulate tissues. The action values for worker exposure are often exceeded. In measurements for a 500 kHz

TABLE 9.8 Exposure Assessment of Workers in Medical Sector

Type of Exposure	Source	Exposure
MRI	1-3 T static	3 T inside the MRI scanner
		1 T close to the MRI scanner
		200 mT at typical working locations close to the MRI scanner
		Less than 1 mT outside the MRI room
MRI	Pulsed gradient magnetic field up to 3 T s^{-1}, usually at around 1 kHz	Depends on spectrum of gradient field
MRI	Pulsed RF 10–400 MHz	Less than 2 W m^{-2} at typical working locations
Electrosurgery	300–600 kHz, 500 W	15 kV m^{-1} close to the cable
		10 A m^{-1} close to the cable

electrosurgery unit, the surgeon's hands were exposed to typical electric and magnetic field values of $15 \, \text{kV m}^{-1}$ and $10 \, \text{A m}^{-1}$ (Ruggera 2005). An earthed metallic shield for the power supply cable could reduce the electric field by a factor of 10. Calculated values for current density are around the exposure limit value at 500 kHz ($5 \, \text{mA m}^{-2}$), whereas calculated SAR values were low (Liljestrand et al. 2003).

References

Aerts S, Calderon C, Valič B, Maslanyj M, Addison D, Mee T, Goiceanu C et al. 2017. Measurements of intermediate-frequency electric and magnetic fields in households. *Environ Res* 154:160–170.

Altpeter ES, Krebs T, Pfluger DH, von Kanel J, Blattmann R, Emmenegger D. 2006. *Study on health effects of the shortwave transmitter station of Schwarzenburg, Q23 Berne, Switzerland.* Major Report. BEW Publication Series No. 05, Bern: Federal Office of Energy.

Bolte JF, Eikelboom T. 2012. Personal radiofrequency electromagnetic field measurements in the Netherlands: Exposure level and variability for everyday activities, times of day and types of area. *Environ Int* 48:133–142.

Bolte J, Pruppers M. 2006. *Electromagnetic fields in the working environment.* Ministry of Social Affairs and Employment (SZW). RIVM report no. 610015001.

Bolte J, Pruppers M, Kramer J, Van der Zande G, Schipper C, Fleurke S. 2008. The Dutch exposimeter study: developing an activity exposure matrix. *Epidemiology* 19:S78–S79.

Bolte J, Van der Zande G, Kamer J. 2011. Calibration and uncertainties in personal exposure measurements of radiofrequency electromagnetic fields. *Bioelectromagnetics* 32:652–663.

Bornkessel C, Schubert M, Wuschek M, Schmidt P. 2007. Determination of the general public exposure around GSM and UMTS base stations. *Radiat Prot Dosimetry* 124:40–47.

Bottura V, Borlino MC, Carta N, Cerise L, Imperial E. 2009. Urban Exposure to ELF Magnetic Field due to High-, Medium- and Low-Voltage Electricity Supply Networks. *Radiat Prot Dosim* 137:214–217.

Boursianis A, Vanias P, Samaras T. 2012. Measurements for assessing the exposure from 3G femtocells. *Radiat Prot Dosimetry* 150:158–167.

Breckenkamp J, Blettner M, Schuz J, Bornkessel C, Schmiedel S, Schlehofer B. 2012. Residential characteristics and radiofrequency electromagnetic field exposures from bedroom measurements in Germany. *Radiat Environ Biophys* 51:85–92.

Brix J, Wettemann H, Scheel O, Feiner F, Matthes R. 2001. Measurement of the individual exposure to 50 and 16 2/3 Hz magnetic fields within the Bavarian population. *Bioelectromagnetics* 22:323–332.

Capstick M, McRobbie M, Hand J. 2007. An investigation into occupational exposure to electromagnetic fields for personnel working with and around medical magnetic resonance imaging equipment. Report VT/2007/017. Brussels, Belgium: European Commission.

Chen J-Y, Gandhi OP, Conover DL. 1991. SAR and Induced Current Distributions for Operator Exposure to RF Dielectric Sealers. *IEEE Trans Electromag Compat EMC* 33(3):252–261.

Cooper TG. 2002. *Occupational Exposure to Electric and Magnetic Fields in the Context of the ICNIRP Guidelines.* Chilton, UK: National Radiological Protection Board, NRPB-W24.

Cooper TG, Mann SM, Khalid M, Blackwell RP. 2006. Public exposure to radio waves 2near GSM microcell and picocell base stations. *J Radiol Prot* 26:199–211.

D'Amore G, Anglesio L, Tasso M, Benedetto A, Roletti S. 2001. Outdoor background ELF magnetic fields in an urban environment. *Radiat Prot Dosim* 94:375–380.

Frei P, Mohler E, Burgi A, Frohlich J, Neubauer G, Braun-Fahrlander C. 2009. A prediction model for personal radio frequency electromagnetic field exposure. *Sci Total Environ* 408:102–108.

Gajšek P, Ravazzani P, Grellier J, Samaras T, Bakos J, Thuróczy G. 2016. Review of Studies Concerning Electromagnetic Field (EMF) Exposure Assessment in Europe: Low Frequency Fields (50 Hz-100 kHz). *Int J Environ Res Public Health* 13(9):875.

Gajšek P, Ravazzani P, Wiart J, Grellier J, Samaras T, Thuróczy G. 2015. Electromagnetic field exposure assessment in Europe radiofrequency fields (10 MHz-6 GHz). *J Exp Sci and Env Epidem* 25:37–44.

Gotsis A, Papanikolaou N, Komnakos D, Yalofas A, Constantinou P. 2008. Non-ionizing electromagnetic radiation monitoring in Greece. *Ann Telecommun* 63:109–123.

Grellier J, Ravazzani P, Cardis E. 2014. Potential health impacts of residential exposures to extremely low frequency magnetic fields in Europe. *Environ Int* 62:55–63.

Hamnerius Y, Persson M. 2005. How to determinate compliance with the directive's exposure limit values (ICNIRP basic restrictions) for electric welding, S5/7–11. In: *Proceedings of the international Electromagnetic Fields in the Workplace workshop* J. Karpowicz, K. Gryz (eds.) Warsaw, Poland.

Heitanen M, Hämäläinen A, von Nandelstadh P. 2002. *Electromagnetic fields in the work environment.* Helsinki: Finnish Institute of Occupational Health.

Hutter HP, Moshammer H, Wallner P, Kundi M. 2006. Subjective symptoms, sleeping problems, and cognitive performance in subjects living near mobile phone base stations. *Occup Environ Med* 63:307–313.

Ilonen K, Markkanen A, Mezei G, Juutilainen J. 2008. Indoor transformer stations as predictors of residential ELF magnetic field exposure. *Bioelectromagnetics* 29:213–218.

International Labour Organization. 1998. Safety in the use of radiofrequency dielectric heaters and sealers. Occupational Safety and Health Series, No. 71. International Labour Office, Geneva.

Joseph W, Frei P, Roosli M, Thuroczy G, Gajšek P, Trček T. 2010. Comparison of personal radio frequency electromagnetic field exposure in different urban areas across Europe. *Environ Res* 110:658–663.

Joseph W, Verloock L, Goeminne F, Vermeeren G, Martens L. 2012a. In situ LTE exposure of the general public: characterization and extrapolation. *Bioelectromagnetics* 33:466–475.

Joseph W, Verloock L, Goeminne F, Vermeeren G, Martens L. 2012b. Assessment of RF exposures from emerging wireless communication technologies in different environments. *Health Phys* 102:161–172.

Khan A, Silva GD. 2010. *Measurement and Analysis of Extremely Low Frequency Magnetic Field exposure in Swedish Residence.* Master's Thesis, Chalmers University of Technology, Gothenburg, Sweden.

Knafl U, Lehmann H, Riederer M. 2008. Electromagnetic field measurements using personal exposimeters. *Bioelectromagnetics* 29:160–162.

Korpinen L, Elovaara J, Kuisti H. 2009. Evaluation of Current Densities and Total Contact Currents in Occupational Exposure at 400 kV Substations and Power Lines. *Bioelectromagnetics* 30:231–240.

Kos B, Valič B, Kotnik T, Gajšek P. 2011. Occupational exposure assessment of magnetic fields generated by induction heating equipment-the role of spatial averaging. *Phys Med Biol* 57:5943–5953.

Kuhn S, Kramer A, Lott U, Kuster N. 2007. Assessment methods for demonstrating compliance with safety limits of wireless devices used in home and office environments. *IEEE Trans Electromagn Compatib* 49:519–525.

Lauer O, Neubauer G, Roosli M, Riederer M, Frei P, Mohler E. 2012. Measurement setup and protocol for characterizing and testing radio frequency personal exposuremeters. *Bioelectromagnetics* 33:75–85.

Leitgeb N. 2014. Limiting electric fields of HVDC overhead power lines. *Radiat Environ Biophys* 53:461. doi:10.1007/s00411-014-0520-2

Leitgeb N, Cech R, Schröttner J, Lehofer P, Schmidpeter U, Rampetsreiter M. 2008. Magnetic Emissions of Electric Appliances. *Int J Hyg Environ Health* 211:69–73.

Liljestrand B, Sandstrom M, Mild KH. 2003. RF Exposure During Use of Electrosurgical Units. *Electromagnetic Biology and Medicine* 22:127–132.

Lindgren M, Gustavsson M, Hamnerius Y, Galt S. 2001. ELF Magnetic Fields in a City Environment. *Bioelectromagnetics* 22:87–90.

Liorni I, Parazzini M, Struchen B, Fiocchi S, Röösli M, Ravazzani P. 2016. Children's Personal Exposure Measurements to Extremely Low Frequency Magnetic Fields in Italy. *Int J Environ Res Public Health* 13:549. doi:10.3390/ijerph13060549.

Mair P. 2006. Effects on the Human Body and Assessment Methods of Exposure to Electro-Magnetic-Fields Caused by Spot Welding. Proceedings of the 4th International Seminar on Advances in Resistance Welding; Wells, November 2006, Available online: https://pdfs.semanticscholar.org/4e34/2f2a52a84342a6ad97d1311578df1d28f993.pdf (accessed on 4 January 2018).

Mann S. 2010. Assessing personal exposures to environmental radiofrequency electromagnetic fields. *Comptes Rendus Physique* 11:541–555.

Martinez MB, Martin A, Anguiano M, Villar R. 2005. On the safety assessment of human exposure in the proximity of cellular communications base-station antennas at 900, 1800 and 2170 MHz. *Physics in Medicine and Biology* 50:4125–4137.

Maslanyi MP, Mee TJ, Allen SG. 2005. Investigation and Identification of Sources of Residential Magnetic Field Exposures in the United Kingdom Childhood Cancer Study (UKCCS). Avalaible online: https://www.gov.uk/government/uploads/system/uploads/attachment_data/file/340205/HpaRpd005.pdf (accessed on 20 May 2017).

Nadeem M, Hamnerius Y, Mild KH, Persson M. 2004. Magnetic field from spot welding equipment–is the basic restriction exceeded? *Bioelectromagnetics* 25(4):278–284.

Paniagua JM, Jiménez A, Rufo M, Gutiérrez JA, Gómez FJ, Antolín A. 2007. Exposure to extremely low frequency magnetic fields in an urban area. *Radiat. Environ Biophys* 46:69–76.

Peyman A, Khalid M, Calderon C, Addison D, Mee T, Maslanyj M. 2011. Assessment of exposure to electromagnetic fields from wireless computer networks (wi-fi) in schools; results of laboratory measurements. *Health Phys* 100:594–612.

Preece AW, Kaune W, Grainger P, Preece S, Golding J. 1997. Magnetic fields from domestic appliances in the UK. *Phys Med Biol* 42:67–76.

Puranen L, Jokela K. 1996. Radiation Hazard Assessment of Pulsed Microwave Radars. *Journal of Microwave Power and Electromagnetic energy* 31(3):165–177.

Röösli M, Jenni D, Kheifets L, Mezei G. 2011. Extremely low frequency magnetic field measurements in buildings with transformer station in Switzerland. *Sci Total Environ* 409:3364–3369.

Rowley J, Joyner K. 2012. Comparative international analysis of radiofrequency exposure surveys of mobile communication radio base stations. *J Expo Sci Environ Epidemiol* 22:304–315.

Rufo MM, Paniagua JM, Jimenez A, Antoln A. 2011. Exposure to high-frequency electromagnetic fields (100 kHz-2 GHz) in Extremadura (Spain). *Health Phys* 101:739–745.

Ruggera PS. 2005. Measurements of emission levels during microwave, Medicines and Healthcare Products Regulatory Agency. High power electrosurgery review update, MHRA-document 05023, Department of Health.

SCENIHR. 2015. Scientific Committee on Emerging and Newly Identified Health Risks, Potential Health Effects of Exposure to Electromagnetic Fields (EMF), European Commission. Available online: http://ec.europa.eu/health/scientific_committees/emerging/docs/scenihr_o_041.pdf (accessed on 19 May 2017).

Schüz J, Grigat JP, Störmer B, Rippin G, Brinkmann K, Michaelis J. 2000. Extremely low frequency magnetic fields in residences in Germany. Distribution of measurements, comparison of two methods for assessing exposure, and predictors for the occurrence of magnetic fields above background level. *Radiat Environ Biophys* 39:233–240.

Skotte JH, Hjollund HI. 1997. Exposure of welders and other metal workers to ELF magnetic fields. *Bioelectromagnetics* 18(7):470–477.

Stam R. 2014. The Revised Electromagnetic Fields Directive and Worker Exposure in Environments With High Magnetic Flux Densities. *Ann Occup Hyg* 58(5):529–541.

Struchen B, Liorni I, Parazzini M, Gängler S, Ravazzani P, Röösli M. 2016. Analysis of children's personal and bedroom exposure to ELF-MFs in Italy and Switzerland. *J Expo Sci Environ Epidemiol* 26:586–596.

Tell RA, Mantiply ED. 1980. Population exposure to VHF and UHF broadcast radiation in the United States. *Proc IEEE* 68:6–12.

Thuróczy G, Jánossy G, Nagy N, Bakos J, Szabó J, Mezei G. 2008. Exposure to 50 Hz magnetic field in apartment buildings with built-in transformer stations in Hungary. *Radiat Prot Dos* 131:469–473.

Thuroczy G, Molnar F, Janossy G, Nagy N, Kubinyi G, Bakos J. 2008. Personal RF exposimetry in urban area. *Ann Telecommun* 63:87–96.

Tomitsch J, Dechant E. 2011. Trends in residential exposure to electromagnetic fields from 2006 to 2009. *Radiat Prot Dosimetry* 149:384–391.

Tomitsch J, Dechant E, Frank W. 2010. Survey of electromagnetic field exposure in bedrooms of residences in lower Austria. *Bioelectromagnetics* 31:200–208.

TransExpo. 2010. International Study of Childhood Leukemia and Residences near Electrical Transformer Rooms. Available online: http://my.epri.com/portal/server.pt?space=CommunityPage&cached=true& (accessed on 19 June 2017).

Trček T, Valic B, Gajsek P. 2007. Measurements of background electromagnetic fields in human environment. IFMBE Proceedings 11th Mediterranean Conference on Medical and Biomedical Engineering and Computing, vol. 16. MEDICON: Ljubljana, Slovenia, pp. 222–225.

Troisi F, Boumis M, Grazioso P. 2008. The Italian national electromagnetic field monitoring network. *Ann Telecommun* 63:97–108.

Valič B, Kos B, Gajšek P. 2012a. Simultaneous Occupational Exposure to FM and UHF Transmitters. *Int J Occup Saf Ergon* 18:161–170.

Valič B, Kos B, Gajšek P. 2012b. Occupational Exposure Assessment on an FM Mast: Electric Field and SAR Values. *Int J Occup Saf Ergon* 18:149–159.

Valič B, Kos B, Gajšek P. 2017. Radiofrequency exposures of workers on low-power FM radio transmitters. *Ann Work Expo Health* 61(4):457–467.

Van Tongeren M, Mee T, Whatmough P, Broad L, Maslanyj M, Allen S, Muir K, McKinney P. 2004. ELF magnetic field exposure in the UK adult brain tumour study: Results of a feasibility study. *Radiat Prot Dosim* 108:227–236.

Viel JF, Cardis E, Moissonnier M, de Seze R, Hours M. 2009. Radiofrequency exposure in the French general population: band, time, location and activity variability. *Environ Int* 35:1150–1154.

Vila J, Bowman JD, Richardson L, Kincl L, Conover DL, McLean D, Mann S, Vecchia P, van Tongeren M, Cardis E. 2016. A Source-based Measurement Database for Occupational Exposure Assessment of Electromagnetic Fields in the INTEROCC Study: A Literature Review Approach. *Ann Occup Hyg* 60(2):184–204.

WHO. 2007. Environmental Health Criteria Monograph No.238. Extremely Low Frequency Fields. Available online: http://www.who.int/peh-emf/publications/elf_ehc/en/ (accessed on 20 May 2017).

Wiart J, Dale C, Bosisio AV, Le Cornec A. 2000. Analysis of the influence of the power control and discontinuous transmission on RF exposure with GSM mobile phones. *IEEE Trans Electromagn Compatib* 42:376–385.

Wilen J, Hornsten R, Sandstrom M, Bjerle P, Wiklund U, Stensson O, Mild KJ. 2004. Electromagnetic field exposure and health among RF plastic sealer operators. *Bioelectromagnetics* 25:5–15.

Yitzhak NM, Hareuveny R, Kandel S, Ruppin R. 2012. Time dependence of 50 Hz magnetic fields in apartment buildings with indoor transformer stations. *Radiat Prot Dosim* 149:191–195.

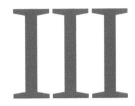

Ionizing Radiation

Mauro Magnoni
Environmental Protection Agency of Piedmont Region, Italy

10

Overview of the Basic Concepts of Radiation Protection

Daniele Giuffrida
*Federal Authority for Nuclear
Regulation*

10.1 Introduction: What is 'Radiation Protection'?

'Radiation' is spontaneously emitted during transformations of atoms, called 'radioactive decays': This radiation has enough energy to strip electrons from the outer shells of atoms, and for this ability it is called 'ionising radiation' in order to distinguish it from, for example, visible radiation, ultra-violet radiation (UV) or microwaves, etc., which are therefore called 'non-ionising radiation'.

Ionising radiation is most commonly encountered in the form of alpha, beta or neutron particles, or as X or gamma rays*. It possesses many valuable properties, for which it is exploited with success and to the benefit of humanity in the medical, industrial, agricultural, research and educational fields.

Ionising radiation, however, may also represent a hazard to human beings, including man: Exposure to ionising radiation can generate a vast range of biological effects, which also comprise sickness, cancer and death of the entire organism.

---

* The old distinction between 'particles' and 'rays' should be dropped, as modern physics proved that, at a microscopic scale, particles can behave as waves (rays), and waves (rays) can also have a corpuscular nature.

After the discovery of ionising radiation (1895–1896), and mainly during the 20th century, a specific branch of physics, encompassing not only radiation physics but also biology, physiology, epidemiology, technology, engineering, genetics, metrology, radioecology, etc., gradually developed and specialised in the assessment of radiation effects and in the protection of humans and other biota from ionising radiation.

Today this discipline is commonly known as 'radiation protection'.

Radiation protection, as a scientific and technical discipline, covers the assessment of radiological risk deriving from exposure to ionising radiation and aims to protect individuals, society and the environment from the harmful effects of ionising radiation without unduly limiting its beneficial effects. These two aspects (assessment of the effects of radiation exposure and protection of organisms from the effects of the exposure) represent two basic pillars of multidisciplinary radiation protection.

Knowledge of the effects of radiation exposure has been accumulated using data from experiments executed on animals and humans and laboratory experiments on cell cultures and tissues; the evaluation of groups of people who have been significantly irradiated (atomic bombs survivors, miners, 'dial painters', cancer therapy patients, etc.) are an integral part of this process. Significant knowledge developments in the relationship dose-effects of radiation exposure came from continuous studies of the effects of exposure of individuals to high doses of radiation, often in conditions which cannot be reproduced (nuclear explosions, radiation accidents, etc.).

Protection of individuals (workers and members of the public) from the risks deriving from exposure to ionising radiation has been sought by developing a paradigm (a model) of how radiological risk is generated, how radiation exposure relates to risks and how risks can be kept under control (principles and practices).

It is worth stressing that radiation protection is not exclusively based on science and objective, scientific data: Its scope, in terms of both paradigm and principles, includes some aspects which are not quantitative, but are based on experience and judgement and other considerations which pertain to social sciences. The current radiation protection paradigm is based on a series of assumptions – reached via consensus among experts – and therefore is more the result of expert judgement than of the output of arithmetic or mathematical calculations. Finally, principles and practices in operational radiation protection have continuously evolved with time, and have been significantly updated and revised whenever new evidence appeared.

It is important to remember that radiation protection recommendations (like the most recent ones, ICRP, 2007) are expressed with reference to a 'standard man' or 'reference person' whose characteristics do not depend on sex, age, race or any peculiar somatic features. Coefficients are calculated from this hypothetical model, which allows us to relate radiation exposure to radiation dose (generated in the model), both from 'external exposure' (i.e. from irradiation of the body from a source which is situated outside the body itself) and from 'internal exposure' (i.e. due to radioactivity which has been introduced within the body and is distributed in organs and tissues according to the chemical biokinetics of the element, while irradiating organs and tissues).

Hence, the ability of the radiation protection system to predict with some precision the magnitude of a specific exposure to ionising radiation depends at best on very specific and limited conditions (e.g. in the case of high dose radiation exposure and well-known radiation fields), and in most cases radiation protection principles serve as guidance rather than a precise instrument for assessing radiation exposure risk.

After more than a century of studies and research, however, the radiation protection community has developed a somewhat complex but comprehensive system to assess radiation exposure, predict the effects of radiation exposure and protect individuals, society and the environment from its negative effects.

Most remarkably, there is wide agreement around the world, in very diversified contexts, about how this should be done:

- At a fundamental level, radiation interacts with cells, tissues and organs, depositing energy which can generate specific effects (see Section 10.2).

- The assessment of radiation exposure is connected to the deposition of energy, and is referred to as radiation 'dose' (see Section 10.5).
- The system of radiation protection correlates doses to risks and uses doses to assess the effects of radiation exposure and as a means of controlling it (see Section 10.7).

10.1.1 Scope of Radiation Protection

The scope of radiation protection can be fully understood by first defining ionising radiation.

The term 'radiation' indicates transfer of energy via a medium in the form of particles or waves. The best example of radiation that we experience in our daily lives is visible light, but radiation of higher and lower wavelengths (energies) exists.

The term 'ionising' indicates a property of a type of radiation (emitted by the nucleus or by the atom or generated by a subatomic particle) whose energy is sufficient to create ionisation in an atom, that is, to strip an electron from an atomic energy level, leaving an ionised atom behind.

This basically limits the definition of ionising radiation to:

1. Electromagnetic radiation with energy above some eV (usually 10 eV, but according to other sources the threshold should be at 30–34 eV).
2. All particles/waves emitted by the atom and the nucleus as the result of a disintegration or a nuclear reaction, i.e. beta particles and electrons, alpha particles, gamma rays, neutrons, heavy ions, pions, neutrinos and other higher energy subatomic particles.

In normal daily practice in the medical, industrial and various other fields, examples of ionising radiation of concern are mainly alpha or beta particles, gamma rays or neutrons, or a combination of these. In some other specialist fields (high energy research accelerators or cosmic ray exposure during space missions, for example), other kinds of ionising radiation may be of concern, but they will not be treated here due to their high specialisation.

The scope of radiation protection, then, is to assess and protect living beings, mainly humans, from the harmful effects of exposure to ionising radiation. Radiation protection is needed in hospitals, universities, factories, airplanes, nuclear power plants, research accelerators and many other workplaces; radiation protection considerations apply to the mining industry, the oil and gas industry, security, border and customs control, cultural heritage and the design and ventilation of some of our dwellings and public places, just to quote a few examples.

10.1.2 Radiation Protection Institutions

The more fundamental scientific basis for the radiation protection discipline is composed (and continuously alimented) by thousands of scientific studies carried out worldwide by scientists and researchers, every year, in universities, research centres, private companies, etc. The most relevant results of those studies are regularly selected, collected, commented upon and presented as coordinated reports by the United Nations Scientific Committee on the Effects of Atomic Radiation (UNSCEAR, website), which was created in the 1950s as an initiative of the General Assembly of the United Nations.

All UNSCEAR reports are published and available online, and they represent an enormous amount of collective state-of-the-art scientific knowledge for radiation studies at any level.

The radiation protection system, also structured upon a series of assumptions and social values that are shared and recognised by most countries in the world, was developed by the prestigious International Commission on Radiological Protection (ICRP, website). The ICRP, an independent non-governmental organisation composed of top international scientists and experts who serve the Commission without receiving any emolument, has issued more than 120 publications since its creation, including various editions of its famous *Recommendations of the ICRP*, which state the most updated and consensual systems for radiological protection in the world. The latest edition of the *Recommendations* (ICRP, 2007) forms the basis for other more technical or legislative documents.

Legislation on radiation protection matters stems from international directives, recommendations and standards: In Europe it is the European Commission's directive known as *EU Basic Safety Standards* (Council Directive 2013/59), which establishes a framework for radiation protection in the European Union's Member Countries. At a more international level, for the member countries of the International Atomic Energy Agency, it is the publication *General Safety Requirements, Part 3*, also previously known as the *IAEA Basic Safety Standards* (IAEA, 2014).

Technical standards are developed at the international and national levels by experts belonging to technical committees (for example, the International Organization for Standardization, ISO, and several national standardisation institutions), and represent so called 'good practices'.

Local and national radiation protection legislation is constituted of national laws, decrees and possibly a series of other acts, issued by the states and usually based largely on international recommendations and good practices. They may differ significantly from country to country.

The professional aspects of the radiation protection profession, which also depend on the technical standards and national legislation in force in a given country, are usually represented by several national professional radiation profession associations (for the medical, industrial and scientific fields and others), which are linked together under the umbrella of the International Radiation Protection Association (IRPA).

IRPA provides updates and guidance in the scientific and professional aspects of operational radiation protection.

10.2 Atoms, Radiation and Interactions

Matter is composed of atoms, basic building blocks for every single material in the universe. Around 110 different atoms, or atomic species, have been identified. Atoms are themselves composed of three particles: electrons, protons and neutrons, the latter bound together in a central nucleus (containing protons and neutrons) and surrounded by orbiting electrons in electronic 'clouds' of defined energy. Particles within the nucleus (also called nucleons) have an internal structure (quarks) and even smaller components (Turner, 2007).

Atoms link together, forming molecules, which then compose visible matter and – in the case of living organisms – cells, tissues, organs and entire organisms. The linkage among atoms is made possible by various mechanisms for sharing electrons within electronic clouds, and keeps atoms more or less closely bound to one another, forming matter in its solid, liquid or gaseous form.

Four basic kinds of energy are present in the interactions within atoms and their components:

- the (attractive) *weak nuclear force*, keeping the components of subatomic particles together in protons and neutrons;
- the (attractive) *strong nuclear force*, keeping neutrons and protons together in the nucleus;
- *Coulombian forces** (of electric nature, therefore attractive for unlike charges and repellant for like charges), keeping the (positively charged) nucleus and the (negatively charged) orbiting electrons together in the atom but tending to push protons apart, for example (and so electrons); and
- *gravitational forces* (attractive), which contribute to maintaining the atom together, but whose impact is usually negligible at this scale.

In nuclear physics and in radiation protection, energies are indicated in the unit electron-volt, eV. One electron-volt is the kinetic energy acquired by an electron when it is accelerated by a difference of potential of 1 V. 1 eV is a very tiny quantity, well apt to express energies at nuclear and atomic scale processes: $1\,eV = 1.6 \cdot 10^{-19}\,J$.

---

* It must be noted that, in contemporary physics, weak and electromagnetic forces have been recognized as being a single, unified, force. However, this distinction will be kept here, for clarity.

While the number of electrons and protons (in equal number, ensuring global atom neutrality), Z, determines the atomic species and the chemical properties of the specific atomic family, the number of neutrons, N, determines the physical properties of a specific atom in a family, identifying the so-called 'isotope' of the chemical family.

The sum of Z and N is called 'mass number', indicated by A, and is usually utilised in radiation protection, together with the chemical element's symbol, to indicate an isotope of interest. As an example, members of the family of sodium chemical species are usually indicated by their chemical symbol, Na, or referred to by their unique atomic number, $Z = 11$. Individual components (isotopes) of that chemical species are indicated as ^{22}Na (or Na-22), ^{23}Na (or Na-23) and ^{24}Na (Na-24), implying that the number of neutrons present in each isotope is, respectively, $N = (22 - 11) = 11$, $N = (23 - 11) = 12$ and $N = (24 - 11) = 13$.

Some isotopes in nature are stable, i.e. they do not change with time. Others, due an instability in the composition of the nucleus (an 'insufficient' or 'excessive' number of neutrons compared to the number of protons, or vice-versa!), are not stable with time and tend to transform into an isotope of another atomic species, with the emission of a particle (or more than one), which is generically indicated as 'radiation', and its accompanying kinetic energy. Those atoms are known as 'radioactive isotopes', 'radioisotopes' or 'radionuclides', and the process they undergo is called 'radioactive disintegration' or 'radioactive decay' and is due to the isotope's property instability, known as called 'radioactivity'.

Radiation is the accompanying result of a spontaneous transformation of an atom or a nucleus, and is naturally present in nature. Natural radioactivity can be enhanced and concentrated by some specific human activities, and due to the effects of this concentration additional harm may derive to workers and to the general public (see Sections 11.1 and 11.2).

Radiation can also be artificially generated, either by using radiation generators (X-ray machines or particle accelerators, for example) or by using concentrated radioactive materials which have been specifically manufactured and commercialised (radioactive sources). The major difference between radiation generators and radioactive sources is that generators can be switched off, usually resulting in a complete halt in the production of radiation beams, but radioactive sources never cease emitting radiation (except when radioactivity has completely disappeared), so they need to be appropriately shielded when they are not in use.

Radiation is emitted by the nucleus (or the atom) with high kinetic energy, compared with the energy of molecular bonds, and can interact with surrounding atoms and molecules and transfer part of its kinetic energy in that process. The interaction with surrounding media ensures that, after a while, all radiation will ultimately be stopped, and its kinetic energy transmitted, in various possible ways, to the medium, also at great distance from the atom originating the radiation beam. It is this process of transmission of energy to surrounding atoms and molecules, which is basically generating modifications within molecules, that may cause harm to living cells, tissues and organs and eventually have an impact on the health of macroscopic organisms (see also Sections 10.5 and 10.6).

From the point of view of radiation protection, radiation emissions of interest are:

- alpha emissions (Section 10.2.1), following alpha radioactive decay;
- beta emissions (Section 10.2.2), either positive (positrons) or negative (negatrons or electrons), following beta radioactive decay;
- gamma emissions (Section 10.2.3), following a gamma transformation in the nucleus, which normally accompanies alpha or beta radioactive decay;
- X emissions (Section 10.2.4), following processes which take place in the atomic electronic cloud as a de-excitation of the atom (for example, the rearrangement of electrons into their energetic levels following the ionisation of the atom) or as a result of the interaction of a light-charged particle with electronic clouds of a medium (bremsstrahlung radiation, the mechanism normally used in the production of X-rays);

- neutrons (Section 10.2.5), usually emitted as the result of a nuclear reaction in the nucleus, or as the result of a nuclear fission process.*

In the next paragraphs, the most salient characteristics of radioactive transformations and the corresponding radiation emitted will be presented, and some indications of radiation interaction with matter will be given. More detailed information can be found in the references (Turner, 2007 and Martin, 2013).

10.2.1 Alpha Decay and Alpha Radiation

Alpha decays and their corresponding emissions, alpha particles, were observed as early as 1899. Alpha particles were used by Ernest Rutherford to probe the nature of the atom itself and prove that the atom is indeed made of a massive, central nucleus, where most of the atom's mass is located. However, it took almost 30 years after this discovery before the physical theory of alpha decay could be formulated, by George Gamow, in 1928. The theory of alpha decay represented one of the greatest achievements in quantum mechanics.

The aggregate which is usually called an 'alpha particle' is composed of two neutrons and two protons tightly bound together, making a very stable aggregate.

In some highly massive isotopes, the instability of the nucleus leads to a profound transformation which culminates with the emission of an alpha particle. Theories do not agree on whether such an aggregate is already formed within the nucleus or not, but all theories agree on the inability of classical mechanics to explain the transformation, that is, the emission of the alpha particle from the nucleus. Binding forces within the nucleus (strong nuclear forces) are so strong that it is not possible, using classical mechanics, to explain how the alpha particles can actually be released from the nucleus! Only with the use of the new and modern concepts of quantum mechanics is it possible to explain the physics of the emission of alpha particles, in a process which was called 'quantum tunneling' and is today widely used in technological applications. Alpha particles, although not energetic enough to overcome the attraction which binds them inside the nucleus, can repetitively bounce against this 'wall' of attraction, and occasionally have a probability to just pass through it, as if a person could, by repeatedly bouncing against a wall, at some point slip through it and found herself on the other side of the wall.

The radioactive decay generating alpha particles, alpha decay, can be written as:

$$_Z^A X \rightarrow {}_{Z-2}^{A-4} Y + {}_2^4 He + Q \qquad (10.1)$$

where X is the initial unstable, alpha emitting radionuclide, Y is its daughter isotope and Q is the energy available in the decay.

The transformation is very dramatic for the original nucleus: Its mass number is decreased by four units and its atomic number is decreased by two units, making alpha decay the only radioactive decay in which both the chemical species and the mass number are varied.

It must also be noted that alpha decay is a two-body process, and the alpha particles generated by a specific radionuclide are all emitted with roughly the same initial kinetic energy and, being originated from the nucleus, usually have initial kinetic energies of the order of some MeV.

Alpha particles travel in a straight line and intensely ionise the material they traverse; the density of ionisation increases with increasing distance (and the decreasing kinetic energy of the alpha particle) up to an ionisation peak (Bragg's peak), after which alpha particles rapidly slow down and eventually stop. The maximum distance that alpha particles travel in a medium is called the 'range' of the alpha particle, and is usually indicated as R. The path is normally linear, and its average length in air, expressed in

---

* Although neutron emission is also possible as a radioactive transformation taking place in some unstable atoms.

TABLE 10.1 Radiation, Typical Energies and Ranges

Radiation	Typical Energy	Typical Range in Air	Typical Range in Water
Alpha particles	3–10 MeV	2–10 cm	20–125 μm
Beta particles	0–3 MeV	0–10 m	<1 cm
Neutrons	0–10 MeV	0–100 m	0–1 m
X-rays	0.1–100 keV[a]	1–10 m	0.1–1 cm
Gamma emission	0.01–10 MeV	0.01–100 m	0.001–1 m

Source: Adapted from Martin, J. E. (2013). *Physics for Radiation Protection*. 3rd ed. Weinheim: Wiley-VCH Verlag & Co. KGaA, Boschstr. 12, 69469 Weinheim, Germany.

[a] This energy range is referred to spontaneous X-rays.

centimetres as a rule of thumb, can be roughly approximated by the initial energy of the alpha particle expressed in MeV: i.e., an alpha particle of 4 MeV has a range of around 4 cm in air.

Alpha particles are massive compared to beta and gamma radiation (they have a mass of around $4000 \text{ MeV}/c^2$), and they transport two positive electric charges. Both these characteristics explain the high ionisation generated by alpha particles along their path in matter, and consequently, the limited distance they can travel.

Table 10.1 indicates typical 'ranges' in air and in water of alpha particles and other radiation.

Alpha particles' energies and branching ratios (probability of occurrence in a decay) for some notable alpha sources are listed in Table 10.2.

10.2.2 Beta Decay and Beta Radiation

Beta radioactive decay and its accompanying particles were also observed very early, by Rutherford himself, in 1899. Studies on the nature of the beta particle soon recognised it as an electron whose

TABLE 10.2 Selected Energies of Typical Alpha Decays

Source	Half-Life	Alpha Particle's Kinetic Energy (in MeV)	Branching Ratio (in %)
^{232}Th	$1.4 \cdot 10^{10}$ y	4.01	77
		3.95	23
^{238}U	$4.5 \cdot 10^{9}$ y	4.20	77
		4.15	23
^{235}U	$7.1 \cdot 10^{8}$ y	4.40	56
		4.37	12
^{239}Pu	$2.4 \cdot 10^{4}$ y	5.15	73
		5.14	15
		5.10	11
^{210}Po	138 d	5.30	99
^{241}Am	433 y	5.49	85
		5.44	13
^{238}Pu	88 y	5.80	76
		5.76	24
^{244}Cm	30 y	5.80	76
		5.76	24

Source: Adapted from Knoll, G. F. (2000). *Radiation Detection and Measurement*. 3rd ed. John Wiley & Sons, Inc. https://doi.org/10.1017/CBO9781107415324.004.

charge-to-mass ratio had been discovered just a few years earlier. However, various aspects of the beta decay process remained mysterious for many years, and it was only in 1933 that Enrico Fermi could publish a comprehensive theory of beta decay, which was so innovative that was at first refused publication. The neutrino, a particle involved in beta decay mechanics whose existence had been postulated by Pauli in 1930, was indeed experimentally detected only in 1956.

Beta decay is one of the most frequent kinds of transformation taking place in unstable isotopes. During beta decay, either a neutron is transformed into a proton, with the emission of a (negative) electron (beta minus emission), or a proton is transformed into a neutron, with the emission of a (positive) anti-electron, also called a positron (beta plus emission). In both cases, a third particle is also emitted, which is a neutrino (or its antiparticle, the antineutrino).

During a beta decay, the isotope's atomic number changes as the number of protons changes, but its atomic mass does not change (as the sum of the numbers of protons and neutrons stays the same): This type of transformation, therefore, allows isotopes to transform into another element while always keeping the same mass number (isotones).

Beta decays can generically be written as:

$$\,^A_Z X_N \rightarrow \,^A_{Z-1} Y_{N+1} + \beta^+ + \nu + Q \qquad (10.2)$$

for a beta plus decay, where β^+ is the positron, ν is the neutrino and Q is the reaction Q-value; or as:

$$\,^A_Z X_N \rightarrow \,^A_{Z+1} Z_{N-1} + \beta^- + \nu^- + Q \qquad (10.3)$$

for beta minus decay, where β^- is the electron, ν^- is the antineutrino and Q is the reaction Q-value.

The presence of the third particle (neutrino or antineutrino) during the beta radioactive decay was postulated due to the characteristic continuous energy spectrum of beta particles emerging from the nucleus, from zero to the maximum energy available in the transformation (Q-value). Neutrinos have mass close to zero and little ability to interact with matter (a very small cross-section); therefore, a complex and specifically designed detector was needed to experimentally prove their existence.

Beta particles interact with the surrounding medium mainly by electric interactions with other electrons bound in surrounding atoms. This allows a continuous progressive loss of the beta particle's energy, which can, during interactions, be deflected in many directions, including being bounced back (as its mass and charge are the same as the atomic electrons). For this reason, the actual path that a beta particle follows while slowing down is very tortuous and complex and cannot be easily predicted, unlike the path of alpha particles. Therefore, no concept of 'range' exists in describing a beta particle's path in matter.

Beta particles (both positive and negative) slow down in the same way, with the notable difference that the positron will eventually disappear while joining an electron and generating two 511 keV gamma rays, in a process called 'annihilation'.

Beta particles can also lose energy by *bremsstrahlung*,* that is, by radiating part of their energy into electromagnetic radiation (X-rays), while being deflected by the electric atomic field: bremsstrahlung energy losses may account for 10%–20% of total beta energy lost, but they are most significant in high-Z materials. For this reason, for example, shielding a beta-emitting radioactive source is usually done using plastic, low-Z materials, instead of lead.

Table 10.3 indicates typical 'ranges' in air and in water of beta particles and other radiation.

Beta particles' maximum energies for some notable pure-beta sources are listed in Table 10.3.

10.2.3 Gamma Transformation and Gamma Radiation

Gamma transformation is the process resulting from an excess of energy in a nucleus. If, as normally happens after a radioactive decay, the daughter nucleus is left in an excited state (i.e., with some excess of energy), the nucleus can get rid of that excess energy by releasing electromagnetic radiation in the form

---

* Bremsstrahlung is the German term usually utilized to indicate this process, and literally means 'braking radiation'.

TABLE 10.3   Selected Maximum Energies of Typical Pure-Beta Decays

Source	Half-Life	Beta Particle's Maximum Kinetic Energy (in MeV)
^{3}H	12.26 y	0.0186
^{14}C	5730 y	0.156
^{32}P	14.28 d	1.71
^{33}P	24.4 d	0.248
^{35}S	87.9 d	0.167
^{36}Cl	$3.1 \cdot 10^5$ y	0.714
^{63}Ni	92 y	0.067
^{90}Sr/^{90}Y	27.7 y/64 h	0.546 and 2.27
^{99}Tc	$2.12 \cdot 10^5$ y	0.292
^{204}Tl	3.81 y	0.766

Source: Adapted from Knoll, G. F. (2000). *Radiation Detection and Measurement*. 3rd ed. John Wiley & Sons, Inc. https://doi.org/10.1017/CBO9781107415324.004.

of what is usually called 'gamma radiation'. The nucleus may go directly to ground, the most stable state, with the emission of a single gamma radiation, or it may need a series of consecutive or simultaneous gamma emissions in order to finally reach the ground state. Gamma radiation is a very common byproduct of any radioactive decay, and it is not a radioactive decay process, strictly speaking, as no change in the nuclear species is involved; it is normally referred to as 'gamma transformation'.

Gamma radiation is electromagnetic radiation in nature, as are visible light, infrared and UV radiation, but it has higher energy, smaller wavelength and higher penetration in matter. Gamma radiation exhibits the properties of a wave (similarly to mechanical waves) and of a particle: generally, it is considered that gamma radiation is emitted in the form of wave packets called 'photons'.

The relationship between photons' energy, frequency and wavelength is expressed as:

$$E = h\nu = hc/\lambda \qquad (10.4)$$

where E is the photon's energy (usually expressed in keV or MeV), h is the Planck constant, ν is the radiation frequency, c is the speed of light in vacuum and λ is the corresponding wavelength.

Gamma radiation does not carry any electric charge; therefore it is classified as 'indirectly ionising radiation', which means that it cannot directly generate ionisations in the medium it transverses, but can knock electrons out of the atom with enough kinetic energy to create ionisation along their paths.

For most of the time, gamma radiation travels in the medium in a straight line without any sort of interaction; at some point, it can strike an atom or a nucleus and deliver a significant fraction of its kinetic energy during this event. Therefore, gamma radiation, like neutrons (another type of indirectly ionising radiation), tends not to lose its energy progressively but during single dramatic events. Moreover, the mean free path between two interactions is quite high and can attain hundreds of meters in the air, for most energetic gamma radiation.

Basically, interaction of gamma radiation with matter (and production of secondary electrons) is based on three distinct phenomena:

- *Photoelectric effect*, which dominates the low energy range and implies the release of an electron from the atom and the absorption of the gamma radiation in the process.
- *Compton scattering*, which predominates in the intermediate energy range: It consists in photons' inelastic scattering from external atomic electrons, to which photons will deliver a fraction of their energy and re-emerge with lower energy (and in a different direction).
- *Pair creation*, an interaction which dominates the higher energy range, during which gamma radiation, under the influence of the atomic or nuclear electric fields, can generate a pair

TABLE 10.4   Selected Energies of Typical Gamma Sources

Source	Half-Life	Gamma Energy (in keV)	Branching Ratio (in %)
^{22}Na	2.6 y	1274.5	100
^{24}Na	15 h	1368.5	100
		2754	100
^{57}Co	272 d	14.4	10
		122.1	86
^{60}Co	5.27 y	1173.2	100
		1332.5	100
^{88}Y	107 d	14.2	53
		1836.1	100
^{131}I	8.02 d	364.5	83
^{134}Cs	2.06 y	604.6	98
^{137}Cs	30 y	32	6
		661.6	85
^{198}Au	2.7 d	411.8	96
^{241}Am	432 y	59.5	36

Source: Adapted from Knoll, G. F. (2000). *Radiation Detection and Measurement*. 3rd ed. John Wiley & Sons, Inc. https://doi.org/10.1017/CBO9781107415324.004.

electron/positron (provided that its original energy exceeds the energetic threshold for the creation of the two particles); gamma radiation delivers to the two particles its original kinetic energy, minus the energy corresponding to the masses of the two electrons (1022 keV).

The attenuation* of a gamma radiation beam in a given material – a stochastic process – is described by a decreasing exponential, whose time constant μ is given by the summation of the three interaction constants of the processes described above:

$$I = I_0 \cdot \exp(-\mu \cdot d) \tag{10.5}$$

where I is the intensity of the beam at the depth d in a given material and I_0 is the beam intensity at the depth conventionally considered 'zero'.

The linear attenuation coefficient μ is given as the sum of the photoelectric (τ), Compton (σ) and pair production (κ) components at a given photon energy:

$$\mu(E) = \tau(E) + \sigma(E) + \kappa(E) \tag{10.6}$$

Table 10.1 indicates typical 'ranges' in air and in water of gamma and other radiation.

Gamma photons' energies and branching ratios (probability of occurrence in a transformation) for some notable gamma sources are listed in Table 10.4.

10.2.4   X-Rays

'X-rays' (as X radiation is commonly and historically called) were discovered by Wilhelm Roentgen in 1895 as an unexpected and unexplainable emission in an experiment with his cathode tube (hence the choice of the letter 'x', indicating the unknown). In just a matter of months, X-rays were well known all around the world and were already successfully being used in the medical practice to treat an infamous wounded patient who had been shot accidentally (Jorgensen, 2016).

---

* Attenuation, as presented here, should be more accurately described in terms of 'good geometry' conditions: 'bad geometry' and 'build-up factors' will not be mentioned here, although represent an important aspect of workplace Radiation Protection.

Like gamma radiation, X-rays are electromagnetic radiation in nature (although they usually have lower energy), and should not be treated separately; however, they have two notable peculiarities that we will highlight here.

X-rays are spontaneously generated by atoms during electrons' transitions from one energetic level to another: Atoms release excess energy via the emission of X-rays, in a similar way that nuclei release excess nuclear energy by emitting gamma radiation. It is important, therefore, to stress that X-rays are generated by atoms and physically generated in electronic level transitions, while gamma radiation is generated by nuclei during nucleons' nuclear level transitions.

Moreover, as the energy difference between some electronic levels and others is peculiar to an atomic species, emissions of atomic X-rays show some peculiar recurring energy lines (characteristic X-rays emissions), from which it is possible to determine the atomic species of the emitting atom (as is done with gamma radiation spectrometry).

The first notable property of X radiation is that it can also be artificially stimulated in atoms by properly illuminating atoms with a suitable beam of radiation (X-rays, electrons, alpha particles, etc.): interaction of radiation with matter creates ionised atoms, which then emit characteristic X-rays.

The second peculiarity of X-rays is that they can also be generated, both naturally and artificially, as a byproduct of charged particles' acceleration: it is the well-known 'bremsstrahlung effect' ('braking radiation' effect), which has been previously mentioned in relation to beta particles. Any charged particle changing direction or velocity irradiates (emits radiation), although the effect is mainly observed in the slowdown of electrons (which are lighter particles and can move and be deflected at very high velocities) or during interactions of high-energy particles (generating so-called 'synchrotron radiation').

Commercially, the slowdown of accelerated electrons in a heavy target via bremsstrahlung is the commercial technique currently still in use to generate X-rays (for medical or industrial purposes, see also Sections 11.4 and 11.5). Theoretically, by using an appropriate beam energy, any desired X-ray energy can be generated. While X-rays generated by atomic transitions in the atom as a result of radioactive decay cannot be controlled (and X radiation is emitted together with other forms of radiation, normally alpha radiation emitted by massive isotopes), bremsstrahlung X-rays can be operated and controlled and can be switched on and off, as they result from the operation of an electron accelerator or a cathode tube.

Table 10.5 indicates typical 'ranges' in air and in water of X and other radiation.

Sources of characteristic low energy X-rays from selected isotopes are listed in Table 10.5.

10.2.5 Neutrons

The existence of neutrons was suggested by Ettore Majorana, and they were first detected experimentally (but not recognised) in 1931, following irradiation of beryllium with alpha radiation. During the 1920s, it was believed that inside the nucleus, together with protons, some 'nuclear electrons' were present, but many experiments and calculations showed that the hypothesis was untenable. In 1932 James Chadwick was able to experimentally discover them, and so to complete the atomic model proposed by Rutherford.

TABLE 10.5   Selected Isotope Sources of X-Rays

Source	Half-Life	X-ray Energy (in keV)	Fluorescence Yield (in %)
^{37}Ar	35.1 d	2.98	8.6
^{41}Ca	$8 \cdot 10^4$ y	3.69	12.9
^{44}Ti	48 y	4.51	17.4
^{49}V	330 d	4.95	20
^{55}Fe	2.60 y	5.90	28.2

Source: Adapted from Knoll, G. F. (2000). *Radiation Detection and Measurement.* 3rd ed. John Wiley & Sons, Inc. https://doi.org/10.1017/CBO9781107415324.004.

Neutrons, which are components of the nucleus tightly bound to protons, are rarely emitted spontane-ously by radioactive decay processes, although some rare examples exist. Most often, neutrons are spon-taneously generated during (spontaneous) fission, or artificially generated as a byproduct of suitable nuclear reactions (that is, reactions induced by particles interacting with the nucleus of an atom, prompt-ing the emission of one or more neutrons as the eventual result of the process).

A typical example of a nuclear reaction producing neutrons is the above-mentioned interaction of alpha particles with the nucleus of the element beryllium, Be. This reaction has a relatively high probability of generating neutrons (around 1 alpha particle in about 10^4 reaches the beryllium nucleus and reacts with it), and can be expressed as:

$$\,^4_2\alpha + \,^9_4\text{Be} \rightarrow \,^{12}_6\text{C} + \,^1_0\text{n} \tag{10.7}$$

This reaction is normally used in neutron-generating radioactive sources, making use of mixtures of Pu-Be, Am-Be, Po-Be or Cm-Be.

Other neutron-producing nuclear reactions include acceleration of a deuterium nucleus onto a target (d-d or d-t, for example); or inducing a photoneutron reaction using energetic gamma rays (able to extract neutrons from a target); and most notably the nuclear fission reaction (both spontaneous and induced fission). The latter example will be developed further in the next chapter.

Neutrons, due to their neutrality and their significant mass (around 1000 MeV/c^2), can easily penetrate the electronic cloud and so are able to interact directly with the nucleus of an atom.

Interactions include scattering (either inelastic or elastic), capture by the nucleus (absorption reaction, or 'neutron capture'), stripping of one or more nucleons from the nucleus and nuclear fission.

The probability of interaction of a neutron with an atom/nucleus is much less than the probability of interaction of a charged radiation, whose Coulombian interaction has a much longer and continuous range of action: while alpha radiation, for example, loses its kinetic energy progressively during its path, eventually coming to a stop, the path of a neutron can be much more complicated, and the proba-bility that it will interact with the nucleus is much smaller (hence its path is comparatively longer).

For these reasons, interactions between neutrons and nuclei are normally described in terms of 'cross sections', indicating (in units of a surface [m^2]) the likelihood that a specific reaction will occur. As these probabilities depend on the energy of the incoming neutron and on the characteristics of the material on which the neutron beam is impacting, cross sections are tabulated in terms of a specific interaction, on a specific material, for a range of energies.

Mechanisms used for the detection of neutrons include detecting protons which are knocked out by neutrons from the nucleus; recoiling nuclei; detecting gamma radiation, usually accompanying neutron capture; or creation of unstable aggregates which produce fission, emitting radiation and heavy ions.

10.3 Radioactivity

'Radioactivity' is the property of unstable isotopes which undergo transformations, during which the isotope is transformed into a different configuration or into a different isotope, and ionising radiation is usually also emitted.

The number of transformations which take place in the unit time is called 'activity', and is usually indi-cated with the letter A.

$$A = \lambda \cdot N = dN/dt \tag{10.8}$$

where A is the activity, N is the number of atoms present in the sample, dN is the number of atoms undergoing transformations in the time dt and λ is the decay constant of the particular transformation.

As radioactive decay is a statistical process, it can be shown that the relative number of transformations taking place in the unit time is a constant, characteristic for a specific radioactive decay: the constant lambda is called 'decay constant' for the specific radioactive decay:

$$\lambda = \frac{dN}{dt} \cdot \frac{1}{N} \tag{10.9}$$

The SI quantity 'activity' is expressed in the special SI unit becquerel [1/s], Bq, as a respectful homage to Henri Becquerel, the French scientist who discovered radioactivity in 1896.

1 Bq corresponds to 1 transformation per second.*

Radioactivity, for any radioactive source, decreases with time, as can be easily shown by integrating the previous Equation 3.7, and follows a statistical behavior, represented by an exponential curve: The number of radioactive atoms remaining in the radioactive sample decreases with time, and reaches zero only at infinite time.

$$A(t) = A_0 \exp(-\lambda \cdot t) \tag{10.10}$$

where A_0 is the sample's activity at the time conventionally taken as zero, and λ is the decay constant of that decay.

For a deeper discussion of the radioactivity concepts see Sections 12.2 and 12.3.

10.4 Radiation Detection

Radiation emitted by radionuclides interact with the surrounding materials (including air, shielding, persons, etc.) according, basically, to three fundamental mechanisms:

- Ionisation, that is, stripping one electron (or more) from an atom, or – less frequently – a nucleon from the nucleus of an atom.
- Excitation, that is, delivering energy to the atom/nucleus and leaving it in an excited state, i.e., in a state of higher energetic content than the fundamental state (and from which, eventually, the atom/nucleus is bound to transition back to the fundamental state, usually by the emission of X/gamma radiation).
- Damage, that is, creating a physical, microscopic irregularity in the structure of the material, for example knocking one atom out of its original position in a crystal lattice and leaving it in an intermediate position (a physical irregularity which can be visible with a microscope).

These three basic interaction mechanisms are exploited in detectors, which need to at least assess the presence of radiation, and in some cases also to quantify it.

There is, however, an important distinction to be made between the behaviour of radiation directly ionising matter via Coulombian forces (electrical interactions among charged particles, as alpha and beta radiation, for example) and those that are not able to directly ionise the atom (neutral radiation, like neutron or gamma radiation).

The first group is called 'directly ionising radiation', and the second 'indirectly ionising radiation'.

As the definition implies, directly ionising radiation creates ions, due to the electrical interaction between charged radiation and the surrounding atoms. As charged particles travel, their electrical fields interact with atomic electric fields, and ionisation (stripping of electrons from atomic shells) may occur. The interaction takes place at a long or short distance, and generates electrons which are more or less energetic and more or less numerous (secondary particles). With electron extraction, an ionised atom is left behind. If the interaction between the incoming radiation and the atom takes place in a suitable detector, these two charged entities (ions) can be easily separated by the application of an electric field. If, for

---

* Which, worth to note, is not necessarily the number of radiation being emitted per second by the radioactive material! That number can be different due to source autoabsorption, especially for low-energy, non-penetrating radiation.

example, the electrons' charge is collected, it can be detected as a current (if the electrons' flow is large enough); if single electrons are collected and appropriately amplified in the detector, they can be detected like individual pulses.

The ability of some radiation detectors to detect and indicate the presence of a single event represents an incredibly powerful characteristic of nuclear measurement, which sets it apart from those in other sciences: consider, for example, that in chemical reactions, reagents can be detected only if their concentrations are on the order of fractions of moles, that is, in a number which is of fractions of 10^{23} atoms, an enormous amount of atoms!

In radiation detection, on the contrary, it is possible to detect one single event, corresponding, for example, to the radioactive decay of one single atom.

Indirectly ionising radiation will dissipate its energy by interacting with one or more atoms and triggering the release of one or more electrons from atomic shells: those electrons, in turn, will dissipate their kinetic energy via the same mechanisms as described above. The effect of the interaction of X and gamma radiation with matter is mostly the result of the interaction of secondary electrons generated during the interaction with target atoms.

Neutrons and gamma radiation, though, can also generate 'nuclear reactions', inducing effects in the nucleus of the atom itself. Nuclear reactions generated by neutrons include scattering of neutrons (either inelastic or elastic), capture of a neutron in the nucleus, stripping of one or more nucleons from the nucleus and nuclear fission, which is the induction of a large instability in the nucleus, eventually concluding with the breaking up of the nucleus into two smaller and more stable units and the emission of large quantities of energy in various forms.

Nuclear reactions generated by gamma radiation also include photofission (fission of the nucleus due to the energy deposited by the gamma quantum).

10.5 Radiation Dose(s)

The concept of 'radiation dose' has been developed to define a quantity able to bridge the physics of exposure to radiation (an interaction taking place at a microscopic scale) with the biological effects of exposure to radiation (which can span several orders of spatial magnitude, from the scale of the DNA to the entire organism).

'Radiation dose', as an ideal quantity, would allow us to assess radiological risks and control them in various natural, occupational and public exposure situations.

Different kinds of radiation, as we have seen, have different interaction mechanisms with matter, generating secondary particles and delivering energy to atoms and molecules: given the same initial kinetic energy, wide differences – in the length of the interaction path and its density of ionisation, for example – exist between, for example, the interaction and energy deposition of alpha radiation (which generates high ionisation, with a high 'linear energy transfer', or LET) and beta radiation (which generates little ionisation, or is said to have a 'low LET').

In the current Radiation Protection paradigm (ICRP, 2007), radiation effects are seen to be generated directly in the atoms/molecules/cells/tissues/organs which have been exposed to radiation*: The deposition of energy, therefore, is considered to be the main initiator of all subsequent radiation effects. The quantity describing the energy deposition is the main physical quantity reflecting the origination of radiological risks.

There is also a wide variety of somatic cells that compose an organism in shape, size, function and susceptibility to be affected by radiation and eventually generate sickness in the tissue/organ or in the whole organism. Indeed, given the same 'amount of radiation exposure', some cells are more radiosensitive and reactive (like cells of the bone marrow) and others are more radio-resistant (like the cells of the heart

---

* It must be mentioned that last decades' experimental studies are showing that a certain complexity exists, in this regard.

or the bones). This cell, tissutal and organic complexity must also be taken into account, the purpose of radiation protection being to understand how radiation exposure generates risk so as to protect people (entire organisms) from exposure to radiation. In the next paragraphs we will examine how the current radiation protection system accounts for those factors.

The deposition of energy by radiation (any kind of radiation, in any kind of medium) is expressed using the quantity Absorbed Dose, *D*, which is defined as

$$D = dE/dm \tag{10.11}$$

where *dE* is the energy which is deposited in the mass volume *dm* by incoming radiation.

The absorbed dose is expressed in [J/kg], or with its special SI Unit Gy (gray).

Absorbed dose can be defined in any volume of any material, has the dignity of a physical quantity and also can be, in specific conditions, experimentally measured. This quantity is used whenever a direct assessment of exposure to radiation is needed, for example in the medical treatment of tumors (see Section 11.4).

Absorbed dose is the fundamental physical quantity in radiation protection and forms the basis for the definition of all other physical quantities that follow.

In order to account for the different mechanisms underlying the deposition of energy in a medium by different types of radiation, a weighting factor, called the 'radiation weighting factor', w_R, has been defined: This factor is obtained by expert judgment and consensus, and allows the definition of Equivalent Dose, *H*:

$$H = \sum_R w_R D_R \tag{10.12}$$

where the summation extends over all the different radiation *R* depositing energy in a medium via respective absorbed doses D_R.

The equivalent dose is expressed in [J/kg], or with its special SI Unit Sv (sievert). Due to the presence of the radiation weighting factor, *H* can no longer be regarded as a physical quantity.

The values of the radiation weighting factors recommended by the ICRP in its Publication 103 are reported in Table 10.6 (ICRP 2007).

Equivalent dose cannot be directly measured. As it can be defined in any medium, it should be specified, and in radiation protection equivalent doses are usually referred to a tissue, an organ or a specific part of the body, so normally it is indicated as 'equivalent dose to the thyroid' or 'equivalent dose to the bone marrow', for example.

Equivalent doses to specific organs are not infrequent in the workplace, for example when radiation sources are operated by hand, under hoods or glove boxes or when external contamination is detected on the skin of a worker during an incident; but they are most frequent in nuclear medicine, where an intake of radioactivity specifically targets an organ or a tissue, where most of the radiation dose is located.

The added value of using equivalent dose instead of absorbed dose is to take into account the ability of different forms of radiation to generate different biological effects, and $W_R s$ are indeed obtained by an

TABLE 10.6 Radiation Weighting Factors w_R Recommended by the ICRP

Radiation Type	Radiation Weighting Factor w_R
Photons	1
Electrons and muons	1
Protons and charged pions	2
Alpha particles, fission fragments, heavy ions	20
Neutrons	From 2 to 20, according to a continuous function of neutron energy

Source: Adapted from ICRP. (2007). The 2007 Recommendations of the International Commission on Radiological Protection. ICRP Publication 103. Ann. ICRP 37(2–4).

TABLE 10.7   Tissue Weighting Factors w_T Recommended by the ICRP

Tissue or Organ	Tissue Weighting Factor w_T
Bone-marrow (red), Colon, Lung, Stomach, Breast, Remainder tissues (Adrenals, Extrathoracic (ET) region, Gall bladder, Heart, Kidneys, Lymphatic nodes, Muscle, Oral mucosa, Pancreas, Prostate (men), Small intestine, Spleen, Thymus, Uterus/cervix (women)	Each 0.12, for a total of 0.72
Gonads	0.08
Bladder, Oesophagus, Liver, Thyroid	Each 0.04, for a total of 0.16
Bone surface, Brain, Salivary glands, Skin	Each 0.01, for a total of 0.04

Source: Adapted from ICRP, 2007. The 2007 Recommendations of the International Commission on Radiological Protection. ICRP Publication 103. Ann. ICRP 37(2–4).

analysis of the relative effectiveness of a type of radiation to generate a specific biological effect in a cell culture (RBE, relative biological effectiveness).

A second weighting factor, taking into account the radiosensitivity of an organ or tissue, and appropriately called 'tissue weighting factor', W_T, has been introduced in order to define the quantity Effective Dose, E:

$$E = \sum_T w_T H_T \tag{10.13}$$

where the summation extends over all the different radiation tissues T, using their corresponding equivalent doses H_T.

The effective dose is also expressed in [J/kg], or with its special SI Unit Sv (sievert). Due to the presence of the radiation weighting factors w_T, E can no longer be regarded as a physical quantity.

The values of the tissue weighting factors recommended by the ICRP in its Publication 103, and the methodology that should be followed in computing E, are reported in Table 10.7.

The effective dose, weighing each equivalent dose component with its relative likeliness to develop sickness in the organism, is a quantity which is related to the 'whole-body' likelihood of being negatively impacted by a given radiation exposure: effective dose can be considered as a 'whole body' dose.

Effective dose cannot be directly measured, and must be used in a priori radiation protection calculations and assessments, not as an a posteriori evaluation of radiological risks to a specific person.

Other, more operational dose quantities have also been developed with the objective of providing radiation protection practitioners with a set of practical quantities that can be measured and used in the practice.

It can be shown that control of exposure to radiation using the operational quantities (for example, the respect of an annual dose limit) ensures compliance with the corresponding effective dose or equivalent dose limit.

Radiation protection instruments can, therefore, be calibrated in operational quantities and used in the practice.

Operational dose quantities include the 'Ambient Dose Equivalent', H^*, and the 'Personal Dose Equivalent', H_p, quantities involved in environmental measurements and personal dosimetry, respectively.

It must be noted that those quantities, although similar in their terminology to equivalent dose, are not to be confused with it, as they differ conceptually, being defined with respect to a quality factor (based on LET) and not the already-mentioned weighting factors (based on RBE). Their potential for use in the radiation protection practice, in which they approximate those quantities that cannot be measured directly, is the main reason they have been introduced.

10.6   Biological Effects of Radiation Exposure

The harmful effects of large radiation exposure were noticed soon after radiation was discovered (Clarke and Valentin, 2009), and, ever since, an enormous number of research studies have been conducted around the world to establish a clear relationship between exposure to radiation and its corresponding biological effects. Many firm conclusions have been drawn over the years: for example, the basic distinction between somatic biological effects (which impact the organism itself) and genetic mutations (which can be transmitted from one generation to another) or the different mechanisms related to biological effects due to high radiation dose exposure and low radiation dose exposure.

A paradigm of the relationship between radiation exposure and biological effects has been reformulated and modified, and may be modified again in the future if the new, compelling and challenging experimental evidence which has been accumulating in the last 20 years will keep suggesting that a revision of the paradigm is needed.

A comprehensive understanding of the fundamental mechanisms of biological effects, and mainly the effects of low-dose exposure, is yet to become fully available.

Radiation deposits energy, and energy breaks bonds between atoms in molecules, including DNA molecules in human cells. This can disrupt the cell's normal functioning.

Human cells are continuously subjected to innumerable other aggressive elements, including age, chemicals and non-ionising radiation, that can harm them and possibly shorten their life-cycles.

Cells are equipped with repair mechanisms, which allow them to recover from a variety of events; cell repair, fully restoring the conditions prior to radiation interaction, is one of three possible outcomes of the exposure to ionising radiation.

Another option is that the cell may not be able to repair the damage inflicted by radiation, or that the cell may be able to repair it in a way which leads to a fatal decline in cell function and the cell may eventually die (apoptosis). If the number of cells which die as a result of exposure to radiation is small, no macroscopic effect is detected in the organism (every day, in every human being, billions of cells die). However, if the number of affected cells in a tissue or organ is large, the result is a macroscopic visible consequence for the organism, a sickness or disease. These kind of effects of radiation exposure are called 'tissue reactions' (previously known as 'deterministic effects', because they always appear, once a specific exposure is met) and include skin redness, swelling, edema, tissue necrosis, nausea, circulatory disease, cataract, sterility (temporary and permanent), central nervous system damage, skin burns and gastrointestinal syndrome.

Each of these effects manifests itself only after a certain dose threshold is exceeded, as reported in Table 10.8.

It can be observed that threshold doses for tissue reactions are very high: Tissue reactions are usually the result of acute, uncontrolled, accidental or incidental large irradiations (atomic bombs, criticality accidents, exposure to high activity orphan sources, etc.) or as a result of radiation therapy (high, acute, fractionated and collimated doses).

The severity of the biological effect increases, once the threshold exceeded, in a linear trend, and at a certain dose threshold the organism collapses due to several simultaneous acute medical conditions (neurological radiation syndrome).

It must also be noted that medical recovery of acutely irradiated subjects may normally be complicated by several additional medical conditions, indirectly generated by radiation exposure, that aggravate the patient's overall medical status (infections, bleeding, burns, etc.).

A third possible outcome of cell irradiation is imperfect cell repair, which may lead to continuous malfunctioning of the cell with no apoptosis. This effect may present itself with a delay, sometimes years or tens of years from the exposure to ionising radiation. This category of biological effects, called 'stochastic effects', as their appearance is not certain, are typical of low-dose radiation exposure, like the exposure to natural radiation or occupational exposure. They include a series of diseases, among which is cancerogenesis.

It is yet not clear how to relate cancer and other low-dose effects to a specific radiation exposure, because the human body is continuously exposed to a series of other confounding factors and aggressive agents, and the causation link is very hard to establish, with the exception of some specific cases.

TABLE 10.8   Acute Radiation Dose Thresholds and Effects

Dose (Gy)	Symptoms	Notes
0–0.25	None	No clinically significant effects.
0.25–1	Mostly none. A few people may exhibit mild prodromal symptoms, such as nausea and anorexia.	Bone marrow damaged; decrease in red and white blood-cell counts and platelet count. Lymph nodes and spleen injured; lymphocyte count decreases.
1–3	Mild to severe nausea, malaise, anorexia, infection.	Hematologic damage more severe. Recovery probable, though not assured.
2	–	Hospital treatment is necessary for survival.
3–6	Severe effects as above, plus hemorrhaging, infection, diarrhea, epilation, temporary sterility.	Fatalities will occur in the range 3.5 Gy without treatment.
4.5	LD_{50}	Lethal dose 50%: half of the subjects are expected not to survive irradiation at 4.5 Gy, despite medical treatment.
>6	Above symptoms plus impairment of central nervous system.	Death expected.
8	Acute digestive syndrome.	Death expected.
>10	Acute neurological syndrome and incapacitation.	Death expected.

Source: Adapted from Turner, J. (2007). *Atoms, Radiation, and Radiation Protection.* 3rd ed. WILEY-VCH Verlag GmbH & Co. KGaA, Weinheim.

Unlikely high-dose biological effects, the probability of incidence of a specific effect, but not its severity, increases with increasing doses in the low-dose range. Only the probability of a an effect appearing, and not effect thresholds, can be computed and assigned.

A huge volume of scientific research has been and is currently being devoted to understanding the mechanisms of the dose-effect relationship at low doses (i.e., below 100 mGy), because that is the range to which most of us are exposed, either as part of the general public or as occupationally exposed workers. Experimental evidence – collected either as results of epidemiologic studies of large populations (the acceptability of the results depends on the statistics, which requires a larger number of subjects while exposures are smaller) or on cell cultures studies – shows a range of possible dose/effect relationships, sometimes contradictory, including the possible existence of a dose threshold for stochastic effects, 'hormesis' (a supposed greater resistance to radiation exposure that some subjects show if they are pre-exposed to radiation) and a supralinear relationship in the lower dose range.

A prudent approach has been chosen by the international scientific community since the 1950s, namely interpreting every single exposure as a small additional (and additive) risk, without considering the existence of a dose threshold. This approach has been called linear–no threshold theory (LNT), and the current system of radiation protection stems from it.

According to this hypothesis, which is prudent and conservative, any exposure to radiation can potentially create additional risk of harm, and therefore should be justified. Moreover, as there would be no 'safe' radiation exposure, the risk approaching zero only with zero radiation exposure, prudence dictates that every radiation exposure should be kept as low as possible. In this approach, the probability that stochastic effects will manifest themselves has been obtained as a linear interpolation of dose-effect data related to high-dose exposures, and is currently evaluated by the ICRP at around 0.05 per sievert (excess relative risk).

It is worth noting some of the latest developments in the field of low-dose exposure, namely the so-called non-targeted effects; abscopal effects, bystander effects and adaptive response, among others (Munira et al., 2013), are challenging the LNT and the entire radiation protection system with it. Contemporary research in the field of low-dose exposure is challenging the 'target theory' paradigm, showing that radiation effects may also manifest themselves in cells which have not been directly exposed to radiation, or in subsequent generations which have not.

10.7 Occupational Radiation Protection

Three basic radiation protection principles have been derived from the previous chapters' considerations and represent the theoretical foundations for operational radiation protection:

- the use of radiation must be previously justified, and should bring more good than harm (Justification Principle);
- according to the LNT, radiation doses should be kept as low as reasonably achievable, taking into account societal and economic factors (Optimisation Principle, or As Low As Reasonable Achievable [ALARA]);
- radiation doses should be limited in order to avoid tissue reactions and the minimise risk of stochastic effects (Limitation Principle).

To put those theoretical principles into practice, several elements are needed: Radiation protection must be conceived holistically as a coordinated result of various components, technical and administrative.

A good and effective nuclear/radiological regulation, providing a clear and applicable set of national rules; a proactive and effective nuclear/radiation regulatory body, providing a clear and applicable system for authorisation, licensing and registration of radiation sources; the existence of professions (RPEs, RPOs, MPs, etc.) and corresponding professional bodies governing duties and interfaces between professionals' roles and responsibilities; training and retraining programmes and opportunities for various actors in the process; a properly drilled and exercised system for emergency preparedness and response; large availability of technical service providers able to measure radiation and radiation doses accurately; and a set of national (or reference to international) technical good practices to benchmark and follow.

A detailed treatment of the strengths of a good regulatory system is beyond the scope of this chapter, but it is important to stress that each one of the mentioned components is not enough by itself to ensure a proper level of radiation protection, but, in the absence of even one of those, the resulting radiation protection system would not be properly coordinated.

Occupational radiation protection's purpose is to protect workers in the workplace, and the general public and the environment as a result of a given work activity.

Protection of workers is composed of several elements, which will be presented and discussed in the following paragraphs.

10.7.1 Prior Analysis of the Work Activity and Assessment of Radiological Risks

Any activity involving the use of radiation should be thoroughly analyzed; its occupational and public radiation doses – including radiation releases, both in terms of discharges to the environment and generation of radioactive waste – should be assessed, including uncertainties.

In the assessment of workers' radiological risks, some assumptions should be made: radiation sources' use and shielding, ambient dose rates and contamination levels, availability of PPEs and their use by workers and time spent in various phases of the activity. The more detailed the analysis of the work process is, the more detailed and useful the radiological assessment will be, and significant conclusions can be drawn.

The prior analysis of the work activity is likely to provide input for various other steps of the process (including regulatory obligations, determination of constraints, etc.) and in the elaboration of the radiological monitoring plan for workers.

10.7.2 Determination of a Radiological Monitoring Plan for Workers

Based on the radiological risks workers are likely to be exposed to in the workplace and on an individual dose assessment for the worker's tasks, a radiological monitoring plan should be set up.

A dose constraint must be set by the employer for all workers or subcategories, and possibly also for radiological impact on the environment, if applicable. A dose constraint is a value of effective (and possibly equivalent) dose, below regulatory limits, which reflects the technological ability to optimise the work process. Doses above these constraints, although they do not trigger any legal consequence, may be an indication to the employer either that the process is not being performed according to international best practices or that a drift in some phase is taking place. Doses above constraints may justify an additional revision of the radiation protection plan or an internal investigation.

Compliance of effective doses (making sure that they respect regulatory limits and possibly also the employer's constraints) is done by controlling the two dose components (external and internal doses), making use of operational dosimetric quantities.

It can be shown that the effective dose can be approximated as:

$$E \approx H_p(10) + E(50) \tag{10.14}$$

where $H_p(10)$ is the personal equivalent dose at 10 mm depth (penetrating radiation) and $E(50)$ is the committed dose calculated over 50 years, which can be expressed as:

$$E(50) = \sum j[e_{(j,inh)}(50) \cdot I_{(j,inh)}] + \sum j[e_{(j,ing)}(50) \cdot I_{(j,ing)}] \tag{10.15}$$

where $e_{(j,inh)}(50)$ is the dose coefficient (50 years) for the inhaled radionuclide j, $I_{(j,inh)}$ is the intake due to inhalation of radionuclide j, and similarly $e_{(j,ing)}(50)$ is the corresponding coefficient for ingestion and $I_{(j,ing)}$ is the intake for ingestion of the generic radionuclide j.

What the two equations show is, basically, that it is possible to express and control effective dose by using the quantity $H_p(10)$, which can be measured using either personal dosimeters or survey meters calibrated in that quantity, and by taking into account internal dose by ingestion and inhalation using dose coefficients, which are tabulated (by the ICRP and by some regulators).

It must also be noted that, in practice, not only effective dose must be subject to control, but also equivalent doses to selected organs (usually the lens of the eye, hands and extremities and the skin), which must then be monitored separately.

The occupational control of personal effective dose is therefore composed of (some of) the following elements:

1. For external exposure: one or more personal, whole body dosimeters calibrated in $H_p(10)$, including passive and active dosimeters.
2. For external exposure (optional): one or more personal dosimeters for specific organs, calibrated in $H_p(0.05)$ or $H_p(3)$, according to the radiological conditions.
3. For external exposure (optional): one or more ambient dosimeters, as needed, where individual monitoring is not deemed necessary, calibrated in $H*(10)$.
4. For internal exposure: routine direct whole-body measurements, with frequency and measurement characteristics depending on the exposure conditions (radionuclides present, PPEs, risks of intake, etc.).
5. For internal exposure (optional): routine urine (or other excreta) sampling measurements, with sampling frequency and measurement characteristics depending on the radionuclides to be assessed, and the exposure conditions.
6. For internal exposure (optional): routine feces sampling measurements, with sampling frequency and measurement characteristics depending on the radionuclides to be assessed and the exposure conditions.

The effectiveness of the individual radiation monitoring plan must be subject to expert judgment. Results of individual monitoring need to be evaluated at least annually, and the plan reassessed; moreover, national obligations on dosimetry communication to the regulator may exist, for example with annual frequency. Making sense of the various radiological data and gathering them into a coherent set of numbers which reflect the exposure conditions is a demanding task and should be assigned to a specialist (see Section 10.7.8).

10.7.3 Control of External Exposure to Radiation

External exposure to radiation can be estimated, in the planning phase, by calculating or simulating dose-rate fields in the workplace (or/and by measuring real doses and dose rates in the workplace) and by estimating timings needed to perform various phases of work activities. Integrated doses can be then summed up, and workload can be adjusted, if needed.

External exposure can be measured in the workplace by using fixed or portable radiation detectors (dose ratemeters and radiation monitors, calibrated in $H^*(10)$ or $H^*(0.07)$) and by measuring individual integrated doses with personal dosimeters (calibrated in $H_p(10)$, $H_p(0,05)$ or $H_p(3)$ as appropriate).

External exposure can be controlled via engineered shielding or by imposing time and/or distance constraints to work activities. As, during planned activities, dose integration performing a task is a continuous, progressive process, continuous monitoring is advisable: the use of electronic dosimeters, which are integrating devices able to provide a live indication related to dose readings, is very useful in this regard.

In some countries' legislation, assigning individual personal dosimeters to workers[*] is a legal obligation, and it is usually accompanied by obligations on ambient monitoring (i.e., ambient monitoring by using ratemeters calibrated in H^*). Ambient monitoring can be considered complementary, when individual doses cannot be evaluated.

Personal dosimeters are passive or active devices which must be worn on the torso (or in other specific positions on the body, according to the dosimeter's calibration), and are calibrated in exposure conditions and with reference to radiation beams which must be appropriately chosen to mimic exposure conditions in the workplace.

During calibration, personal dosimeters are placed on appropriate anthropomorphic phantoms, which simulate the human body and its radiation backscattering properties, and exposed to well-known radiation fields for well-known exposure periods. At the end of the calibration exposure, the information[†] in dosimeters is retrieved (in the jargon, dosimeters 'are read'), and a proper calibration factor, relying dosimeter readings to radiation exposure, is defined. When the dosimeters are returned for monthly readings after use in the workplace, such factors are needed to establish a 'dose reading'. According to the dosimeter's characteristics, dosimeter 'reading' can be a destructive process for the information stored in the dosimeter.[‡]

The process taking place in the calibration laboratory is, of course, very different from the real dosimeter exposure in the workplace: In the latter, many factors influence the actual radiation reaching the detector, including the position of the dosimeter on the worker's body (which may not be the same as on the calibration phantom), the position of the worker with respect to the radiation source (which is continuously varying, hence the position of the dosimeter is varying with respect to the field, and there is a real possibility of auto-shielding by the worker's body) and the composition of the radiation field (which is usually far from ideal, both in terms of radiation – mixed fields may be present – energies and direction, etc.).

A good dosimetric system can estimate, with precision of $\pm 30\%$–50%, external doses to which workers are exposed in practice.

---

[*] Or only to some categories of workers, where categorization exists.

[†] Radiation exposure is stored in the dosimeter in different ways, depending on the technology of the specific dosimeter. Film dosimeters (photographic plates) store information as latent images with varying levels of darkening (density). Thermoluminescent dosimeters (TLDs) store electric charges in semiconductors' energy band gaps, which are retrieved as light pulses when TLDs are heated. Optically stimulated luminescent dosimeters (OSLs) release light glows when illuminated by a light beam of a certain frequency (laser light).

[‡] For example, in film dosimeters, the development process makes the latent image become permanent, and radiation exposure can be read and assessed several times, if needed; in TLDs, retrieving the exposure information involves reheating the crystal and detecting its light glow, which is basically destroying the information stored in the dosimeter: OSLs have the advantage that can be re-read several times.

10.7.4 Control of Internal Exposure to Radiation

Internal exposure can be modeled and estimated in the planning phase, and can be inferred by measuring radioactivity in the workplace (surface or air contamination) and making assumptions about the timing of various phases of work activities.

Internal occupational exposure cannot be directly measured, but can only be assessed by measuring the human body or its excreta and by reconstructing radioactivity intakes and dosimetry.

Control of internal exposure can be achieved with engineered controls, ventilation, filtration and the use of appropriate personal protective equipment (PPE).

Internal exposure takes place whenever radiation is emitted from a source which is inside the human body. This can be due to the introduction in the body of a medical source for therapeutic purposes, or to the incidental or accidental intake of radioactivity during a planned or incidental work condition. The amount of radioactivity introduced in the human body is called 'intake', and the most common pathways for intake are inhalation and ingestion (with direct skin absorption and wound intake being more specific paths).

Once radioactive material is introduced in the body, it is transferred by body fluids according to the material's chemical properties, and may be accumulated in one or more organs or eventually excreted from the body in urine, feces, sweat, etc. Radioactivity in the excreta can be detected and quantified. Excretion of radioactivity from the body can also be forced by using specialised chemicals for some radionuclides and in specific exposure conditions. Those highly specialised chelating agents can facilitate clearance of certain elements or molecules for which they have been designed, by selectively bonding and sequestrating them.*

As it travels through the body, the material will keep decaying, emitting radiation and irradiating tissues with which the material is in contact and additional tissues. The time needed to clear a specific chemical element from the body can be considered as an additional (biological) half-life, and the clearance of the radioactive material from the body, taking into account both radioactive decay and biological clearance, can be expressed as:

$$\frac{1}{t_{eff}} = \frac{1}{t_{phy}} + \frac{1}{t_{bio}} \tag{10.16}$$

where $t_{1/2eff}$ is the effective half-life, $t_{1/2phy}$ is the radioactive decay (physical) half-life and $t_{1/2bio}$ is the biological clearance half-life.

If the physical half-life is much smaller than the biological half-life (for example, a short-lived radiopharmaceutical), the physical half-life predominates and the radioactive material will disappear into its daughter product (radioactive or not) faster than being excreted from the body; similarly, if the biological half-life is much smaller and predominates over the radioactive decay (for example, a long-lived alpha emitter), the total half-life will be very close to the biological clearance half-life, and the material will be cleared from the body faster than its radioactive decay.

The assessment of the dose resulting from radioactivity intakes follows a two-step process:

1. Assessment of the path and transit time of radioactivity inside various tissues and organs in the body (biokinetics of the radioactive material in the body).
2. Assessment of the radiation doses to the various tissues in which radiation is interacting (dosimetry).

This complex analysis in occupational radiation protection is made simpler with the use of tabulated coefficients relating intakes to doses, a work undertaken by ICRP Committee 2.

The dose due to the intake of radioactivity, usually computed over the 70 years following the intake, is called the 'committed dose' and is used in the control of occupational exposure.

---

* For example, DTPA, a chelating agent most suited for alpha-emitting transuranics, which would deliver high-LET doses for a long period of the subject's life.

In the operational practice, this process is followed:

1. Quantification of radioactivity in excreta (radioactivity excretion) or in the whole body (radioactivity retention), if possible.
2. Use of various information and assumptions, including estimation of time of intake, radionuclide chemical form, AMAD, biokinetic modeling, etc.
3. Once a reasonable intake has been established, use of (dosimetric models or) dose coefficients to assess the committed dose resulting from the radioactivity intake.*

Internal dose assessment methodology is reported in IAEA, 1999; ISO 20553, 2006; NCRP, 2010.

Detection of existing radioactivity in the body can be done be directly measuring the radiation emitted by the body of the individual (where this is possible, mainly for gamma-emitting radionuclides) in conditions of low radiation background. This kind of assessment is called the 'direct method' and is usually performed with the use of a 'whole body counter', a detection system composed of a shielded room, a fixed geometry in which the individual must sit/lie for a period of time and a gamma detector enabling an accurate determination of the radiation emitted by the subject (usually a NaI or Ge detector). Measurements usually last tens of minutes, if low activity and high accuracy are needed, which is frequently the case. The calibration of such a system is performed using specifically developed anthropomorphic calibration phantoms, which recreate tiny details of the human body and in which radioactive sources can be precisely positioned.

The main challenge of these systems is to achieve reduced radiation background and good calibration with different sources, which allows some flexibility in the interpretation of measurement data. A clear limitation of direct systems is the ability to measure only gamma-emitting radionuclides (in some cases, also X-ray emitters), and the impact that the subject's body has in the attenuation of the source.

Another methodology, called the 'indirect method'[†] for the assessment of radionuclides' intakes, which is most indicated for non–gamma-emitting isotopes, is the analysis of excreta: In this case, a human sample (most frequently urine, feces or mucus) is measured. This usually requires sample preparation and the use of sophisticated laboratory detection instrumentation, such as alpha spectrometry, beta spectrometry, mass spectrometry, liquid scintillation, etc., in addition to more popular gamma spectrometry and total alpha/beta activity determination.

It is possible to detect very low levels with a high degree of accuracy, although the need for sample pre-treatment and the dependency on very careful laboratory practice for final accuracy have a tremendous impact on the quality of the result.[‡] Hence, the highest care and proper laboratory techniques should always be used when manipulating and analyzing human samples used for internal dose assessments.

Once the radionuclides' activity in the subject's body has been determined using direct methods or the radionuclides' (concentration of) activity in the subject's excreta samples has been determined using indirect methods, appropriate assumptions and excretion models, together with biokinetics and dosimetric models, the subject's committed doses can be estimated. The usual practice is to rely on tabulated dose coefficients, retention and excretion rates for various radionuclides which have been previously calculated for more generic exposure conditions (see ICRP Publications). However, determination of internal exposure (once the exposure has taken place) is a complex task, and doses are normally associated with large uncertainties.

---

* Computer software and codes exist, which support these calculations, especially when a higher level of detail and personalisation is needed (for example, in retrospective dosimetry, or when a potentially higher internal dose needs to be accurately assessed). See (IDEAS. (n.d.), NIRS. (n.d.), PHE. (n.d.)), for example.
† Indirect methods can also be used in addition to direct methods.
‡ Even a tiny cross contamination of a urine sample, coming, for example, from just a few droplets of water with a significant natural uranium content (not an infrequent condition) can significantly alter the sample analysis results, and be erroneously interpreted as an intake of natural uranium, leading to high committed dose.

The psychological aspects of internal contamination cannot be underestimated, and workers should be informed and trained beforehand to better understand the meaning of such exposures. For these reasons, monitoring programmes tend to limit the likelihood and extent of internal contamination: for those work activities where internal contamination risks cannot be avoided in normal operating conditions (tritium absorption through the skin and ^{60}Co inhalation in nuclear reactors, for example), or when incidental condition occur, workers' internal exposures need to be monitored.

Protection from internal contamination and internal exposure due to inhalation risks in the workplace can also be obtained by:

1. Reducing, limiting or avoiding the spread of contamination in the air by controlling the source (engineering actions at the source or in the workplace).
2. Confining the contamination source, for example using plastic tents and ventilation systems with adequate filtration (engineering actions at the source or in the workplace).
3. Protecting workers' respiratory systems with masks and filters, respirators and/or respiratory suits (PPEs).

No prevention or protection system, however, can ever provide 100% assurance, hence monitoring systems should be in place in order to detect, as soon as possible, conditions that could cause internal contamination.

10.7.5 Radiation Monitoring in the Workplace

Radiation dose rates and contamination levels must be kept under control in the workplace: this is done via fixed radiation detection instrumentation (fixed ionisation chambers for dose-rate monitoring, for example), portable instrumentation (dose-rate or surface contamination monitors, for example) and laboratory instrumentation (liquid scintillation counter for assessing tritium in the air, or a gamma spectroscopy system, for example).

External irradiation is usually easier to detect and assess than the presence of contamination.

Gamma and neutron dose rates in some plants or laboratories are detected with fixed instrumentation located in key positions in rooms or corridors or close to processes, and may be linked to access control systems (interlocks). Modern dose-rate monitoring systems are usually interconnected using a dedicated ethernet network, which centralises signals and displays them locally or remotely: therefore, indications of instant, average and historical dose-rate values and their changes are available to radiation protection operators online.

Gamma detectors in use are usually ionisation chambers, proportional counters and, less frequently, Geiger Muller counters and scintillation detectors; proportional counters are usually used for neutron detection.

Portable instruments used for gamma dose-rate monitoring purposes are usually ionisation chambers, Geiger-Muller detectors, scintillation detectors, semiconductors, etc.; neutron monitoring with portable detectors is usually done using proportional counters, Bonner spheres and others.

Surface contamination is usually assessed with the use of scintillation detectors (ZnS is a popular choice) of various portability, shape and size. Other, more specialised types of surface contamination detector also exist when detection of low-energy beta emitters (e.g.: ^{3}H or ^{14}C) is needed: butane windowless proportional counters, for example.

Surface contamination is defined as 'removable surface contamination' if it is possible to spread or remove the radioactivity from the contaminated matrix, and as 'fixed contamination' if removal is normally not possible; 'total surface contamination' is indicated as the sum of the two. The presence of contamination is usually a signal that radioactivity leakage took place during a process and may require a corrective action. The presence of removable contamination, though, also indicates the potential for contamination dispersion in the air, hence a potential risks of workers' internal contamination. When removable contamination is present, some action (confinement, decontamination, etc.) is usually needed, making prompt contamination detection crucial in workplace monitoring.

Direct surface contamination measurements are possible only in an area in which existing workplace background radiation is low enough to allow proper contamination detection below a prescribed level (i.e. in high dose-rate areas, it is usually not possible to detect surface contamination nearing clearance levels).

Most frequently, an indirect measure of removable contamination is therefore also needed, either as a complement to the direct-total contamination measurement (in order to separate the removable component from the fixed one), or as the only way to assess the presence of (removable) contamination in the workplace. In those cases, a swipe, called a 'smear test', must be performed and subsequently measured using a fixed laboratory instrument (or the suitable arrangement of a portable instrument in an area of lower background radiation).

The purpose of the smear test is to transfer, by an appropriate combination of motion and pressure onto the surface, a fraction of the existing removable surface contamination, and to measure the smear test with a detector of known efficiency (geometry is fixed by the smear test size). The fraction being transferred depends on many factors (whether the smear test is dry or with solvent/liquid, whether the surface is flat or wrinkled, what the surface material is, etc.): Usually, for a normal smear test on a normal surface, it is customary to choose a transfer coefficient of 0.1. The path that should be followed during a smear test performance is S-shaped, spanning a surface of about 100 cm^2, and the operator should be careful not to wipe the same surface again and to use a constant pressure and constant velocity. This is not easily done in practice, but it must also be said that the operator's ability to take smear test samples improves with practice. A rigid canvas can help to make sure that the wiped surface is correct, especially when significant precision and repeatability of the measurement are needed.

Concentration of surface concentration of activity C (usually expressed in Bq/cm^2) can be obtained by the following equation:

$$C = S/(\varepsilon \cdot A \cdot f \cdot T) \qquad (10.17)$$

where C [Bq/cm^2] is the surface contamination, ε is the appropriate efficiency of the fixed detector, A is the surface which has been wiped during the smear test, R is the removal factor and f is a safety factor (which may not be present, depending on local regulations).

For example, if 3 cps (counts per second) in the beta channel and 0.3 cps in the alpha channel are being measured on a smear sample of 150 cm^2 of a floor surface, using a detector whose efficiency (4 pi) is 43% for alpha radiation and 29% for beta radiation of similar energies, then the removable contamination on the floor can be calculated as follows:

$$C_{\text{beta}} = S/(\varepsilon \cdot A \cdot f) = 3[\text{cps}]/(0.29[\text{cps/Bq}] * 150[\text{cm}^2] * 1 * 0.1) = 3/(0.29*150*0.1)\text{Bq/cm}^2$$
$$= 0.69 \text{ Bq/cm}^2$$

$$C_{\text{alpha}} = S/(\varepsilon \cdot A \cdot f) = 0.3[\text{cps}]/(0.43[\text{cps/Bq}] * 150[\text{cm}^2] * 1 * 0.1) = 0.3/(0.43*150*0.1) \text{ Bq/cm}^2$$
$$= 0.047 \text{ Bq/cm}^2$$

In both cases, thresholds for contamination levels (which may have been fixed locally as 0.37 and 0.037 Bq/cm^2) are exceeded.

10.7.6 Sampling and Measuring of Radioactivity in the Laboratory

In several cases, assessment of radioactivity cannot be properly performed in the workplace. The cases include high dose-rate areas, determination of activity concentrations which are extremely low and analysis of human samples.

Some examples of analyses which may be performed in the laboratory are determination of total activity, total alpha and total beta activity, total gamma activity, gamma spectrometry, alpha spectrometry, radon concentration in air, liquid scintillation, ICP-MS, etc.

A wide variety of laboratory instruments, whose description is beyond the scope of this chapter, are currently available in the market. Continuous technological improvements in materials, detectors and their performance has become a standard in this field, and this requires continuous follow-up and update from radiation protection practitioners.

More than in other types of measurements, though, and regardless of the high quality of any laboratory instrument, it is the overall quality of the process which drives the final outcome of a measurement. The operator's competence, training and ability to strictly follow technical laboratory procedures are of utmost importance and ultimately determine the quality of a laboratory measurement result.

10.7.7 Assessment and Accounting for Individual Doses

Keeping occupational doses under control is the final scope of occupational radiation protection. As we have seen, this entails several aspects, including assessing radiation dose rates and contamination in the workplace and measuring personal integrated doses.

The regular assessment of workers' effective doses, registration of these doses in a personal file and periodic review of dosimetry is good practice. Periodic reviews may highlight trends in work practices and trigger remedial action before investigation levels are exceeded, or show that re-evaluation of a worker's position is needed. Periodic reviews may be part of national legal obligations.

In the assessment of effective dose, internal doses must be assessed with the determination of the inhalation and ingestion (and other, if present) committed dose components. Action usually needs to be taken in the presence of significant internal contamination. External doses must be also assessed, the starting point being personal dosimeter readings, taking into account both regulatory and operational dosimeters readings, exposure on other facilities, if present. Annual external doses must be calculated by assembling various dosimetric information according to established procedures.

Records of medical examinations, training and other significant events (e.g. pregnancy, long absence from work or radiotherapy) should also be recorded in workers' personal files, in line with the existing privacy policy.

If requested by national regulation, dosimetric files may need to be shared or transmitted to national authorities.

10.7.8 Professionals in Radiation Protection

Operational occupational radiation protection is a complex series of tasks encompassing legal obligations, specialised technical knowledge, use of radiation detection instrumentation and the personal ability to train workers and instill a culture of radiological safety.

Many national legislations around the world have introduced the need for specialised professions to carry out some of the tasks outlined in these paragraphs.

Many differences in various legislations exist, of course, but the similarities, especially given the harmonisation role of the IAEA, are numerous.

10.7.8.1 The RPE – The Radiation Protection Expert

The occupational radiation protection system in the workplace must be set up by a competent professional who has knowledge of theoretical radiation protection principles, the effects of exposure to radiation and operational radiation protection and its applications. This individual should be able to determine and assess radiological risks in various phases of the work activity and develop a series of internal procedures and rules, which must ensure compliance with the laws in effect and be aligned with the employer's dose constraints. This individual must also be able to choose the most appropriate radiation detection instrumentation, advise on their calibration, use and limitations, be able to interpret complex results of direct and indirect radiation measurements and correspondingly advise the employer. He or she must be knowledgeable about dosimetry and its interpretation, be able to calculate doses, including internal doses and support the employer in the unlikely case of a radiation emergency.

This professional must be an expert in the radiological field, and should take responsibility for the design of the company's radiation protection system.

He or she may be called the 'qualified expert', 'radiation protection expert', 'radiation protection adviser' or some other title, depending on the applicable national legislation, but this person's role is advisory in nature, as he or she must assist the employer/operator/licencee to determine the best structure and organisation for radiological protection in the workplace (Lambert et al., 2016).

10.7.8.2 The RPO – The Radiation Protection Officer

Another professional in occupational radiation protection has operational duties in the workplace. This individual is required to take measurements in the workplace, take samples and analyse them and more generally to execute the radiation protection system that has been developed by the radiation protection expert.

This individual is responsible for a series of operational tasks, and he or she should report to and assist the expert in several daily tasks like routinely checking instrumentation, taking routine environmental samples and checking radioactive sources' leakage.

This individual's qualifications do not need to be as high as the those of the radiation protection expert, as complex calculations are not required, but he or she must possess some operational skills, including the ability to liaise with workers and be motivational, as he or she is responsible for daily operational tasks in the workplace.

10.7.8.3 The MP – The Medical Physicist

In the medical field, special attention must be given to the radiological safety of the patient: radiation is used, in the medical field (see Chapter 12), to obtain diagnostic information or a therapeutic result, and in both cases the benefit to the patient must largely overcome the additional radiological risks of the exposure to radiation.

While the choice to perform a radiological diagnostic/therapeutic procedure is the responsibility of the doctor, an expert professional in the radiological aspects of the medical use of radiation must support the doctor in every phase of the subsequent procedures.

This professional must be competent to ensure that the medical instruments are functioning properly and that unnecessary irradiation of tissues not involved in the procedure is kept to a minimum; that shielding is appropriate for patients and family members; that radiotherapy treatment plans are developed according to the patient's specific needs; and various other tasks connected with the medical radiation protection of patients.

This professional, whose competence and specialisation should be very high, is usually recognised as a trained specialist in the legislation of many countries, and is usually known as a 'medical physicist'.

A medical physicist's duties are different from those of a radiation protection expert; the medical physicist focuses on the patient's dose and the equipment needed to best perform medical procedures, while the radiation protection expert has no responsibility for the patient's dosimetry and is focused on the medical facility's personnel occupational doses and on doses to the general public and the environment.

References

Clarke, R. H., Valentin, J. (2009). The History of ICRP and the Evolution of its Policies. *Annals of the ICRP*, 39(1), 75–110. https://doi.org/10.1016/j.icrp.2009.07.009.

Council Directive 2013/59/Euratom of 5 December 2013 laying down basic safety standards for protection against the dangers arising from exposure to ionising radiation, and repealing Directives 89/618/Euratom, 90/641/Euratom, 96/29/Euratom, 97/43/Euratom and 2003/122/Euratom, http://eur-lex.europa.eu/legal-content/EN/TXT/?qid=1495976809093&uri=CELEX:32013L0059 (consulted on 28-MAY-2017).

IAEA. (1999). IAEA RS-G-1.2 – Assessment of Occupational Exposure Due to Intakes of Radionuclides. Vienna. Retrieved from http://www-pub.iaea.org/MTCD/publications/PDF/P077_scr.pdf

IAEA. (2014). Radiation Protection and Safety of Radiation Sources: International Basic Safety Standards, General Safety Requirements Part 3 INTERNATIONAL ATOMIC ENERGY AGENCY VIENNA ISBN 978–92–0–135310–8 ISSN 1020–525X No. GSR Part 3, http://www-pub.iaea.org/MTCD/publications/PDF/Pub1578_web-57265295.pdf (consulted on 28-MAY-2017).

ICRP. (2007). The 2007 Recommendations of the International Commission on Radiological Protection. ICRP Publication 103. Ann. ICRP 37(2–4).

IDEAS. (n.d.). IDEA System Expert System for Internal Dosimetry. Retrieved July 1, 2017, from https://www.idea-system.com/ideas-guidelines/introduction/.

International Commission on Radiological Protection (ICRP) website: http://www.icrp.org/ (consulted on 28-MAY-2017).

International Organization for Standardization website (ISO): https://www.iso.org/home.html (consulted on 28-MAY-2017).

International Radiation Protection Association website (IRPA): http://www.irpa.net/ (consulted on 28-MAY-2017).

ISO. (2006). ISO 20553 – Radiation protection – Monitoring of workers occupationally exposed to a risk of internal contamination with radioactive material.

Jorgensen, T. J. (2016). *Strange Glow: The Story of Radiation*. Princeton, New Jersey: Princeton University Press.

Knoll, G. F. (2000). *Radiation Detection and Measurement*. 3rd ed. John Wiley & Sons, Inc. https://doi.org/10.1017/CBO9781107415324.004.

Lambert, K., Alhaj, A., Arabia, S., Brandl, A., Kase, K., Gallego, E., Coates, R. (2016). Working Group on Radiation Protection Certification and Qualification. Retrieved from http://www.irpa.net/docs/IRPA Guidance on Certification of a RP Expert (2016).pdf

Martin, J. E. (2013). *Physics for Radiation Protection*. 3rd ed. Weinheim: Wiley-VCH Verlag & Co. KGaA, Boschstr. 12, 69469 Weinheim, Germany.

Munira, K. et al. (2013). Non-targeted effects of ionizing radiation–implications for low dose risk, *Mutation Research*, 752(1), 84–98. doi:10.1016/j.mrrev.2012.12.001. https://www.ncbi.nlm.nih.gov/pmc/articles/PMC4091999/pdf/nihms579724.pdf (consulted on 30-MAY-2017).

NCRP. (2010). NCRP Report n. 164 - Uncertainties in internal radiation dose assessment. National Council on Radiation Protection and Measurements. Retrieved from https://www.ncrppublications.org/Reports/164.

NIRS. (n.d.). Mondal 3. Retrieved July 1, 2017, from http://www.nirs.qst.go.jp/db/anzendb/RPD/mondal3.php.

PHE. (n.d.). IMBA® Professional Plus Internal Dosimetry Software – Home. Retrieved July 1, 2017, from https://www.phe-protectionservices.org.uk/imba.

Turner, J. (2007). *Atoms, Radiation, and Radiation Protection*. 3rd ed. WILEY-VCH Verlag GmbH & Co. KGaA, Weinheim.

United Nations Scientific Committee on the Effects of Atomic Radiation (UNSCEAR) website: http://www.unscear.org/unscear/en/index.html (consulted on 28-MAY-2017).

Source of Radiation Exposure in the Workplace: Nuclear, Medical and Industrial Sources

Daniele Giuffrida
Federal Authority for Nuclear Regulation

11.1 Introduction: Annual Average Exposure to Radiation

We are all exposed to ionising radiation throughout our lives. Exposure is usually split into *natural radiation exposure*, which will be the object of the next two chapters, and *artificial radiation exposure*, which is the result of human activities, whose occupational aspects will be treated in this chapter.

During exposure to artificial radiation, either radioactive sources are purposely used for their properties (nuclear reactors, security X-ray scans, radiopharmaceuticals, etc.) or radiation is incidentally present as a byproduct of another activity (increased cosmic irradiation rays during air travel, for example).

TABLE 11.1 World Average Annual Exposure Due to Natural and Artificial Sources

Source	Annual Average (mSv)	Typical Range (mSv)
Inhalation of radon gas	1.26	0–10
External gamma irradiation	0.48	0.3–1
Ingestion of radioactivity	0.29	0.2–1
Cosmic radiation	0.39	0.3–1
All natural radiation sources (total)	2.4	1–13
Medical diagnosis (excluding therapy)	0.6	0–tens
Fallout from atomic bomb tests	0.005	–
Occupational exposure	0.005	0–20
Fallout due to the Chernobyl accident	0.002	–
Nuclear fuel cycle	0.0002	0–0.02
All artificial radiation sources (total)	0.6	0–tens

Source: Adapted from United Nations Scientific Committee on the Effects of Atomic Radiation, UNSCEAR 2008 Report to the General Assembly, with scientific annexes, Volume I: (Sources), Scientific Annexes A and B.

It has been recently estimated (NCRP Report 160, UNSCEAR 2008, Annexes A and B) that annual natural world average doses are about 3 mSv per capita. This average dose has large variations (up to 50 mSv/y and more) due to specific geographical and geological conditions, and is mainly dependent on the magnitude of radon gas exposure.

A recent analysis of public radiation exposure data in the United States indicates that artificial radiation exposure also accounts for an annual average of around 3 mSv, the main source of exposure being the use of medical radiation in society[*], which has kept increasing since the 1980s.

Table 11.1 presents average worldwide public exposure data, and dose ranges in which they typically lie.

Within the medical uses of radiation, which represent a significant portion of public exposure data in western countries (see Table 11.2), half of the average annual dose is due to the use of computerised tomography (CT) scans, one quarter to nuclear medicine, and the rest to interventional fluoroscopy and radiology. Unlike natural exposure, which cannot be switched on and off, actual annual medical doses depend on the number and type of medical procedures that an individual undergoes during a specific year.

Within exposure to consumer products, average annual doses are split equally among cigarette smoking (inhalation of ^{210}Po), air travel (increased exposure to cosmic rays) and the effect of building materials.

Within other various uses of radiation in the industrial, medical, security and research sectors, most of the annual average exposure is due to caring for medical patients; doses due to the operation of nuclear power plants account for only 15% of the subtotal.

Within occupational exposure, the highest annual average exposures are incurred by airline crews and medical doctors, with nuclear, industrial and research sectors each accounting for less than 10% of the annual average dose of 0.005 mSv.

11.2 Artificial Radiation Sources

Radiation sources (as radiation generators and radioactive sources are generically called) are widely used in the medical, industrial, research and academic sectors.

The first category, *radiation generators*, is composed of electrical devices that can produce ionising radiation and that can be controlled, so that the production of radiation can be initiated and terminated when

---

[*] This average exposure has been calculated excluding radiotherapy, whose doses to the patients are much higher than average doses during diagnostic procedures.

TABLE 11.2   United States Population's Average Annual Exposure to Artificial Sources

Source	Annual USA Average (mSv/y)	Fraction (%)[a]
Medical radiation to patients (excluding radiotherapy)	3.00	–
Computed tomography (CT scans)	–	49
Nuclear medicine	–	26
Radiography	–	11
Interventional radiology	–	14
Consumer products	0.13	–
Cigarette smoking	–	35
Building materials	–	27
Air travel	–	26
Industrial, security, educational uses	0.003	–
Caring for nuclear medicine patients	–	72
Operating nuclear reactors	–	15
Occupational exposure	0.005	–
Medical occupational exposure	–	39
Aviation occupational exposure	–	38
Nuclear power occupational exposure	–	8
Industrial and commercial occupational exposure	–	8
Research, military and governmental occupational exposure	–	7
Total average annual USA exposure	3.14	–

Source: Adapted from National Council on Radiation Protection and Measurements, 2009. NCRP Report 160: Ionizing radiation exposure of the population of the United States. National Council on Radiation Protection and Measurements.

[a] The percentage in the column 'Fraction' refers to the sub-total of that category reported in the column 'Annual USA average': for example, CT scans account for 49% of 3 mSv/y, the subtotal of the 'Medical radiation to patients' category, so they account for around 1.5 mSv/y (not indicated in the table).

the generator is powered on or off. Examples of radiation generators are industrial X-ray machines, CT scanners and linear accelerators.

The second category, *radioactive sources,* comprises manufactured materials containing one or more radioactive isotopes, specifically selected for their radiation emission, half-life, particles' energy, intensity, etc.

Radioactive sources can be characterised as *sealed (radioactive) sources* if the radioactive material is securely fixed in a solid matrix and is able to resist to some level of mechanical shock, exposure to heat or water submersion, according to the requirements of technical standards (ISO 2919:2012). These kinds of sources are usually used in industrial activities (for example, nucleonic gauges, NDT, well logging, etc.). If the conditions for the definition of 'sealed source' are not met, then the source is characterised as an *unsealed (radioactive) source*; this kind of radioactive source, which can release radioactivity, is normally used in laboratories, for example, to provide for the limited quantities of radioactivity required during experiments or analyses, or in nuclear medicine departments to prepare radiopharmaceutical doses for patients.

National radiation regulations define the lower limits for a radiation generator to be subject to the provisions of the regulation: below those limits, no action is required from the user or manufacturer of the radiation generator. This is the case with cathode tubes, extensively used in the past for TVs, which were not subject to any radiation protection provisions because the intensity, quality and energy of the emitted radiation were below the limits for the classification of that equipment as a radiation generator.

While natural radioactive materials are present in all materials, including food (and even occur inside the body), natural or artificial radioactive materials may be intentionally added to some consumer products, which results in additional public exposure to radiation; hence a clear boundary is needed to identify the quantity of radioactive materials that are subject to radiation protection regulations. As with radiation generators, national radiation regulations define if a certain quantity of radioactivity must be considered a radioactive source and when it is considered exempt from the regulation. Usually, the definition is expressed in terms of the radionuclides' activity and concentration of activity (EU Council Directive 2013/59/Euratom, IAEA, 2014: Radiation Protection and Safety of Radiation Sources: International Basic Safety Standards).

Once a device or a material exceeds the boundaries to be classified as a radiation source, rules and obligations regarding its manufacturing, importation, commercialisation, purchase, transportation, use, storage, accountability, transfer and waste management, including disposal, may be present in the national legislation. The set of provisions that must be met for the possession and use of radiation sources is called 'radiation sources licensing' and is an important element of any national radiation protection organisation.

11.3 Radiation Sources in the Nuclear Field

A nuclear power plant* is, basically, a big water heater.

Heat is generated in the reactor core by a continuous, enormous number of nuclear fission reactions; each fission event, by itself, releases only a tiny amount of energy compared to the overall power of the reactor[†].

The heat released by each single fuel element[‡] requires that fuel elements be continuously cooled down by a recirculating fluid (water, in a light water reactor) in order avoid meltdown; the water is then used to create steam, which is sent to turbines connected to electricity generators, much like the principle of a bicycle's dynamo.

Thousands of megawatts (MW) of electricity can be generated by a single unit in a modern nuclear reactor, while more than three times that power is being generated in the reactor's core as heat (thermal power), and – in large part – is lost in the electricity generation process or is discharged into the environment as heat.

Nuclear energy is an essential component of the energy mix in most modern advanced countries, and is the only viable energy source for effectively combating climate change and replacing fossil fuels in the short term. There are almost 450 nuclear reactors in operation around the world in 2017 (IAEA, 2017), and they generate a significant fraction of all the electricity needed by the world's population. More than 60 reactors are currently under construction, 20 of which are in China, 7 in Russia, and 4 in the UAE and in the USA.

While nuclear energy is for many a controversial subject, several viable technological alternatives already exist for the storage of irradiated nuclear fuel and radioactive waste, in rational and centralised deposits which have a minimal impact on populations and the environment.

Nuclear accidents are likely to happen again, as they are in every industrial activity; technological research and efforts must be used to reduce the likelihood and the consequences of nuclear accidents. It must be stressed, though, that the impact of modern reactor accidents is limited, although more effective techniques for emergency response and decisions regarding population relocation are needed, as shown during the recent Fukushima accident. Moreover, continuous provision of information and education on health issues connected with nuclear power, and building long-lasting trust in the relationship between the public and government, seem to be important elements in supporting national nuclear power policies.

---

* Many different reactor designs are worldwide available for commercial production: for sake of clarity and simplicity, a Light Water Reactor (either a Pressurized or a Boiling Water Reactor) will be used as a reference in this chapter, as this design category represents the majority of currently operating world reactors.
† Around 200 MeV are released in a fission event of ^{235}U, a much larger Q-value than other nuclear reactions' energies and radioactive decays, or the few eV of an oxidation chemical reaction (like coal burning, for example).
‡ There are around 200 fuel elements in a typical PWR reactor, each element being composed of around 260 fuel rods of around 4 m height, for a total of around 51000 rods in the reactor core (Buongiorno, 2010).

During a nuclear fission event (on ^{235}U, for example), a neutron reaches the uranium's nucleus without any electrical interaction, and can interact with and/or be absorbed by the nucleus. The latter event generates an instability in the nucleus, which will eventually split into two (or, more rarely, three or more) fragments. The fragments are called *fission fragments* and are pushed apart during the fission process with around 170 MeV of kinetic energy. Fission fragments quickly dissipate their kinetic energy by interacting with other materials' nuclei, and then continue their radioactive decay processes until they reach a stable nucleus configuration.

Other particles are also emitted during fission, including two or three more neutrons, which may induce fission on other uranium nuclei in what is called a *nuclear chain reaction*. Gamma emissions, alpha and beta particles contribute to heat released in the fission process.

Neutrons can also be absorbed by materials in a process which is called *neutron capture* and does not lead to nuclear fission, but to the transformation of the original (stable) nucleus into a radioactive, unstable new isotope. Take, for example, the reaction between a neutron and a stable cobalt nucleus, a component of any steel alloy:

$$^{59}Co + n \rightarrow Co\text{-}60$$

The nucleus created as the product of neutron irradiation is ^{60}Co, a radioactive beta-gamma emitter, whose half-life is around 5 years: The irradiation of stable cobalt nuclei in reactor-shielding materials, or in any metallic piping, or molecules in the coolant water itself, can create new radioactive nuclei which may decay into isotopes of different chemical species.

This process is one of the two main processes generating radioactivity in a nuclear reactor: The main process is the *nuclear fission reaction* itself, with the splitting of U atoms into radioactive fission fragments, usually contained inside fuel elements' pellets (*fission products*). The second process is the result of neutron irradiation and capture and is called *neutron activation*, which leads to the creation of *activation products*.

Within a nuclear reactor, in addition to the reactor core and its activation and fission products, other radiation sources may also be present.

For example, intense neutron sources are needed to start the reactors. Those sources are introduced in the core and create a neutron population to help its initial startup.

Other radiation sources may be used during reactor construction to check the integrity of the welding and the quality of critical components (non-destructive testing, NDT), similar to what happens in other industrial activities (see also Section 11.5); a nuclear reactor construction site is, indeed, a very large industrial site. Those radiation sources are usually removed once testing of the structures and components is complete.

Other radioactive sources of lower activity may also be used during reactor operation for the extensive radiation-measuring and radiation-protection equipment.

By design, nuclear reactors' radiation sources are mostly confined within the reactor containment building, and largely inside the reactor core. Radioactivity is usually present, in small quantities, inside the containment building, mainly due to activation processes; in some cases, irradiation and contamination can simultaneously be present, with a predominant gamma and a significant neutron component. In a liquid spillage, for example, alpha, beta and gamma isotopes may be found at the same time.

Reactor coolant water is usually very radioactive due to neutron irradiation, contact with fuel elements (which usually have cracks and crevices from which small quantities of fuel may leak) and the recirculation of metallic, irradiated and activated residues in the core (for example, from the creation of sludge, oxidation particles, etc.).

The chemistry of the primary water needs to be perfectly controlled for reasons of both nuclear safety and radiation protection, hence huge amounts of cooling water are continuously filtered by several purification systems, including demineralisers, filters and resin columns to reduce its radioactive content (Prince, R., 2012). Those systems, after some time, accumulate significant activity of fission fragments and activation products, and may become highly irradiated (in some cases, dose rates in excess of some

Sv/h are measured around resin tanks). Operations to remove and change resin columns need to be properly planned and executed in order to keep integrated dosimetry under control; there is a risk of workers being externally and internally irradiated.

Another significant source of dosimetry to workers is fuel exchange operations during outages. As nuclear fuel in a nuclear reactor slowly 'burns down' and becomes less profitable to use after some years, due to the accumulation of fission products and other neutron absorbers*, the planned exchange of a portion of the existing fuel elements with new elements and the reshuffling of the other fuel elements in the core is carried out every 3–5 years. All operations take place under water, with the reactor core open and completely submerged; the water provides shielding, cooling and a transparent medium in which to operate. 'Exhausted' irradiated fuel elements, which are highly radioactive, are moved into the reactor fuel pond, where they are allowed to cool down for some period of time (usually years), and new fuel elements are moved into the reactor core. All these operations, which are carefully planned to control workers' doses and minimise unproductive plant shutdown time, require the coordinated presence of a large number of additional contractors and usually represent a significant fraction of the annual integrated dose for the reactor's staff. As a reference, in France a workers' dose during outages represents 81% of the total dose, which means that doses received during normal operation of the reactors represent only 19% of total workers' doses (ISOE, 2014).

For many years, large operators around the world have been working on initiatives to keep nuclear power plants' occupational doses under control and to benchmark them, with the objective of overall integrated dose reduction, to conform to the ALARA principle. ISOE n.d, an international OECD/NEA sharing and benchmarking initiative started in 1992, aims to spread good practices in reducing plant workers' doses. As an example of the results of these efforts, which are also shared at the international level by many national operators and regulators, integrated occupational 3 years' average doses have gone from around 2.6 man-Sv/reactor in 1992–1994 to 0.9 man-Sv/reactor in 2012–2014.

When decontamination is needed, i.e. to remove radioactive materials resulting from, for example, liquid spillage or leakage, simultaneous risks of external and internal irradiation exist. Proper work procedures and appropriate personal protective equipment, including appropriate suits, gloves, overshoes and masks fitted with appropriate filters, if not respirators or more advanced full-body protections, must be used. Dosimeters, an integral part of workers' clothing, must be worn at all times, and specific dosimeters may be needed to monitor, for example, doses to the extremities. Regular contamination checks on clothing and on the skin surface must be also performed; whole-body checks at the exit of the reactor containment building are routinely performed (using hands and feet monitors or portal monitors). Checks on equipment and tools must also be executed to limit the potential spread of contamination outside the controlled areas of the reactor.

Usually, a certain baseline level of internal contamination in nuclear power plant workers is possible, for example due to ^{60}Co and ^{3}H; intake of radioactivity must be regularly checked and monitored, and corresponding internal dosimetry calculations and assessments routinely performed.

An operational nuclear power plant will also routinely emit small amounts radioactivity into the environment. Activation products, mainly tritium and noble gases, ^{131}I, ^{14}C and particulates from irradiated air in the reactor's building and slightly activated or contaminated waters, after assessment of radioactivity content, are allowed to be discharged into authorised discharge paths, including reactor off-gas (for gaseous effluents) and river flows (for liquid effluents), for example. These radionuclides contribute to the radiation dose to the public living near a nuclear power plant, as radioactivity may reach the population directly or indirectly (via the food chain). Prior to the construction of a nuclear power plant, evaluation of the impact of radioactivity release to the environment is required. Actual radioactivity releases are measured with a series of detectors positioned around the plant, using a coordinated sampling and measurement plan of environmental matrixes like air, water, depositions, fruit, fish, milk, etc.

---

* Nuclides able to absorb neutrons, but do not create additional fissions, therefore representing a loss in the neutron population balance (and are therefore called 'neutron poisons').

Doses to populations due to routine discharges have been assessed as very low, and in most cases negligible.

With more and more reactors nearing original project end-of-life, if it is technically feasible to continue to operate a plant beyond its project life, operators tend to request authorisation to extend its operational life in order to maximise return on investment.

A plant's life extension may include a series of additional correlated activities (e.g. steam generator exchange) which are intermediate between ordinary plant operation and decommissioning, and may have a significant impact on dosimetry (replacement of a steam generator, for example).

A nuclear power plant remains radioactive after its operational life cycle is over, due to the presence of activation and fission products in various parts of the reactor and ancillary buildings. The final removal of radioactivity and activated components from the plant and segregation, the demolition of structures and management according to their activity content and the final release of the former nuclear site to unrestricted uses represent a complex administrative and technical process called *decommissioning* (IAEA, 1999). A reactor's decommissioning may take place immediately after shutdown, but most often is delayed by some years in order for the radioactivity to decrease, thereby reducing radiological risks and occupational doses.

Radioactive waste management is a key factor during decommissioning, as it is during operation. The proper identification, quantification, segregation and disposal of radioactive wastes according to their waste stream makes a decommissioning project more successful or economical. Radionuclides due to activation which are usually encountered during decommissioning include ^{3}H, ^{14}C, ^{39}Ar, ^{41}Ca, ^{54}Mn, ^{55}Fe, ^{59}Ni, ^{63}Ni, ^{60}Co, ^{65}Zn, ^{93}Zr, ^{108m}Ag, ^{110m}Ag, ^{133}Ba, ^{133}Cs, ^{137}Cs, ^{152}Eu, ^{154}Eu, ^{155}Eu, etc., with radionuclide inventories which depend on integrated neutron flux, composition of materials and time (IAEA, 1998).

Many nuclear reactors have been successfully decommissioned around the world to date, and significant experience is growing in this relatively new field.

11.4 Radiation Sources in the Medical Field

Several radiation sources are used in the medical field, both for diagnosis and for therapy (IAEA, website). As mentioned above, medical exposure to radiation is currently by far the largest artificial source of annual dose for the populations of the EU and the USA.

A summary of radiation sources used in medicine will be presented briefly here, separating radiation generators and radioactive sources and their respective uses in medical diagnosis and therapy and leaving to the interested reader a number of excellent additional references References.

11.4.1 Radiation Generators Used for Diagnosis

The first use of a radiation generator in medicine took place just a few weeks after the discovery, by Wilhelm Roentgen, of the existence of so-called X-rays at the end of 1895; in the beginning of 1896, X-rays were used on a wounded patient in Canada to locate a bullet in his leg (Jorgensen, 2016).

Since then, X-rays have been extensively used in patient diagnosis, and enormous progress in the quality of medical X-ray imaging has accompanied this development.

X-ray machines are basically tubes in which an electron beam current of hundreds of mA is generated by a heated filament (called a cathode, and usually made of tungsten), then accelerated by a difference of potential of some tens of kV against a metallic target (called an anode and usually made of tungsten, rhenium, molybdenum or graphite).

The anode may be stationary or rotating, and X radiation is produced by *bremsstrahlung*, although more than 99% of the energy of the electron beam is dissipated as heat in the anode, requiring heat removal.

Radiation is filtered out of the X-ray tube housing by using a set of absorbers (inherent and additional/ added filtration) to selectively absorb lower energy photons and an unwanted part of the energy spectrum, and is collimated towards the patient's target organ to reduce scattered radiation and limit the irradiation of non-targeted tissues (Bushberg et al., 2002).

Electromagnetic radiation can penetrate a subject's body and is more attenuated by hard tissues and bones than by soft tissues: If the resulting image is collected (by means of traditional photographic film cassettes, or with modern electronic CCDs), then it is possible to appreciate an internal vision of the patient's body, possibly deriving a medical diagnosis. An entire, very sophisticated branch of medical physics is devoted to improving the quality (definition, contrast, resolution, noise, blur, distortion, arti-facts, etc.) of the medical image, while keeping under control radiation doses to the patient's targeted and non-targeted tissues (Hendee and Riternour, 2002; Sprawls, 1995).

An evolution of X-ray machines which was largely popular at the beginning of the 20th century, and whose applications are still in use nowadays – although under stricter regulation and control – is *fluoros-copy*. Fluoroscopy relates to X-ray imaging as cinematography relates to photography; fluoroscopy allows the doctor a real-time visualisation of moving anatomic functions.

Today's fluoroscopy makes use of image intensifiers, screen displays and video recording systems, and is significantly improved since the original design of a fluorescent screen over which a dim live image would appear to the practitioner standing right behind it (direct fluoroscopy without image intensifiers has been banned in many countries).

Fluoroscopy imparts significant radiation doses to the patient, but is required in specific conditions, when detailed live information on the functioning of an organ (for example, the digestive tract) is needed or during interventional radiology (Balter, 2001).

According to recent research, there are large variations in dose levels for fluoroscopy procedures, which are the result of various differences in local practice, equipment and staff.

Interventional procedures have experienced a dramatic increase in recent decades, due to several ben-efits, in particular related to outpatient procedures (Wagner and Archer, 2004).

Interventional procedures may lead to significantly higher doses to the patient (and also to the practi-tioner and other medical staff) (ICRP, 2013a,b), hence they have been the object of close scrutiny and guidance (ICRP, 2000, ICRP 2010, NCRP 2010).

A third, very sophisticated application of X-ray imaging is the computerised tomography scan, or *CT scan*. This technique allows one to visualise, with enhanced precision, detailed slices of a patient's organ. Unlike X-ray images and fluoroscopy, which are obtained by illuminating the patient with a single, short flash of X-rays (with a long, persistent beam of X-rays) and collecting the resulting image (video) as pro-jected on a CCD/film (on a fluoroscopic screen or imaging device), tomographic scans are obtained by detecting and recording the output of several pulses of X-ray thin slices' beams, each of which is generated at a different angle with respect to the target, the machine being rotated around the subject. The term 'com-puterised' comes from the fact that the final image must be reconstructed by taking account of the infor-mation stored in all the digital images which have been produced and recorded, and this task is achieved with the aid of a dedicated computer.

Due to the fact that CT scans are the result of a series of beam exposures, this technique also implies larger doses to the patient than traditional X-rays.

According to international surveys, it appears that large differences exist among medical imaging pro-cedures in different countries, in terms of the availability of medical imaging, the quality of the images and the radiological safety to patients. New technologies have generated significant improvements in the qual-ity of diagnostic medical imaging and in the corresponding reduction of doses delivered to patients, although the pace of adoption of new technology is not always fast (especially in countries with less devel-oped healthcare systems).

Table 11.3 presents typical average exposure to the patient from selected diagnostic medical imaging procedures (performed in countries with more advanced national medical systems) and their evolution over time.

TABLE 11.3   Average Effective Doses Due to Selected Imaging Procedures

Examination	1970s	1980s	1991–1996	1997–2007
Chest radiography	0.25	0.14	0.14	0.07
Abdomen X-ray	1.9	1.1	0.53	0.82
Mammography	1.8	1	0.51	0.26
CT scan	1.3	4.4	8.8	7.4
Angiography	9.2	6.8	12	9.3

Source: United Nations Scientific Committee on the Effects of Atomic Radiation, UNSCEAR 2008 Report to the General Assembly, with scientific annexes, Volume I: (Sources), Scientific Annexes A and B.

11.4.2 Radioactive Sources Used for Diagnosis

'Nuclear medicine' is the branch of medicine making use of radioisotopes for diagnosis (unsealed radioactive materials) and therapy (sealed and unsealed radioactive materials).

The radiopharmaceutical is usually administered to the patient intravenously, orally or by inhalation. The radioisotopes contained in the radiopharmaceutical, subject to the biokinetics of the molecular compound (appropriately chosen for its chemical properties), are distributed in the patient's body – which becomes slightly radioactive – and concentrated in specific volumes, allowing detection of the position and extent of the radioisotopes in the body and organs with a suitable radiation detector.

The radioisotopes will be cleared from the patient's body with a time constant depending on the excretion time of the molecule and the physical decay of the radioactive nuclide (see Section 10.7.4). In nuclear medicine, molecules and isotopes are chosen for a specific diagnostic purpose, and the half-life and the excretion path of the labelled molecule are optimised to obtain the best medical imaging while keeping under control both radiation doses to the patient and exposure time.

Several radiopharmaceuticals have been developed, and usually are characterised by half-lives which are a compromise between shorter half-lives (to limit irradiation to the patient and family members after the exam) and longer half-lives (allowing enough time to manufacture the radiopharmaceutical, transport it, inject it and obtain the requested medical information): typical radiopharmaceuticals' half-lives are in the order of tens of minutes.

Unlike radiation generators, which can be switched on (and off) at any time, radiopharmaceuticals need to be manufactured, dispensed, transported, measured, prepared and injected, and their effectiveness is dependent on a chain of several variables, including (but not limited to) the availability of advanced, specialised and complex radiation imaging detectors. Medical use of radioisotopes requires diverse and advanced equipment and competency, and is currently not available in all healthcare facilities.

Medical radioisotopes for diagnosis are usually produced in nuclear reactors or particle accelerators (most frequently, small cyclotrons) or in radionuclide generator systems (i.e. systems containing a long-lived precursor isotope which continuously generates the shorter half-life radionuclide, which is then extracted from the generator, like ^{99}Mo-^{99}T).

Preparing radiopharmaceuticals, i.e. labelling a specific drug with the intended radioisotope, usually takes place in highly specialised commercial radiopharmaceutical facilities, with the notable exception of several hospital radiopharmacies that can produce isotopes and radiopharmaceuticals for their own use.

Radioisotopes used in nuclear medicine diagnosis belong to one of three categories: ^{99m}Tc-labelled compounds (which are the most popular, and are used for various medical procedures), radioiodinated compounds (containing isotopes of iodine, usually ^{123}I, ^{125}I or ^{131}I) or beta$^+$-emitting radiopharmaceuticals, used in PET (positron emission tomography) systems (usually ^{11}C ^{13}N, ^{15}O and ^{18}F).

Radiation detectors used in nuclear medicine comprise some of the most advanced radiation detection medical systems: They range from simpler detection systems, such as NaI probes used for thyroid uptakes studies, to very complex detector systems like CT-PET coupled machines.

TABLE 11.4   List of Selected Radionuclides Used in Nuclear Medicine for Diagnosis

Radionuclide	Use of Radionuclide
^{99m}Tc	RE, regional perfusion, kidney function, bone metabolism, lung emboli, blood volume, cerebral blood flow
^{123}I	Thyroid uptake
^{18}F	Whole-body localisation of tumors
^{57}Co, ^{58}Co	Vitamin B12 deficiency
^{111}In	Blood element imaging
^{201}Tl	Cardiac blood flow and cellular metabolism
^{67}Ga	Detection of soft tissue malignancies
^{125}I	Radioimmunoassay, blood volume
^{51}Cr	Kidney clearance, blood volume
^{59}Fe	Ferrokinetic studies
^{3}H	Body water content
^{82}Rb	Cardiac blood flow and cellular metabolism

Source: Adapted from United Nations Scientific Committee on the Effects of Atomic Radiation, UNSCEAR 2008 Report to the General Assembly, with scientific annexes, Volume I: (Sources), Scientific Annexes A and B.

Gamma cameras (composed of an array of NaI or similar scintillation detectors, and used to determine the exact distribution of the uptake in the organs) often come in dual or triple configurations, and can be used for static, dynamic or tomographic imaging studies with great flexibility.

Improved spatial resolution is achieved using PET scanners – complex 3D coincidence detection devices – using a unique property of beta^{+} emissions annihilation, during which two simultaneous, opposite 511 keV gamma rays are emitted.

A list of radionuclides which are used in diagnostic nuclear medical practice is reported in Table 11.4.

Radiation doses from different nuclear medicine procedures may vary significantly, and among similar facilities in the most advanced healthcare systems. Various patients' effective doses for selected examinations, resulting from an international survey, are reported in Table 11.5.

11.4.3 Radiation Generators Used for Therapy

Today, radiotherapy is one of the main treatments for cancer , although it can also be used to treat other minor diseases. It is assumed that 50%–60% of cancer patients benefit from radiotherapy

TABLE 11.5   Mean Patient Effective Dose for Selected Nuclear Medicine Diagnostic Examinations

Examination	Mean Effective Dose (mSv)
Renal	0.4–3.8
Gastroenterology	0.1–5.7
Brain	2.0–7.5
PET	5.6–10.8
PET-CT combined	6.0–10.8

Source: Adapted from United Nations Scientific Committee on the Effects of Atomic Radiation, UNSCEAR 2008 Report to the General Assembly, with scientific annexes, Volume I: (Sources), Scientific Annexes A and B.

(Khan et al., 2014). Radiotherapy can be used as the main modality for cure, as a palliative treatment or as an adjuvant treatment in conjunction with surgery.

The potential use of X-ray machines for therapy was recognised at the very beginning of the history of X-rays, with the first accounts of therapeutic use of X-ray machines in 1896.

Radiation is used in therapy for its ability to microscopically deliver huge amounts of energy to the patient's tissues, hence to induce a response against a specific disease. In this case, radiation is not used to identify the status and function of a patient's body region but to kill cancer cells. Large radiation doses (on the order of tens of grays*) are involved in this process, and usually dose fractionation is needed[†] in order to minimise unwanted deleterious effects and to help the body recover from an exposure. These procedures, involving high radiation doses and dose rates, may create the potential for tissue reactions and accidents if not managed properly. An extremely accurate dosimetry of radiation delivered to the tissues is required. Irradiation of tissues outside the region of interest must be minimised, and an individual treatment plan, based on the patient's anatomy, must be prepared, with specific collimators manufactured on the body's morphology[‡].

Radiation generators used in radiotherapy usually fall into these categories:

1. X-ray machines,
2. Electrons' linear accelerators,
3. Light or heavy ions particle accelerators.

X-ray machines in use for therapeutic purposes are usually subdivided into *kilo-voltage X-ray generators* (contact therapy machines producing X-rays of 25 to 40 kVp and superficial therapy machines producing X-rays of 40 to 120 kVp) and *ortho-voltage X-ray generators* (machines producing X-rays of 150 to 300 kVp).

Medical *linear accelerators* (also called *linacs*), usually with energies between 4–25 MV, are used to produce intense electron beams (also called *megavoltage therapy*) with an isocentric setup, i.e. allowing rotation of the accelerator around the patient's body – and the tumor location – and irradiation of the tumor from different angles, in order to achieve more accurate and predictable dose distribution in the volume. To achieve electrons' energies exceeding 4 MeV, electrons' acceleration with microwaves is used, in a complex setup of a series of waveguides with bending magnets to redirect the beam out of the tube housing. Wedges, blocks and multileaf collimators are used to conform the electrons' beam as required.

Superficial therapy is normally used for treatment of skin lesions, at a limited distance between the unit's focus and the patient and using small radiation field sizes: Radiation beams are collimated on the patient's skin with the use of cones. Orthovoltage therapy is used for palliative treatment of bone lesions close to the surface: Cones are used to set up the beam on the patient's skin. Megavoltage therapy is deeply penetrating (focus-skin distance is on the order of 1 m), and is skin-sparing; therefore, it is used for the treatment of deeper tumors and tumor beds.

Light ion and *heavy particle accelerators* are promising alternatives to traditional therapy, and are being developed in a limited number of specialised facilities around the world. For specific tumors (e.g. melanoma of the eye, deep-seated tumors such as brain, lung and prostate tumors), proton beams already represent the treatment of choice.

Prescribed doses during radiotherapy treatments are in the 40–60 Gy range for most treatments, with lower doses used to treat some specific tumors (leukemia, testis tumors and some pediatric tumors) (Williams et al., 2000).

---

* It must be noted that in radiotherapy, radiation doses are usually referred to the irradiated tissue/organ, and therefore are expressed in terms of *equivalent doses* to that volume (expressed in grays).

† For example, 2 Gy per day over 30 consecutive days, for a total dose of 60 Gy.

‡ These considerations are applicable both to radiation therapy obtained using radiation generators (X-ray machines and accelerators) and radionuclides, although preparation of the radiopharmaceutical and its management, including radioactive waste management, is additionally required in the latter case.

TABLE 11.6   Characterists of Brachytherapy

Type of Brachytherapy	Radioactive Sources	Dose Rates at the Tissue	Duration of the Treatment
Low-dose-rate	^{137}Cs	0.4–2.0 Gy/h	A period of several days in one fraction, or more often two
Medium-dose-rate	Higher-activity ^{137}Cs sources	Up to 12 Gy/h	A period of several days in one fraction, or more often two
High-dose-rate	^{192}Ir or ^{60}Co	2–5 Gy/min	Minutes or less and the treatment generally delivered through several fractions

Source: Adapted from United Nations Scientific Committee on the Effects of Atomic Radiation, UNSCEAR 2008 Report to the General Assembly, with scientific annexes, Volume I: (Sources), Scientific Annexes A and B.

11.4.4 Radioactive Sources Used for Therapy

Both sealed and unsealed radioactive sources can be used in radiation therapy.

Sealed sources are used in radiotherapy, either at a small distance from the tissue to be irradiated (*brachytherapy*, usually using beta radiation) or at a larger distance from the tissue and aimed at irradiating larger volumes (*teletherapy*, usually using gamma radiation).

Most popular brachytherapy* sources were made of ^{226}Ra, but recently they have been replaced by ^{137}Cs, ^{192}Ir and ^{60}Co. Sources are applied in contact to the tissues, normally in the form of seeds or wires, and can also be implanted (temporarily or permanently) in the treatment of specific kind of tumors.

Table 11.6 shows some of the characteristics of brachytherapy treatments, including radioactive sources used and the typical range of dose rates generated in the tissue of interest.

Teletherapy[†] sources[‡] used to be made of ^{137}Cs, but are now almost exclusively made of ^{60}Co, an isotope which decays with a half-life of 5.26 years to stable ^{60}Ni, producing two gamma rays of 1.17 MeV and 1.33 MeV. For this reason, radiation from this kind of source is also called megavoltage radiation[§]. This kind of radiation has the ability to irradiate deeper tissues than superficial and orthovoltage radiation, and is therefore also used in isocentric setups, similarly to medical linacs.

As in diagnosis, the use of unsealed radioactive sources in therapeutic nuclear medicine relies on the ability of the chemical compound used in the procedure to concentrate in a specific part of the patient's body. Unlike in diagnosis, the intended use of radiation in therapy is not to obtain information from outside the body of the patient but to deposit a large radiation dose, predominantly in selected tissues in the volume that must be treated.

Therapeutic nuclear medicine procedures are used very rarely compared to diagnostic nuclear medicine procedures.

Table 11.7 provides some information on radionuclides used during nuclear medicine radiotherapy.

11.5 Radiation Sources in the Industrial Field

There are several uses of radioactive sources and radiation generators in the industrial field: Each one of them takes advantage of some peculiarity of the radiation emitted by the source.

Many applications developed in the early 20th century, mainly in consumer products, are now obsolete or no longer acceptable (radium fountains, thorium cosmetics, uranium ice creams, etc.), and will not be described here.

---

* From the Greek term for 'short' (distance).

† From the Greek term 'far' (distance).

‡ Also called 'telecurie units'.

§ It must be noted that 'megavoltage radiation' can also be obtained with the use of medical linear accelerators, in a process in which high-energy electrons are directed against a metallic target and produce *bremsstrahlung* X radiation.

TABLE 11.7   List of Selected Radionuclides Used in Nuclear Medicine for Therapy

Radionuclide	Form	Administration
^{123}I	MIBG	Intravenous
^{131}I	Iodide or MIBG	Oral or intravenous
^{32}P	Phosphate	Oral or intravenous
^{89}Sr	Chloride	Intravenous
^{90}Y	Colloid and spheres	Intra-articular and intra-cavitary
^{169}Er	Colloid	Intra-articular
^{186}Re	Colloid	Intra-articular

Source: Adapted from IAEA website on the Radiation Protection of Patients (https://rpop.iaea. org/RPOP/RPoP/Content/AdditionalResources/Training/1_TrainingMaterial/index.htm, accessed on 18-JUN-2017), featuring plenty of training and reference material in the medical applications of radiation sources.

11.5.1 Non-Destructive Testing

The field of non-destructive testing (normally indicated as NDT and sometimes as 'industrial radiography') is very wide and well developed.

In this specific application, X-ray generators (usually 250 kVp), gamma-emitting radioactive sources (typically, a few TBq of ^{192}Ir, ^{60}Co or ^{75}Se) or electron accelerators (of higher energies, typically dual-energy systems, also reaching 15 MeV) are used to test the quality of welds, for example, or the integrity of metallic components.

X and gamma radiation is more attenuated by thicker layers of metallic materials, and produces a different level of shading on a photographic film which has previously been wrapped behind the object to be analyzed.

Radiation sources used in NDT are usually very intense and produce radiation fields which can generate tissue reactions in a few minutes/hours: For this reason, the manufacture, transport and use of those sources is strictly regulated and must be subject to utmost care (IAEA, 2011).

Radiation sources must be operated remotely, using electrical remote controls for radiation generators and systems of source-positioning guides and remote 'crank-out' for radioactive sources.

A new, promising application of NDT may be to verify the integrity of buildings following earthquakes and other natural disasters. A recent use of NDT during the earthquake in Nepal showed the feasibility of this approach: As in usual metallic plates or welding inspections, X-rays can provide insight into invisible flaws in standing buildings (IAEA, 2017).

11.5.2 Nucleonic Gauges

Radioisotopes are widely used all over the world as gauges for density, moisture, level, thickness, etc. This technique (also called *nucleonic gauging*) is most suited to extreme environmental conditions or to high-speed production lines. One hundred thousand nucleonic gauges are operated worldwide (IAEA, n.d. Nucleonic gauging).

Level gauging in large tanks (oil or coal, for example) is obtained by coordinately moving a radioactive gamma source and a detector up the opposite sides of the tank and measuring the transmitted radiation: When the measurement abruptly changes, this indicates that the level of the material in the tank has been reached. These gauges are most useful where direct contact gauges cannot be used because of heat, pressure or aggressive or corrosive substances.

Thickness is usually measured with the use of intense, linear beta-emitting sources, the source and the detector on opposite sides of the material to be measured, providing a signal which operates thickness control feedback. The material, usually plastic film, textiles, paper, etc. some meters wide, may move along at speeds of 400 m/s.

Similarly, large linear beta-emitting sources are used to inactivate accumulated tribology electricity in the newspaper industry, avoiding dangerous electric discharges to operators.

11.5.3 Materials Identification

Identification of materials, including the ability to detect minute quantities of chemical elements, is one of the most powerful applications of nuclear techniques and can be applied to the most varied subjects, including artistic manufacts and cultural heritage, elemental composition analysis for consumer protection, smuggling detection and other security applications.

Backscattering radiation can provide an indication of the composition of a material.

X-ray fluorescence emissions can be stimulated, and small, portable units (XRF) are available for process control, security, food analysis, scrap sorting and artistic studies.

Neutron activation analysis, in which a neutron source induces activation of elements in a target, which become radioactive and can be detected with gamma spectrometry, is widely used in archeology, forensics, history of art, etc.

Available neutron sources for this technique are nuclear reactors, spontaneous neutron emitters (such as ^{252}Cf), alpha-berillium sources (Am-Be, Pu-Be, etc.), d-t or d-d accelerators, etc. (IAEA, 2012).

11.5.4 Smoke Detectors

Some types of smoke detectors are equipped with an alpha-emitting radioactive source (typically, ^{241}Am) facing an electrode whose electric potential difference allows alpha particles' secondary to be collected and create a steady electric current. When smoke gets inside the detector, it alters the alpha flow and the electric current generated by alpha particles, which triggers the alarm.

These detectors are being gradually phased out in many countries and replaced with non-radioactive detectors, mainly due to the need for collection and disposal of the radioactive source after use and concerns about gamma dose rates to the public (^{241}Am also emits 59 keV gamma radiation).

11.5.5 Well Logging

The use of radiation sources is well established in the oil and gas industry: When drilling a new field, radiation sources can help in determining if a well has encountered a certain rock, mineral, oil, gas, water or other substance of interest and to detect the extent of the field. For this purpose, gamma-emitting or neutron-emitting radioactive sources (am-Be or Pu-Be) or radiation generators (usually d-d or d-t accelerators, producing neutron flux) can be used.

Expert interpretation of detected scattered radiation can provide information about the material present in the well. Information obtained during well logging includes *lithology* (U, Th, K), by measuring natural radiation in the well; *density*, by using ^{137}Cs sealed sources (usually of 5–75 GBq activity) and *porosity*, using neutron sources.

11.5.6 Industrial Sterilisation

Industrial irradiators are used to sterilise consumer products (various medical supplies such as bandages, hypodermic syringes and surgical instruments; bottle corks, food containers, cosmetics and cosmetics containers, tissues, toothpicks, etc.), to kill germs and bacteria without modifying the properties of the material being irradiated (Morehouse and Komolprasert, 2004).

Very high activity ^{60}Co sources (or electron accelerators) are usually used in this process.

Cobalt sources, usually in the form of rods, are loaded into racks inside a heavily shielded bunker, and are kept in a pit – often under water – for shielding. Once the material to be irradiated is positioned in the bunker (or is moved into the bunker on a conveyor), the rack is extracted from the pool and allowed to irradiate the product. An irradiation can last tens of minutes, during which integrated doses of hundreds of kGy can be reached, exterminating germs. Special dosimeters are used to ensure that the required sterilisation dose has actually been achieved.

11.5.7 Food Irradiators

Some industrial irradiators are used in a similar way to irradiate food and reduce microbial load, control foodborne pathogens, inhibit the germination of root crops and extend the shelf life of perishable produce.

Usually, food irradiation takes place with gamma rays (^{60}Co or ^{137}Cs), X-rays (less than 5 MeV) and electrons (less than 10 MeV).

This practice has been the object of close scrutiny and is performed according to the directives of the International Atomic Energy Agency (IAEA), the World Health Organisation (WHO), and the Food and Agricultural Organisation (FAO) of the United Nations (IAEA, 2015).

Most foods can be irradiated: spices, herbs, strawberries, raspberries, bananas, mangoes, onions, garlic, etc., extending shelf life from days to weeks. The quantity of food that is irradiated is growing each year, mainly in the Asia and Pacific region and in the Americas.

11.5.8 Industrial Irradiators to Change a Material's Properties

Industrial irradiators are used to induce specific material effects or improvements, including glass darkening, jewel coloring, polymer cross-linking, paint drying, natural rubber vulcanisation, etc.

Many applications are in the field of plastics: Non-stick cookware is treated with the use of gamma radiation to avoid food sticking to the metal surface; commercial polymers are irradiated to turn them into high-performance plastics, improving the mechanical strength or resistance to heat of cables, tubes, wheels, etc.

In this technique, both gamma sources and electron accelerators can be used, although accelerators, with electrons energies of 1–10 MeV, are the most popular radiation source.

11.5.9 Industrial Tracers

Radioisotopes are used to monitor fluid flow in industrial systems, to detect corrosion and leakage and to break in engines.

Small radioactive sources (tracers – usually short-lived, gamma-emitting isotopes) are mixed in the process, and samples containing the tracer are collected at specific phases of the industrial process and measured. The radioactivity detected and its relation to the original tracer can provide insights on the practice.

11.5.10 Security Applications

In the last decades, significant effort has been devoted to improving security in many aspects of people's lives.

X-ray screening is widely available at airports, public buildings and large gatherings of people to check luggage the for presence of explosives or other weapons.

Several similar and more advanced techniques have been developed, including neutron detection and quick-scanning techniques for larger objects, like cargo containers.

Passive radiation systems for the detection of existing radioactive materials in transported goods have been available for many years; more recently, more advanced techniques using industrial irradiators and generators (interlaced, dual-energy accelerators) have appeared.

TABLE 11.8   Radionuclides and Their Application in the Industrial Sector

Radionuclide and Half-Life	Applications
^{241}Am (half-life: 432 y)	Used in backscatter gauges, smoke detectors and fill height detectors and in measuring ash content of coal
^{137}Caesium (30.17 y)	Used for radiotracer technique for identification of sources of soil erosion and deposition, as well as in density and fill height level switches. Also for low-intensity gamma sterilisation
^{51}Cr (27.7 y)	Used to label sand to study coastal erosion, also a tracer in study of blood
^{60}Co (5.27 y), ^{140}La (1.68 d), ^{46}Sc (83.8 d), ^{110m}Ag (250 d), ^{198}Au (2.7 d)	Used together in blast furnaces to determine resident times and to quantify yields to measure furnace performance
^{60}Co (5.27 y)	Widely used for gamma sterilisation, industrial radiography, density and fill height switches
^{198}Au (2.7 d) and ^{99m}Tc (6 h)	Used to study sewage and liquid waste movements, as well as tracing factory waste causing ocean pollution, and to trace sand movement in river beds and ocean floors
^{198}Au (2.7 d)	Used to label sand to study coastal erosion
^{3}H (in tritiated water) (12.3 y)	Used as a tracer to study sewage and liquid wastes
^{192}Ir (73.8 d)	Used in gamma radiography to locate flaws in metal components
^{85}Kr (10.756 y)	Used for industrial gauging
^{54}Mn (312.5 d)	Used to predict the behaviour of heavy metal components in effluents from mining waste water
^{63}Ni (100 y)	Used in light sensors in cameras and plasma displays, also in electronic discharge prevention and electron capture detectors for thickness gauges. Also for long-life beta-voltaic batteries. Made from nickel-62 by neutron capture

Source: Adapted from WNA, 2017. Radioisotopes in Industry|Industrial Uses of Radioisotopes–World Nuclear Association [WWW Document]. URL http://www.world-nuclear.org/information-library/non-power-nuclear-applications/radioisotopes-research/radioisotopes-in-industry.aspx (accessed 6.28.17).

These scanning X-ray systems, equipped with appropriate detection, reconstruction and imaging software, can scan an entire cargo container in less than a minute and are able to identify atomic species present in the materials, hence facilitating smuggling prevention.

Table 11.8 (adapted from WNA, 2017, Radioisotopes in Industry) summarises the various uses of radioisotopes in the industry.

11.6 Radiation Sources in Education and Research

Many applications of radioisotopes and radiation generators exist in the research, education and laboratory sectors. Universities, colleges, high schools and other academic and scientific institutions use radioactive materials and generators during courses, laboratory demonstrations, experimental research, etc. Radioisotopes are used in archeology, environmental sciences, agriculture, pest control, etc.

A selected number of applications will be briefly presented here.

11.6.1 Radiolabelling

Biological, ecological and environmental studies are conducted with the use of radiotracers, tracking dispersion of contaminants, deposition in soil, uptake in plants and other organisms or oceanic currents.

Some typical radionuclides used are ^{3}H, ^{14}C, ^{32}P, ^{33}P, ^{35}S and ^{125}I.

11.6.2 Carbon Dating

Dating archeological organisms with the use of ^{14}C is a technique that has been in use for many decades and has represented an enormous leap in the quality of archeological dating. This technique is based on the equilibrium between carbon in living organisms' bodies and carbon in the environment. ^{14}C, a naturally occurring radioactive isotope of carbon, is continuously generated by cosmic rays in the atmosphere, and is kept in equilibrium, in its isotopic proportion, in living organisms. Once the organism dies, the carbon exchange with the environment stops, and ^{14}C keeps decaying without being any more exchanged. By measuring the residual ^{14}C in the sample, it is possible to relate it to the concentration of non-radioactive carbon and determine the time at which the organism died.

A major limitation of this technique is represented by ^{14}C half-life, which is 5730 y, hence making it impossible to date samples that are 100,000 years old or older. Other, more sophisticated techniques, based on the ratio ^{235}U/^{238}U, must be used in those cases.

11.6.3 Agriculture

Radiation techniques are widely used in agriculture: Plant seeds, for example, have been exposed to ionising radiation for decades in order to develop more resistant and productive types of plants, and more than 1800 new crops have been generated (Fried, n.d).

Radiation uptake studies can help to determine how much of a pesticide is dispersed in the environment and how much is used by the plant: In these studies, fertilisers radiolabelled with ^{15}O or ^{32}P are used.

Other isotopes that can be used in agriculture studies are isotopes of C, N, Mg, Si, Cl, Fe, Zn, Se, Br, Hg and Pb (IAEA, 2001).

11.6.4 Pest Control

Besides making plants stronger, radiation can be used to control insect populations, thereby decreasing the use of dangerous pesticides.

As early as the 1930s, it was understood that it was possible to rear, sterilise and release enough sterile males of an insect species to overwhelm the natural population and thus control or eradicate an infestation (Fried, n.d).

This method, called the 'sterile insect technique' (IAEA, n.d. Sterile Insect Technique), is used worldwide against insect species of agricultural importance, such as the Mexican fruit fly and the Mediterranean fruit fly, and has been tested for the control of tsetse flies, mosquitoes, the melon fly, the oriental fruit fly, the onion fly and other important agricultural pests.

11.6.5 Gas Chromatography

Laboratory analyses using gas chromatographs, devices containing small, radioactive, pure beta-emitting sources (usually ^{3}H or ^{63}Ni), can help in determining the chemical components of any kind of sample, including air samples, cigarette smoke, petroleum, etc.

11.6.6 Education

Several generators and radioactive sources may be used in colleges and universities (NCRP, 2007).

Some examples are *diffractometers* (analytical X-ray machines that irradiate a sample of material with X-rays to study the scattered pattern produced by its crystalline lattice), *electron-beam microscopes, X-ray fluorescence systems* (X-rays used to excite orbital electrons in a sample and then measure the spectrum of emitted characteristic X-rays) and *particle accelerators* (Van de Graff, linear accelerators, cyclotrons, etc.).

Some common radioactive sources include sealed sources of ^{54}Mn, ^{133}Ba, ^{109}Cd, ^{57}Co, ^{60}Co, ^{90}Sr, ^{137}Cs, ^{152}Eu, ^{22}Na, ^{210}Po, ^{204}Tl, ^{65}Zn, etc.

11.6.7 Radioisotope Power Sources

Some radioisotopes are used as batteries, where special applications require the steady emission of a lot of energy for a long period of time. This is the case with pacemakers, which are frequently powered by ^{238}Pu sources, navigation beacons and satellites.

Radioisotopes batteries (radioisotope thermoelectric generators, which use thermocouples to exploit the difference in heat between a ^{238}Pu source and the cold environment of space) have been successfully used in space vehicles, in the investigation of the planet Saturn made by the Cassini space probe (which had 82 radioisotope heater units and 3 radioisotope thermoelectric generators on board), (NASA, 2008, n.d) and the rover *Curiosity* in the Mars Science Laboratory (using the multi-mission radioisotope thermoelectric generator).

References

Balter, S., 2001. *Interventional fluoroscopy: physics, technology, safety.* (ed.) S Balter, pp. xiii+284, Wiley-Liss, New York, NY.

Buongiorno, J., 2010. PWR Description. Retrieved June 30, 2017, from https://ocw.mit.edu/courses-/nuclear-engineering/22-06-engineering-of-nuclear-systems-fall-2010/lectures-and-readings/MIT22_06F10_lec06a.pdf

Bushberg, J., Seibert, J., Leidholdt, E., Boone, J., 2002. *The Essential Physics of Medical Imaging*, 3rd ed. Medical Physics. Philadelphia, PA, USA. doi:ISBN: 978-0-7817-8057-5.

Council Directive 2013/59/Euratom of 5 December 2013 laying down basic safety standards for protection against the dangers arising from exposure to ionising radiation, and repealing Directives 89/618/Euratom, 90/641/Euratom, 96/29/Euratom, 97/43/Euratom and 2003/122/Euratom, http://eur-lex.europa.eu/legal-content/EN/TXT/?qid=1495976809093&uri=CELEX:32013L0 059 (consulted on 28-MAY-2017)

Fried, M.I., n.d. Historical Introduction to the Use of Nuclear Techniques for Food and Agriculture. Retrieved from https://www.iaea.org/sites/default/files/18005480406su_fr.pdf

Hendee, W.R, Riternour, E.R, eds., 2002. *Medical Imaging physics*, 4th ed. Wiley-Liss, Inc., New York.

IAEA, 1998. *Radiological Characterization of Shut Down Nuclear Reactors for Decommissioning Purposes, Vienna.*

IAEA, 1999. *Decommissioning of Nuclear Power Plants and Research Reactors: Safety Guide*, Vienna, 19–24.

IAEA, 2001. Use of isotope and radiation methods in soil and water management and crop nutrition. Vienna. https://doi.org/10.1017/CBO9781107415324.004

IAEA, 2011. Specific Safety Guide No. SSG-11–Radiation Safety in Industrial Radiography 128.

IAEA, 2012. Neutron generators for analytical purposes. International Atomic Energy Agency.

IAEA, 2014: Radiation Protection and Safety of Radiation Sources: International Basic Safety Standards, General Safety Requirements Part 3 INTERNATIONAL ATOMIC ENERGY AGENCY VIENNA ISBN 978–92 –0–135310–8 ISSN 1020–525X No. GSR Part 3, http://www-pub.iaea.org/MTCD/publications/PDF/Pub1578_web-57265295.pdf (consulted on 28-MAY-2017)

IAEA, 2015. *Manual of good practice in food irradiation*. Vienna. Retrieved from http://www-pub.iaea.org/MTCD/Publications/PDF/trs481web-98290059.pdf.

IAEA, 2017. Japan to Support Use of NDT Technology for Recovery from Earthquakes, Floods in Asia and the Pacific [WWW Document]. Jun–17. URL https://www.iaea.org/newscenter/news/japan-to-support-use-of-ndt-technology-for-recovery-from-earthquakes-floods-in-asia-and-the-pacific (accessed 6.28.17).

IAEA, 2017. REFERENCE DATA SERIES No. 2 2017 Edition–NUCLEAR POWER REACTORS IN THE WORLD.

IAEA, n.d. Nucleonic gauging [WWW Document]. URL https://www.iaea.org/topics/nucleonic-gauging (accessed 6.28.17).

IAEA, n.d. Sterile Insect Technique, Insect Pest Control–NAFA. Retrieved June 30, 2017, from http://www-naweb.iaea.org/nafa/ipc/

IAEA, website on the Radiation Protection of Patients (https://rpop.iaea.org/RPOP/RPoP/Content/AdditionalResources/Training/1_TrainingMaterial/index.htm, accessed on 18-JUN-2017), featuring plenty of training and reference material in the medical applications of radiation sources

ICRP, 2000. Avoidance of Radiation Injuries from Medical Interventional Procedures. ICRP Publication 85. *Ann. ICRP* 30(2)

ICRP, 2010. Radiological Protection in Fluoroscopically Guided Procedures Performed Outside the Imaging Department. ICRP Publication 117. *Ann. ICRP* 40(6)

ICRP, 2013a. Radiological protection in cardiology. ICRP Publication 120. *Ann. ICRP* 42(1)

ICRP, 2013b. Radiological protection in paediatric diagnostic and interventional radiology. ICRP Publication 121. *Ann. ICRP* 42(2)

ISO 2919:2012 – Radiological protection – Sealed radioactive sources – General requirements and classification, 3rd edition

ISOE, 2014. *Occupational Exposures at Nuclear Power Plants Twenty-fourth Annual Report of the ISOE Programme, 2014*, 24th ed.

ISOE, n.d. ISOE Network–Home [WWW Document]. URL http://www.isoe-network.net/ (accessed 6.28.17).

Jorgensen, Timothy J., 2016. *Strange Glow: The Story of Radiation*. Princeton University Press. Princeton, New Jersey.

Khan, F.M., Gibbons, J.P., Pine, J.W., 2014. *The physics of radiation therapy*, 5th ed. Philadelphia, Wolters Kluwert/Lippincott Williams and Wilkins.

Morehouse, K.M., Komolprasert, V., 2004. Irradiation of Food and Packaging: An Overview. (accessed 6.28.17) https://www.fda.gov/food/ingredientspackaginglabeling/irradiatedfoodpackaging/ucm081050.htm.

NASA, 2008. Space Radioisotope Power Systems Multi-Mission Radioisotope Thermoelectric Generator. Retrieved from https://mars.nasa.gov/msl//files/mep/MMRTG_Jan2008.pdf

NASA, n.d. Cassini: The Grand Finale: Radioisotope Thermoelectric Generators (RTGs). Retrieved June 30, 2017, from https://saturn.jpl.nasa.gov/radioisotope-thermoelectric-generator/

National Council on Radiation Protection and Measurements, 2009. *NCRP Report 160: Ionizing radiation exposure of the population of the United States*. National Council on Radiation Protection and Measurements

NCRP, 2007. *NCRP report n. 157: Radiation protection in educational institutions*. NCRP report.

Prince, R., 2012. *Radiation protection at light water reactors*. Springer.

Radiation Dose Management for Fluoroscopically-Guided Interventional Medical Procedures, 2010. NCRP Report No. 168, National Council on Radiation Protection and Measurement. Bethesda, MD

Sprawls, P., 1995. *Physical principles of medical imaging*. Medical Physics Pub.

United Nations Scientific Committee on the Effects of Atomic Radiation, UNSCEAR 2008 Report to the General Assembly, with scientific annexes, Volume I: (Sources), Scientific Annexes A and B

Wagner, L.K., Archer, B.R., 2004. *Minimizing risks from fluoroscopic x-rays; bioeffects, instrumentation and examination, a credentialing program for physicians*, 4th edn. Partners in Radiation Management. Houston.

Williams, J.R., Jerry R., Thwaites, D.I., 2000. *Radiotherapy physics–in practice*. Oxford University Press, Oxford.

WNA, 2017. Radioisotopes in Industry | Industrial Uses of Radioisotopes–World Nuclear Association [WWW Document]. URL http://www.world-nuclear.org/information-library/non-power-nuclear-applications/radioisotopes-research/radioisotopes-in-industry.aspx (accessed 6.28.17).

Environmental Radioactivity and Radioecology

Mauro Magnoni

*Environmental Protection
Agency of Piedmont Region*

12.1 Introduction

The birth of the environmental radioactivity as an independent scientific discipline can be dated to around the 1950s. In fact, it goes back to those years when there was widespread diffusion all over the world of the radioactive dusts injected into the upper atmosphere by the nuclear explosions of a large number of bombs detonated by the USA and the USSR during the early stages of the Cold War from 1945 to 1963.

As a consequence of that, enhanced levels of environmental radioactivity were measured by scientists in many different countries, especially in the northern hemisphere, where there were two major nuclear polygons involved in weapons testing: the Nevada Test Site (USA) and the Semipalatisk Test Site (Kazakhstan, USSR). This situation caused great concern in public opinion, leading to protests in many countries and to the birth of the peace and disarmament movements. The famous Einstein-Russell Manifesto, published on 9 July 1955 and signed by 11 top scientists, was an important step towards the public awareness of the threat posed by the tests and the use of such nuclear weapons; this document pointed out not only the terrific destructive capability of the hydrogen bomb (H bomb), but also the risk to mankind of a global radioactive contamination of the Earth (https://pugwash.org, 1955).

The Limited Test Ban Treaty, signed in Moscow in 1963 by the USSR and the USA, was the first, albeit partial, response of the politicians to public concerns: The two most important nuclear powers, the USA and the USSR, decided to stop testing nuclear weapons in the atmosphere, sea and space. This treaty was very important, and not only politically; it was, in fact, the first important step in stopping the nuclear arms race, but also in protection of the environment. The prohibition on performing nuclear detonation in the atmosphere was quite effective and limited in a substantial way the release of new artificial radioactivity into the environment; however, initially, radioactivity levels in the environment were affected only slightly by this decision. The long and, in some cases, very long half-life of many radionuclides allows radioactive contamination to persist for many decades. Nowadays it is still possible to detect in the environment some artificial radioisotopes coming from the nuclear tests (see Section 12.5.2).

The widespread radioactive contamination of the world pushed the researchers to investigate the presence of artificial radioactivity in the terrestrial and aquatic environments, as well as in foodstuffs. Over the following years, a great number of studies were conducted in many countries in order to understand the fate of the different radionuclides introduced into the biosphere. The atomic explosions produced a huge quantity of different radioisotopes that behave in many and very complex ways, according to their chemical characteristics. These studies gave rise to a new discipline, *radioecology*: a new science that combines the basic concepts of physics, chemistry and biology with ecology and radiation protection.

Another important date for the study of environmental radioactivity is 1986: On 26 April of that year a catastrophic nuclear accident involved the Chernobyl nuclear power plant in Ukraine, in the former Soviet Union. One of the four 1000 MWe reactor units of the plant went completely out of control and was totally destroyed by a huge steam explosion due to an uncontrolled increase of the temperature in the reactor, following a series of errors by reactor operators who had deliberately turned off the safety systems. The big fire in the graphite moderator generated by the explosion brought radioactive dust up to a height of about 2 km, thus allowing the dispersion of the radioactive cloud over long distances: as a result, all of Europe was interested in an important radioactive fallout whose intensity varied greatly from one place to another, being influenced essentially by precipitation (OECD NEA Report, 2002).

The Chernobyl accident greatly boosted the studies of environmental radioactivity in all European countries. In some areas the deposition of fission products reached very high levels, one order of magnitude, or even more, greater than those of the fallout from bombs.

The objectives of these studies were initially focused on a simple radioprotection perspective: the main problem was to evaluate the dose released to human beings due to the radioactive contamination of the environment and foodstuffs. Soon after, however, it was clear that the distribution of different radionuclides into the biosphere allowed the investigation of many complex environmental dynamics. The study of the fate of the artificial radionuclides can thus be used as a powerful tool for the understanding of the complex mechanisms of various ecosystems. In these kinds of studies the radioactive elements are not primarily considered a source of radiological risk, but tracers of the complex dynamics and interactions between the different environmental compartments.

In the following sections, after a brief introduction to the basic concepts of the radioactive phenomenon, we will present the main features of environmental radioactivity, discussing in some detail the characteristics of radioactive pollution and presenting the theoretical and experimental methods used to investigate and evaluate the behaviour of the radioactive elements in the environment.

12.2 Radioactive Decay

The discovery of radioactivity was due to a French physicist named Henri Becquerel. In 1896, he noticed that an unknown radiation, emitted by some uranium salts (potassium uranylsulfate) was able to induce phosphorescence on a photographic emulsion taken in a dark place. The penetrating radiation responsible for this strange phenomenon, different from X-rays, was soon recognised as a new kind of radiation due to uranium, the heaviest natural element in the Mendeleev Table.

The word 'radioactivity' was used for the first time by Marie Curie two years later, when she was studying the properties of another radioactive element, radium. In 1903, Rutherford and Soddy reached a comprehensive understanding of the new phenomenon, still valid nowadays: that radioactivity is the property of certain elements to spontaneously transform themselves into other elements by means of the emission of a particles radiation whose origin is subatomic.

Rutherford also established experimentally the so-called radioactive decay law, i.e. the mathematical relationship that describes the decrease in time of the quantity of every radioactive substance:

$$N(t) = N_0 \cdot e^{-\lambda \cdot t} \tag{12.1}$$

where N_0 is the number of atoms in the radioactive substance at a given time ($t = 0$), $N(t)$ is the number of atoms at the generic time $t > 0$ and λ is the decay constant, the key parameter of the radioactive decay law, whose numerical value is typical for each radioactive element.

The physical meaning of the decay constant λ can be explained considering the probabilistic nature of the radioactive phenomenon: It is in fact the probability per unit of time that a given atom belonging to a radioactive substance decays.

However, instead of the decay constant, it is convenient to indicate, for practical use, the time needed for an initial arbitrary number of radioactive atoms to transform into half that of the original value, which is conventionally known as the 'half-life' of the transformation. The half-life has a unit of time, and is usually expressed in seconds, hours, days, years or their multiples, as half-lives can span many decades, from billions of years to femto-seconds.

The half-life gives an idea of how quickly a radioactive material is undergoing transformations and will disappear. It is a more intuitive and simple concept, although strictly related to the decay constant, as we can see in the following. Starting from the mathematical expression of the decay law, we can write, by the definition of half-life:

$$\frac{N_0}{2} = N_0 \cdot e^{-\lambda \cdot t_{1/2}}$$

Making, then, the natural logarithm of the two members of the above equation, we get:

$$\ln\left(\frac{N_0}{2}\right) = \ln(N_0) - \lambda \cdot t_{1/2}$$

Finally, simplifying and rearranging the expression, we obtain the fundamental relationship:

$$t_{1/2} = \frac{\ln 2}{\lambda} \tag{12.2}$$

that relates the half-life of a radioactive element to the corresponding decay constant value.

It must be pointed out that any experimental measurement of radioactivity is performed by means of the detection of the ionising radiation emitted by the radioactive atoms: The measurable quantity is therefore the disintegration rate of the radioactive substance rather than the real number N of radioactive atoms. This quantity is called *activity* and, by definition, is proportional to the number of radioactive atoms, the proportionality constant being its decay constant:

$$A = \lambda \cdot N \tag{12.3}$$

The radioactivity decay law can also be written in terms of activity:

$$A(t) = A_0 \cdot e^{-\lambda \cdot t} \tag{12.4}$$

where A_0 is the activity at time zero ($t = 0$), $A(t)$ is the activity at time $t > 0$ and λ is the decay constant. In Figure 12.1 a typical decay curve of a generic radioisotope is shown.

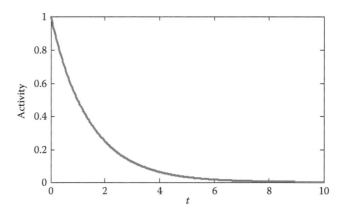

FIGURE 12.1   Exponential decay of a generic radioactive element: The time in the abscissa is reported in half-life units.

The SI unit of measure of the activity corresponds to 1 disintegration per second and is called Becquerel (symbol: Bq), after the name of the discoverer of radioactivity. However, it is still used as the historical unit of measure of the activity, the Curie (symbol; Ci), after Marie Sklodovska Curie. This unit of measure is much larger than the Becquerel. It corresponds to the activity of 1 gram of radium (radioisotope ^{226}Ra). The following relationship holds between the two units: $1 \text{ Ci} = 3.7 \cdot 10^{10} \text{ Bq}$.

12.3 Alpha, Beta and Gamma Emissions

Radioactive decay takes place when the nuclear structure of an isotope is unstable. The reason for this instability is strictly related to the composition of the nucleus, i.e. the number of protons and neutrons. The stability of a nucleus is the result of the complex balance between the attractive short-range nuclear forces, acting indifferently among the nucleons (protons and neutrons), and the Coulombian repulsion forces due to the electrical charge of the protons. A detailed treatment of this issue is beyond the scope of this book and can be found in any introductory nuclear physics textbook (see for example, Johns and Cunningham, 1983). Let's discuss briefly only the most important outcomes of this instability. When a nucleus is unstable, the radioisotope spontaneously emits a corpuscular radiation in order to gain stability. There are two kinds of corpuscular radiation emitted by the radioactive elements: α radiation and β radiation (Lieser, 1997).

α radiation: α radiation is composed of very massive particles, ionised helium atoms (^{4}He^{++}) emitted by unstable nuclei. It is typical of some heavy elements: the larger size of the atoms limits the effectiveness of the attractive short-range nuclear forces. Thus, the relative importance of the Coulombian repulsion forces increases and the α particles are therefore able to escape from the nucleus with a given energy, on the order of 5 MeV. Each α emitter has its own typical α energy: this interesting property allows the identification of an unknown α radioisotope measuring the energy of its emission by means of instruments called α spectrometers. For the typical values of the α energies and some more details on α radiation, see Section 10.2.1.

The nuclear reaction for the α decay of a generic element X can be written as follows:

$$^A_Z X \rightarrow {}^{A-4}_{Z-2} Y + \alpha$$

in which the generic radioactive nucleus parent X, characterised by the atomic weight A (sum of the number of protons and neutrons) and the atomic number Z, is transformed into the nucleus Y, characterised by the atomic weight A-4 and atomic number Z-2.

The range of α radiation is quite short, a few centimetres, in air: In fact, it is easily stopped by a sheet of paper and, for that reason, external α irradiation poses no significant health threat. The typical ranges of α particles in biological tissues do not exceed 70 μm, so they stop in the dead skin layer without causing any harm. By contrast, if radioactive elements enter the body by ingestion, inhalation, wounds or transcutaneous absorption, α radiation is much more dangerous that the other types of radiation (for instance β and γ); in this case the α particles are in fact able to directly irradiate the living tissues and, being high LET radiation (linear energy transfer, see Section 10.5), i.e. a type of radiation that releases its energy in a short range, they induce greater biological damage to the living matter.

β radiation: β rays are a corpuscular radiation composed of high energy electrons (e^-) emitted by the nucleus of a radioactive element. The physical process of β decay is very different from that of α decay: in fact, the electron emitted comes from the transformation into a proton of one of the neutrons belonging to the nucleus. Therefore, the radioactive nucleus parent X, characterised by a mass number A and an atomic number Z, is transformed into the nucleus Y, characterised by the same mass number A and an increased atomic number, Z+1. Another important difference is that β decay is a three-body decay: Besides the electron, another particle is emitted, the antineutrino. The antineutrino ($\bar{\nu}$) and the neutrino (ν) are particles with no electric charge and very small masses (<2.2 eV). Therefore, the reaction for β decay can be written as follows:

$$^A_Z X \rightarrow \, ^A_{Z+1} Y + e^- + \bar{\nu}$$

This fact has a very important consequence on the energy of the electrons emitted during β decay. The total energy E of the decay is split into two parts, between the electron and the antineutrino, according to a probability distribution law: The energy spectrum of β radiation is therefore continuous, ranging from *0* to a maximum value, E_{max}, also called the *end point energy* of β decay. As a result of this, no β emitter isotope can be identified by measuring the energy of its β particles. The detection of a pure β emitter radioisotope* is, in fact, quite a complex task: A careful chemical separation is needed before measuring, and identification is usually made by experimentally checking the half-life of the decay.

The β particles slow down in matter by complex electronic interactions with the atomic electrons, and they are usually stopped by low Z material shields, the thickness of which rarely exceeds 1 cm (see also Section 10.2.2).

The β decay described above is more precisely indicated as β^- decay (beta minus decay), in order to distinguish it from β^+ (beta plus decay), a very similar process in which the corpuscular particle emitted is the positron, a fundamental particle positively charged and with the same mass of the electron.

For this type of decay we will have:

$$^A_Z X \rightarrow \, ^A_{Z-1} Y + e^+ + \nu$$

In this case the radioactive nucleus parent X is transformed into the nucleus Y, characterised by the same mass number A, while the atomic number Z is decreased by one to Z-1. The other particle emitted in this case is the neutrino (ν): in this process, one of the protons belonging to the nucleus is transformed into a neutron.

There is also a third type of β decay: so-called electron capture (EC) decay. This decay occurs when the wave function of one of the atomic electrons spinning around the nucleus has a value significantly different from zero also in the neighbourhood of the nucleus: The atomic electron is then so close to the nucleus that it can be captured by the nucleus itself. It is absorbed by one of the nuclear protons that is transformed into

---

* A pure β (or α) emitter is a radioisotope that emits only a β (or α) radiation: the absence of γ emissions makes these radioisotopes more diffucult to be detected

a neutron with a simultaneous emission of a neutrino (v):

$$^A_Z X + e^- \rightarrow \,_{Z-1}^{\;\;A} Y + v$$

This transformation is analogous to β^+, decay, giving birth to a daughter with Z-1 protons and A nucleons.

γ radiation: γ radiation belongs to the highest energy part of the electromagnetic spectrum (from a few tens of keV (kilo electronVolt) to several MeV (Mega electronVolt). This type of radiation is generally emitted by the nuclei of radioactive elements that are left in an excited state as a consequence of an alpha or beta decay. The excess energy in the decayed nuclei, due to the rearranging of the nucleons inside the nuclear structure, is released by means of the emission of one or more photons, whose energy is characteristic of the radioactive element. This is a very important property from an experimental point of view, because it makes it possible to identify most of the radioactive elements simply by measuring the energies of their γ emissions. These measurements are performed by means of special detectors, the γ spectrometers.

The energies of γ radiations emitted by radioactive elements depend on the structure of the energy levels of the nucleus, which are arranged in a very similar way with respect to the atomic levels of the atoms; the only (but relevant) difference is the much higher energy values of nuclear transitions – hundreds of keV or even more, while the typical energies of atomic transitions are generally lower by order of magnitude.

γ radiation is a highly penetrating radiation. It can be stopped only using shields made of 'high Z material', lead in most cases. A more detailed description of the interactions of γ radiation with matter can be found in Section 10.3.2.

In most cases, γ emissions follow very quickly (on the order of picoseconds) the alpha or beta emissions of the radioactive nucleus and, for that reason, they can be considered as being emitted simultaneously with the corpuscular radiation. However, sometimes, after the alpha or beta emission, the nucleus can survive for a longer time in its excited state: minutes, hours or even days. The daughter radionuclide is then considered in a particular physical state, different from its ground state, called a metastable state: in these cases the letter *m* is added as a superscript beside the number indicating the mass number of the element. For example, the isotope Tecnetium-99, an artificial radioisotope frequently used in nuclear medicine and produced in a metastable state from ^{99}Mo by a β^- emission, decays into its ground state with a half-life of about 6 hours, emitting a 140 keV gamma radiation: It is therefore indicated as ^{99m}Tc to make it distinguishable from its ground state, Tecnetium-99 (^{99}Tc), a β^- emitter with a long half-life, 211,100 years, that decays into the stable isotope ^{99}Ru.

12.4 Nuclear Fission, Enriched and Depleted Uranium

Nuclear fission was discovered in 1938 by chemists Otto Hahn and Fritz Strassmann, with the fundamental contribution of the physicist Lise Meitner, who first understood the physical nature of the new phenomenon and correctly interpreted the experimental data (Meitner and Frisch, 1939). They observed that uranium, if bombarded with slow neutrons, can split in two fragments plus two or three neutrons. The fission fragments are highly radioactive and they are all β^- emitters. Fission releases an enormous amount of energy, about 200 MeV per fissionated nucleus. This is the physical process on which the atomic bomb and nuclear power reactors are based.

Only a small fraction of natural uranium can undergo fission when bombarded by slow neutrons: The ^{235}U isotope, which represents only 0.72% by of natural uranium, which is in fact composed almost exclusively of the isotope ^{238}U. For that reason, in order to employ uranium for its fissile properties, it is necessary to enrich the percentage of ^{235}U: an enrichment up to 2%–3% of ^{235}U is generally enough for the fabrication of the fuel for nuclear reactors, while the percentage of ^{235}U necessary for building nuclear weapons is about 90%.

During the Manhattan Project, considerable efforts were devoted to the uranium enrichment process, but only two of the first three atomic bombs (the first bomb, detonated at Alamogordo, and the Hiroshima bomb) were built using ^{235}U as the nuclear explosive. The third bomb, dropped on Nagasaki, was a plutonium bomb: The fissionable element used was ^{239}Pu, an artificial radioisotope generated from ^{238}U by neutron capture and subsequent β^- decay.

Nowadays, the enrichment of natural uranium is still a critical and difficult industrial process. Gas centrifuge systems are now used to enrich uranium. This method utilises a much more effective technology than that available in the early stages of the nuclear era (gas diffusion), but the enrichment process still requires much time to produce large quantities of weapons-grade enriched material necessary to build an atomic bomb (90% ^{235}U). In some cases, it is therefore probably easier, from the military point of view, to use the fissionable isotope ^{239}Pu as a nuclear explosive instead of trying to enrich uranium. Plutonium, a transuranic radioisotope, is in fact produced in large quantities during the normal operation of any nuclear power plant, especially with a low burn-up of the fuel. Therefore, it could be easier to produce weapons-grade plutonium (pure ^{239}Pu) in larger quantities than enriched uranium. Plutonium is found in the spent fuel rods and can be separated by chemical processes from the other nuclear radioactive wastes, the highly radioactive fission fragments. These operations can take place in fuel reprocessing plants, the nuclear installations devoted to the treatment and conditioning of the spent fuel coming from the power plants.

An interesting byproduct of enriched uranium is depleted uranium, often indicated as DU; it is uranium with a very low ^{235}U content, typically less than 0.2%, the natural level being 0.72%. The manufacturing of nuclear fuel for nuclear power plants has produced large quantities of depleted uranium, which initially was treated as low radioactive waste to be disposed of. Very soon, however, it became clear that this special material, which has a very high density (about 19 g/cm^3) and a high atomic number Z ($Z = 92$, number of protons), could have very interesting re-uses. Depleted uranium was utilised to make a large variety of peculiar technological items: very effective radiation γ shields for high activity sources, stabilisers for sailboat keels and, especially in the past, counterweights in the wings of some aircraft (the Boeing 747 and others).

The depleted uranium issue became very popular among the general public in 1991, after the first Gulf War. In that conflict, American and British troops for the first time widely used munitions with depleted uranium tips: these munitions were able to easily penetrate the armour of the enemy's tanks. The extensive use of these weapons on the battlefield released large quantities of dust containing high DU concentrations and caused great concern among the veterans exposed to the dust and the people living in the affected areas.

Despite its military utilisation, DU must not be considered a nuclear weapon because the properties for which is used are not nuclear ones: The very high density of uranium and the pyrophoric effects, i.e. the capability of the uranium bullets to fire when they hit the targets, allowing them to easily penetrate the armour of tanks, are characteristics that are independent of the fissile properties of uranium.

Depleted uranium was also massively used by US Air Force planes during the Balkan Wars of the Nineties: the Bosnia-Herzegovina War, the Serbia and Montenegro War and the Kosovo War. Studies on the environmental contamination of DU in Iraq and in the Balkans showed detectable DU contamination only surrounding the hit targets and unexploded DU penetrators (Danesi et al., 2003; UNEP, 2001, 2003); however, it is worth noting that the detection of DU traces in the environment is a very difficult task due to the relatively high natural uranium background. The presence of traces of DU in environmental samples needs the evaluation of the isotopic ratios of the uranium isotopes: ^{235}U/^{238}U and ^{234}U/^{238}U (Magnoni et al., 2001).

Many other studies were also conducted on the health effects of the exposure of human beings to depleted uranium, but none of them were able to demonstrate a clear relationship between DU exposure and various diseases. However, it must be pointed out that the great difficulties of these kind of studies – in particular, the very large uncertainties on the quantification of the real exposure to DU of the people supposed to be contaminated – have up to now prevented reaching a conclusion on this matter (see, for example, SCHER, European Commission, 2009).

Notwithstanding the limited public awareness and the limited media coverage, the most important and critical problem regarding nuclear applications remains the diffusion of classic nuclear weapons and atomic bombs (SIPRI, 2017). In particular, the proliferation of nuclear weapons, i.e. the spread of nuclear weapons capabilities in different countries all over the world, is one of the most serious and important issues in international politics. The Non-Proliferation Treaty (NPT), the political instrument that the United Nations promoted to tackle this problem, was signed in 1968 and aimed to prevent the diffusion of nuclear weapons among countries other than the 'historical nuclear powers' – the US, Russia, China, France and that United Kingdom, all of which developed their atomic weapons before 1967. The International Atomic Energy Agency (IAEA) is the UN body charged with the controls and inspections needed to check for the accomplishment of the bonds of the Treaty by the United Nation Members States who signed it.

12.5 Radioactivity in the Environment

As already mentioned in the Introduction (Section 12.1), the main origin of the widespread diffusion of artificial radionuclides into the environment were the nuclear weapons tests during the Cold War. Also, major nuclear accidents that occurred at some nuclear power plants and installations contributed significantly to the dispersion of artificial radioactivity: Windscale, UK in 1957 (UNSCEAR, 1993); Majak, Russia in 1957 (UNSCEAR, 1993); Chernobyl, Ukraine in 1986 (UNSCEAR, 2000, 2008); Fukushima, Japan in 2011 (UNSCEAR, 2013).

However, in spite of the local relevance of man-made contamination, at a global level artificial radionuclides represent only a small fraction of the total inventory of environmental radioactivity. In fact, the inventory of all the natural radionuclides that can be found in the Earth's crust, in the seas and in the atmosphere is much greater than the artificial radioactivity released into the environment globally. It was estimated that through 1962 (the year before the signature of the limited Test Ban Treaty), nuclear tests produced about 740 PBq (1 PBq $= 10^{15}$ Bq) of ^{90}Sr and a similar amount of ^{137}Cs (NCRP, 1987). The most severe accident that occurred since then (Chernobyl, 1986) released into the environment activities on the order of 10 PBq for ^{90}Sr and about 85 for ^{137}Cs.

By contrast, it can be easily calculated that the inventory of the all the natural radionuclides, considering only a thin layer of the Earth's crust, exceeds these values by an order of magnitude. This fact is confirmed by the estimation of the exposure to ionising radiation of the world population: The natural contribution largely dominates the total effective dose released to human beings (Section 11.1). A more detailed discussion of the characteristics of the two main components of environmental radioactivity, the natural and the artificial ones, will be made in the next two subsections.

12.5.1 Natural Radioactivity

The natural radionuclides present in the environment can be divided into three broad categories:

    a. Radionuclides belonging to the natural series
    b. Primordial radionuclides
    c. Cosmogenic radionuclides

    a. *The natural series radionuclides*: The radionuclides belonging to the natural series (also known as the natural families) are characterised by the fact that their existence depends on a radionuclide, the parent of the series, having the half-life on the order of the age of the Universe, billions of years or more. This physical property allowed the survival of all the elements belonging to the families from heavy nuclei nucleosynthesis times.

       The most important natural series is the so-called 'uranium series', the parent of which is the isotope ^{238}U of uranium, with a long half-life: $t_{1/2} = 4.47 \cdot 10^9$ years. Uranium is abundant in the Earth's crust: It's estimated that in the upper crust the average level is around 2.8 ppm (parts per million), corresponding to about 35 Bq/kg. ^{238}U decays with the emission of an α particle, transforming itself into another radioactive isotope, the β^- emitter ^{234}Th (thorium), which has a quite short half-life, only 24 days. This radioisotope decays into another unstable element, the β^- emitter protoactinium (^{234}Pa), which has a very short half-life – only 1.17 minutes – and in turn decays into another radioactive element, and so on. The radioactive chain continues with many α and β transformations and finally ends in a stable isotope of lead, ^{206}Pb (see Figure 12.2).

The uranium series is also called the 4n+2 series, because the atomic mass A of all its members can be expressed with the formula 4n+2, where n is an integer.

An important characteristic of this family is the presence, approximately in the middle of the chain, of a particular radionuclide, a noble gas called radon (^{222}Rn). This isotope, produced from radium (^{226}Ra, $t_{1/2} = 1600$ years) by α decay, is an α emitter too, with a half-life of 3.82 days. It is very important from the radioprotection point of view: A large percentage of the dose released to human beings is in fact due to radon. The radon issue deserves special attention: We will discuss it in more detail in Chapter 13.

Another very important natural series is the thorium series, Thorium (^{232}Th, $t_{1/2} = 1.40 \cdot 10^{10}$ years), is the parent of this family, also known as the 4n series because the atomic mass of all its members can be expressed by means of this formula. Thorium is quite abundant in the Earth's crust. The average value in the upper crust is estimated to be 10.7 ppm, corresponding to about 42 Bq/kg. Therefore ^{232}Th, together with all its progeny, contributes significantly to natural radioactivity. The series ends with another stable isotope of lead, ^{208}Pb (Figure 12.3).

The third natural series is the 4n+3 series, also known as the actinium series. The parent of this family is ^{235}U, $t_{1/2} = 7.038 \cdot 10^8$ years, another uranium isotope that is much less abundant than ^{238}U: As already mentioned, only 0.72% by weight of natural uranium is composed by this isotope. The actinium series ends with another stable isotope of lead, ^{207}Pb. In Figure 12.4 a simplified version of the actinium series in shown.

In normal situations, the radioprotective relevance of this series is very limited: The very low natural abundance of the parent of the series, ^{235}U, generally prevents the release of significant doses to

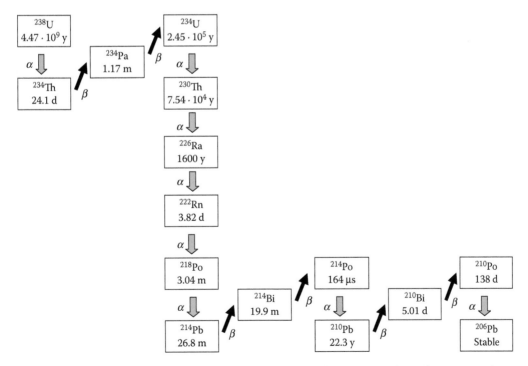

FIGURE 12.2 The uranium series. (Adapted from W.W. Nazaroff and A.V. Nero, *Radon and Its Decay Products in Indoor Air*, A Wiley Interscience Publication, John Wiley & Sons, 1988.)

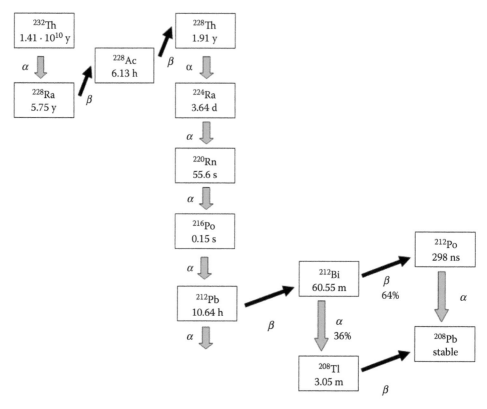

FIGURE 12.3   The thorium series. (Adapted from W.W. Nazaroff and A.V. Nero, *Radon and Its Decay Products in Indoor Air*, A Wiley Interscience Publication, John Wiley & Sons, 1988.)

human beings. Significant exposure due to the isotopes belonging to the actinium series are possible only for workers employed in facilities involved in the manipulation and storage of large quantities of enriched uranium – nuclear fuel factories, uranium enrichment installations, nuclear weapons storage sites, etc.

The radionuclides belonging to natural series release a considerable amount of the total effective dose normally received by the general public, the major part being attributable to the inhalation of radon and its progeny (see Table 11.1, Section 11.1).

The large variety of the chemical species of the natural series radioisotopes and their very different half-lives, spanning several orders of magnitude, from a few microseconds to billions of years, makes the distribution of these radionuclides in the different environmental compartments very complex. For example, the radon isotope ^{222}Rn, being a noble gas, escapes easily from the rocks and enters into the atmosphere, where its short-lived radon daughters, namely ^{218}Po, ^{214}Pb, ^{214}Bi and ^{214}Po, after being produced, are attached to fine and ultra-fine atmospheric particulate. The fate of radon and its daughters is thus separated: radon remains suspended in the atmosphere and follows the motion of the air masses, while the daughters are more easily removed, especially in an indoor environment. Similar mechanisms, driven by the different chemical behaviours of the radioactive components of the natural series, also apply in many other terrestrial and aquatic environments. For those reasons, in real cases, a general disequilibrium appears among the various components of the radioactive series. The chains are often 'broken' in two or more parts and the relationships between the activity concentrations of the radionuclides are very complex and difficult to model. However, there is a special, very interesting case that allows a simple evaluation

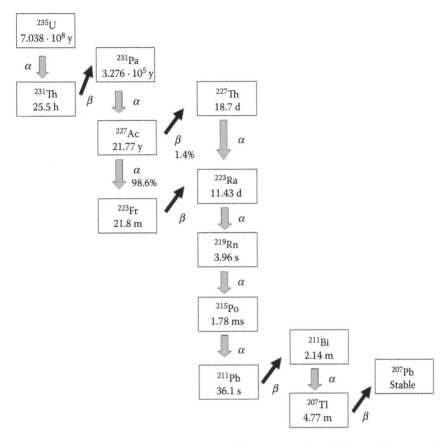

FIGURE 12.4 The actinium series. (Adapted from ICRP Publication 38, *Radionuclides Transformation – Energy and Intensity of Emissions*, 1983.)

of the relationship existing between two or more members of a radioactive chain: the *secular equilibrium condition*.

Secular equilibrium can take place when the half-life of the parent radionuclide is much larger than the half-life of the daughter radionuclide and the system (radionuclide parent + progeny) is completely isolated, i.e. no exchange with the outside environment can occur. If both these conditions are met, the activity concentration of the parent radionuclide eventually, after a brief transitory time, equals the activity concentration of the daughter. This result can be mathematically demonstrated solving the differential equations that describe the system.

Let's thus consider, for the sake of simplicity, a radioactive chain composed of only two generic radioactive substances, the parent A and the daughter B, characterised respectively by their decay constants: λ_A and λ_B.

The differential equations of this system are:

$$\frac{dN_A}{dt} + \lambda_A \cdot N_A = 0$$

$$\frac{dN_B}{dt} + \lambda_B \cdot N_B - \lambda_A \cdot N_A = 0$$

(12.5)

where N_A and N_B are, respectively, the numbers of the atoms of the radioactive species A and B. If, at time zero, the element B is absent and the number of the atoms of the parent A is $N_{A°}$, the

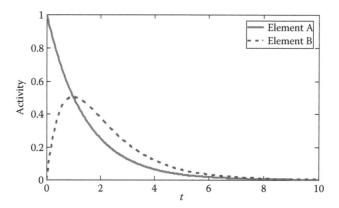

FIGURE 12.5   The time evolution of the activities of the radioactive elements A and B.

boundary conditions of the system may be written as follows: $N_A(0) = N_{A°}$ and $N_B(0) = 0$. The solutions are therefore given by:

$$N_A(t) = N_{A°} \cdot e^{-\lambda_A \cdot t}$$

$$N_B(t) = \frac{\lambda_A N_{A°}}{\lambda_B - \lambda_A} \cdot \left(e^{-\lambda_A \cdot t} - e^{-\lambda_B \cdot t}\right) \tag{12.6}$$

These equations may be rewritten in terms of the activity, A, that is the experimentally measurable quantity (the disintegrations per unit of time) and that is strictly related to the number of radioactive atoms N by the fundamental relationship $A = \lambda \cdot N$. We have therefore:

$$A_A(t) = \lambda_A N_{A°} \cdot e^{-\lambda_A \cdot t}$$

$$A_B(t) = \frac{\lambda_B \lambda_A N_{A°}}{\lambda_B - \lambda_A} \cdot \left(e^{-\lambda_A \cdot t} - e^{-\lambda_B \cdot t}\right) \tag{12.7}$$

In Figure 12.5, a typical graphical representation of the solutions, when the half-lives of the two elements are of the same order of magnitude, is shown.

As we can see, the activities of the species A and B change dramatically over time: The relative contribution of the two radioactive elements is not constant and no simple relationship can be found. However, the picture looks completely different if the half-life of the parent A is much greater than the half-life of the daughter B, $t_{1/2A} >> t_{1/2B}$ and than the typical observation time t, $t_{1/2A} >> t$. The corresponding condition for the decay constants is thus $\lambda_A << \lambda_B$ and the solutions (12.7) may be drastically simplified, because the following expressions can be approximated thereby: $e^{-\lambda_A} \cdot t \approx 1$ and $\lambda_B/(\lambda_B - \lambda_A) \approx 1$. We have thus:

$$A_A = \lambda_A N_{A°}$$

$$A_B(t) = \lambda_A N_{A°} \cdot \left(1 - e^{-\lambda_B \cdot t}\right) \tag{12.8}$$

The second expression (12.8) gives asymptotically (i.e. when $t \to \infty$, practically 5–6 times the half-life $t_{1/2B}$): $A_B(t) = \lambda_{A°} \cdot N_A$; we have, therefore, a very interesting result: $A_A = A_B$. The activity of element B equals the activity of element A! In Figure 12.6 the growth of radioactive species B approaching secular equilibrium with element A is shown.

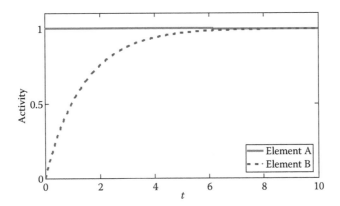

FIGURE 12.6 The element B, produced by the decay of the element A, reaches secular equilibrium after a time corresponding to 5–6 its half-life.

Equations and solutions similar to (12.5) and (12.7) also can be written for chains composed of a large number of radionuclides: $A, B, \ldots, Z$. The mathematical expressions are, of course, much more complicated, but the result is the same: If the system is isolated and the half-life of the parent of the chain is much larger than the half-lives of all its components, the secular equilibrium condition holds for all the members of the family, and we will have: $A_A = A_B = \cdots = A_Z$.

The secular equilibrium condition is very important in those environmental radioactivity studies involving chains with many radionuclides. In fact, it can drastically simplify the treatment of very complex problems. A typical use of the secular equilibrium condition is the quantitative evaluation of radionuclides that are difficult to measure because of their particular physical and chemical characteristics. For example, the isotope of radium ^{226}Ra is quite difficult to detect directly; in fact, it emits α particles, very easily absorbed by matter, and it has a relatively weak γ emission, which is not easily and properly resolved by most spectrometric systems. Therefore, if secular equilibrium can be assumed for the following system: ^{226}Ra, ^{222}Rn, ^{218}Po, ^{214}Pb, ^{214}Bi, the strong γ emissions of ^{214}Pb and ^{214}Bi can be used for a much easier evaluation of the activity of radium.

Figure 12.7, for example, shows the growth of ^{222}Rn in a closed system containing at time t = 0 only ^{226}Ra: It can be seen that secular equilibrium is reached after about 20 days, which is approximately 5–6 times the half-life of ^{222}Rn.

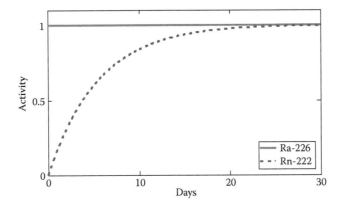

FIGURE 12.7 Approach to secular equilibrium for the system ^{226}Ra / ^{222}Rn: The activity of ^{222}Rn equals the radium activity after approximately 20 days.

In very special cases, the secular equilibrium condition can be applied to all the members of the natural families; however, much more common are the situations, as in the example discussed above, where secular equilibrium holds only for part of the chain.

b. *Primordial radionuclides*: Besides ^{238}U, ^{232}Th and ^{235}U, there are other primordial radionuclides with very long half-lives (billions of year): all these radioisotopes were generated during stellar and supernovae nucleosynthesis. They didn't give birth to a chain because they decayed directly into stable elements. The most important of these radionuclides is potassium-40, ^{40}K: it has a half-life of $1.25 \cdot 10^9$ years and makes up approximately 0.012% of the potassium found in nature, the other stable isotopes being ^{39}K (93.26%) and ^{41}K (6.73%). It decays β^- to ^{40}Ca about 89.3% of the time and by electron capture to ^{40}Ar about 10.7% of the time: the occurrence of traces of argon in the Earth's atmosphere is due to the decay of ^{40}K. It can be easily detected by γ spectrometry because of a strong γ radiation (1460.75 keV), emitted during the EC decay. ^{40}K is widely present in the environment and its activity concentration can vary greatly on the Earth's crust: The average value in the upper crust is estimated to be 630 Bq/kg, but a high variability is observed depending on the type of rock. Activity concentrations up to a few thousand Bq per kilogram are not uncommon in some types of rocks or soils. Another very important characteristic of ^{40}K is related to its peculiar behaviour in living organisms, particularly in humans. The potassium concentration is in fact under homeostatic control in the human body: While the potassium content varies significantly between the different organs, ^{40}K concentrations are maintained nearly constant within the organs by the cellular metabolism in order to allow the correct behaviour of cellular processes. For that reason, no huge accumulation or total depletion of ^{40}K is possible, and the activity concentration of ^{40}K in the human body remains approximately constant at about 60 Bq/kg on average, so the total amount of ^{40}K for a standard man (70 kg) is on the order of 4000–4500 Bq.

Due to its ubiquitous presence in the environment and in the human body, ^{40}K delivers to humans a significant fraction of the total effective dose: A value of 0.3 mSv/y can be estimated, that is, about 10% of the total average value for the world population.

The other relevant primordial radionuclide is rubidium-87, ^{87}Rb, which has a very long half-life: $t_{1/2} = 4.80 \cdot 10^{10}$ years. ^{87}Rb is a pure β emitter (no γ radiation is emitted during the decay process) and, for that reason, it is not easily detectable. It is distributed worldwide and its average concentration in the Earth's crust is evaluated as 70 Bq/kg. Its content in the human body is estimated to be about 8 Bq/kg. However, due to its lower concentration, its radioprotective relevance is very limited compared to that of ^{40}K.

The other primordial radionuclides detectable in the Earth's crust all have very long half-lives, including Samarium-147 (^{147}Sm, $t_{1/2} = 1.05 \cdot 10^{11}$ years), Lutecium-176 (^{176}Lu, $t_{1/2} = 2.2 \cdot 10^{10}$ years), Lanthanium-138 (^{138}La, $t_{1/2} = 1.12 \cdot 10^{11}$ years) and many others. However, none of them are interesting from the radioprotective point of view due to their low or very low concentrations in the Earth's crust: 0.7 Bq/kg for ^{147}Sm, 0.04 Bq/kg for ^{176}Lu and 0.02 for ^{138}La.

c. *Cosmogenic radionuclides*: The cosmogenic radionuclides are a group of natural radionuclides generated by the interaction of cosmic rays with the upper part of the Earth's atmosphere. The most common interactions involve the cosmic nucleons and result in spallation or neutron capture reactions (Lal and Peters, 1967). A spallation reaction occurs when a heavy component of the cosmic radiation hits an atmospheric atom, knocking nucleons off the target nucleus, usually O or N. As a result, a nucleus with the same or lower mass is produced. Neutron capture, on the other hand, consisting in the absorption of a neutron by the target atom, produces a radionuclide that is one mass heavier than the target.

There are a great number of cosmogenic radionuclides. However, only four are produced in considerable quantities and can be considered relevant for environmental radioactivity and radioprotection: tritium (^{3}H), radiocarbon (^{14}C), sodium-22 (^{22}Na) and beryllium-7 (^{7}Be).

Tritium is the radioactive isotope of hydrogen; its nucleus is composed of one proton and two neutrons. Its half-life is 12.23 years. It's a pure, low-energy β^- emitter: The end-point energy of the

β^- particles emitted by tritium is only about 18 keV, the lowest energy value among all the β emitters. Because of this, tritium is quite hard to detect in most samples. The measurement methods normally employed require a careful pre-treatment that usually involves the distillation of the sample before radioactive counting. Tritium is extracted from the environmental matrices and converted into tritiated water, which is more easily measurable.

In the environment, two types of tritium can be distinguished, depending on the different molecular bonds of the radioisotope:

- HTO (tritiated water)
- OBT (organically bound Tritium)

These two different chemical forms have their own specific fates in the environment and in the biosphere. OBT moves according to the chemical and biological properties of the organic molecules to which is bonded, while tritiated water usually follows the normal water cycle. For that reason, tritium is often used as a tracer for aquifer dating and for other groundwater and geological studies. Typical levels of tritium in surface water are usually of the order of a few Bq/l. Greater concentration can be found locally near the liquid waste discharge points of operating nuclear power plants. In fact, in spite of its natural origin, a large quantity of tritium can also be produced artificially, in particular in nuclear power plants: The intense neutron flux generated by nuclear fission leads to the activation of the water circulating near the core of the reactor, and the neutrons are captured by the hydrogen atoms that are transformed into tritium. For that reason, the liquid and gaseous wastes released during routine periodical discharges from a nuclear power plant may contain enhanced levels of tritium with respect to the natural background. However, the relatively limited amount of such discharges makes the dose contribution delivered to humans generally negligible (Section 11.3).

A more important artificial tritium source were the nuclear weapons tests in the atmosphere. Tritium was produced and released into the atmosphere in huge quantities during the early stages of the Cold War (1945–1963). It was estimated that before the signature of the Test Ban Treaty in 1963 the tritium inventory in the northern hemisphere due to the bomb tests reached the considerable level of 70 EBq (1 EBq = 10^{18} Bq), completely overshadowing the natural production rate, which would lead to a much lower equilibrium inventory, about 1.3 EBq in each hemisphere. Fortunately, the threat to human health due to this huge amount of radioactivity released into the environment was limited by the very low energy of the β particles emitted by tritium, which makes it one of the least radiotoxic radioisotopes. But the very large quantity of tritium injected in the atmosphere did provide a small but significant contribution to the total delivered dose.

It was estimated that in 1970 the dose contribution to human beings due to tritium was on average about 20 µSv per year in the northern hemisphere (NCRP, 1987). This value was calculated assuming that the water present in the environment was in equilibrium with the body water. Nowadays (2017), the tritium dose is much lower, certainly well below 10 µSv; radioactive decay has considerably reduced the level of tritium caused by the bomb tests, but tritium is still present in the environment in a larger quantity (about 3,3 EBq) than would have existed solely from natural production.

Another very important cosmogenic radionuclide is radiocarbon, ^{14}C. It, too, is a pure β emitter with a relatively weak β emission (150 keV). The radionuclide is produced by spallation reaction. ^{14}C decays into the stable isotope ^{14}N. The half-life of ^{14}C is quite long: $t_{1/2} = 5730$ years. As with tritium, its inventory was affected by a considerable artificial production rate during the nuclear weapons tests in the atmosphere; it was estimated that the atmospheric levels of ^{14}C were increased by a factor of two in the northern hemisphere. However, the very effective exchange of carbon between the atmosphere and the ocean led to a rapid decline of atmospheric ^{14}C concentration, which has now returned to the natural level.

^{14}C is also produced in nuclear reactors via neutron capture by the nitrogen, carbon and oxygen present in the fuel and the other components of the reactor. It is usually released into the atmosphere in small quantities as gaseous waste. The discharges are estimated to be two orders of magnitude lower than the natural production rate. The radioprotection relevance of ^{14}C is thus generally very limited.

Its concentration in the body can be considered to be in equilibrium with the environmental ^{14}C and, because of its long half-life and low removal rate to the deep oceans, the dose commitment due to radiocarbon is spanned over many years.

Certainly, the importance of radiocarbon is mainly due to its very interesting scientific applications in the dating of paleontological samples. Its long half-life, and the fact that the fraction ^{14}C/^{12}C in living organisms is constant (approximately 1 atom of ^{14}C per 10^{12} atoms of ^{12}C), allows the use of radiocarbon for the dating of organisms that lived up to tens of thousands of years ago. In 1955, Libby was awarded the Nobel Prize for having established the carbon dating method.

^{22}Na is a radionuclide produced by cosmic ray spallation reactions on argon as a target nucleus. It is a β^+ emitter (end-point energies 545 keV and 1.82 MeV) and some very strong γ emissions (511 keV and 1275 keV). The half-life of ^{22}Na is $t_{1/2} = 2.60$ years. The distribution of ^{22}N in the environment is highly affected by the mixing and deposition atmospheric processes. Due to its quite long half-life, once it is washed out by precipitation, it takes time to return to its equilibrium values in the atmosphere. ^{22}N is transferred from rain to flora: activity concentration of the order 0.1–0.2 Bq/kg (wet weight) were measured in lichens, bryophytes and other plants. Traces of ^{22}N (0.1–3 Bq/kg wet weight) were also detected in the flesh of wild animal (elk, deer, caribou). Also, the levels of ^{22}N in the environment were highly affected by the nuclear weapons tests. It was estimated that in 1962 the activity concentrations of ^{22}N increased about 50 times the natural cosmogenic concentration levels: In 1966–1968 activity concentrations in air (at ground level) were measured in the range of 10–50 mBq/m^3, values much greater than the current ones, which are now generally well below the sensitivity of the standard detection monitoring systems (<0.1 mBq/m^3). Nowadays radioactive decay has, in fact, returned the ^{22}N levels to natural equilibrium values.

Due to the low concentration levels found in the biosphere, ^{22}N has a very limited radioprotective relevance. The average effective dose released to people is estimated as only 3 nSv/year, a very negligible level. However, the measurements of ^{22}N in the different environmental compartments can be very important for the study of terrestrial ecosystems.

The last cosmogenic radionuclide worth mentioning here is ^{7}Be. It decays by electron capture. It has also a strong γ emission (477.6 keV), which makes it easily detectable by means of γ spectrometry in many environmental matrices: atmospheric particulate samples, rain, surface water, pine needles, etc. The half-life of ^{7}Be is $t_{1/2} = 53.29$ days. Even the activity concentrations of ^{7}Be were increased during the time of the weapons testing, but this additional contribution has now completely disappeared because of the very short half-life of this radioisotope.

The dose delivered to human beings by ^{7}Be is very low: an effective dose of the order of 30 nSv/year is estimated.

It is, however, very interesting to study this radionuclide as a tracer for the movement of air masses in order to investigate atmospheric dynamics and, in particular, the exchanges between the troposphere and the stratosphere (Vieeze and Singh, 1980; Dutkiewicz and Husain, 1985; Yoshimori 2005; Usoskin and Kovaltsov, 2008). ^{7}Be is much more abundant in the stratosphere than it is in the troposphere: the typical levels of ^{7}Be at mid-latitudes are of the order of a few mBq/m^3 (at ground level), while in the stratosphere they are orders of magnitude greater. Therefore, a sudden increase of ^{7}Be concentrations at ground level can be interpreted as the consequence of the intrusion of air masses of stratospheric origin. It was observed that the levels of ^{7}Be are also slightly affected by solar activity. An inverse correlation with the 11-year solar cycle was found: The solar activity affects the cosmic ray flux on the Earth and, for that reason, also the production rate of most cosmogenic radionuclides.

12.5.2 Artificial Radioactivity in the Environment

While the discovery of artificial radioactivity can be dated to 1934, when Irène Joliot-Curie and Frederic Joliot, who were awarded the Nobel Prize one year later, synthesised the first man-made radionuclide, phosphorus-30, ^{30}P, bombarding ^{27}Al with alpha particles, the relevance of artificial radioactivity rose dramatically only after World War II, during the Cold War: a great number of atmospheric nuclear explosions produced and spread all over the world a large quantity of fission fragments, neutron-activated products and transuranic elements (plutonium, americium, etc.), producing worldwide contamination.

Soon after the detonation of a nuclear weapon, the radioactive load is mainly due to highly radioactive fission products, such as ^{133}Xe, ^{85}Kr, ^{132}Te, ^{132}I, ^{131}I, ^{103}Ru, ^{140}Ba, ^{140}La, ^{141}Ce, ^{144}Ce, ^{95}Nb and ^{95}Zr. After a few weeks, these short-lived radionuclides have almost all disappeared, and only a few long-lived fission products, namely ^{137}Cs ($t_{1/2} = 30.17$ years), ^{90}Sr ($t_{1/2} = 28.79$ years), ^{134}Cs ($t_{1/2} = 2.06$ years) and ^{106}Ru ($t_{1/2} = 373.6$ days) remain, eventually together with smaller quantities of some transuranic elements (isotopes of Pu, Am, Np, Cm) and activation products (for example, ^{110m}Ag, $t_{1/2} = 249.9$ days). The fallout process, i.e. the deposition of radioactive dust on the ground, can last for months, depending on the yield of the nuclear explosion; the more powerful the weapon, the higher the height attained by the fireball generated by the explosion. The fireball floats on the atmosphere and continues to rise until it cools, reaching thermal equilibrium with the surrounding air. If the power of a nuclear weapon is beyond a certain level (about 20 kton), the fireball can pass the tropopause and the radioactive dusts are injected into the stratosphere, settling on the ground very slowly.

Table 12.1 shows some data on the relationship existing between the yield of a nuclear weapons and the dimensions of the characteristic mushroom cloud.

Therefore, the time needed for the total depletion of the stratospheric radioactive dust load could be several weeks or even months, allowing a uniform radioactivity deposition on the ground over the entire hemisphere. For that reason the deposition levels of the long-lived fission products (^{137}Cs, $t_{1/2} = 30.17$ years, ^{90}Sr, $t_{1/2} = 28.79$ years) and some transuranic elements (^{239}Pu, $t_{1/2} = 24000$ years) coming from this global fallout are remarkably similar all over the Northern and Southern hemispheres, with higher values in the North because of the greater number of nuclear tests performed. It is presently estimated (2017) that in the Northern hemisphere at mid-latitudes (40°–50°), a deposition, about 2000 Bq/m^2 of ^{137}Cs, can be still attributed to nuclear weapons fallout (Facchinelli et al., 2002; UNSCEAR, 2008). In the Southern hemisphere the levels are much lower, about 500 Bq/m^2.

Nevertheless, the dose contribution to the world population from global fallout is now very small, compared to that from other radiation sources (see Table 11.1, Section 11.1).

A radioactive contamination very similar to that from nuclear weapons could also arise as a consequence of a major nuclear power plant accident. The isotope composition of the radioactive clouds released during a severe accident at an operating nuclear reactor is quite similar to that deriving from a nuclear explosion: At the early stages of the accident the activity is dominated by short-lived fission

TABLE 12.1 Nuclear Weapon Yields and Dimensions of the Radioactive Mushroom Cloud

Yield of the Atomic Bomb (kilotons)	Mushroom Cloud Radius (km)	Maximum Cloud Height (km)
1	0.7	2.1–2.7
10	2.3	6.7–8.2
20	3.1	8.2–10
200	8.0	11–17
5000	29	16–27

Source: Adapted from L.T. Matveev, *Fundamentals of General Meteorology Physics of the Atmosphere*, Israel Program for Scientific Translations, Jerusalem, 1967.

products (noble gases, isotopes of iodine, lanthanides), while the relative contributions of the different radioactive elements could be very different.

In Tables 12.2 and 12.3, the estimated radioactive releases due to the Chernobyl and Fukushima accidents are reported.

However, in the medium and long term, the consequences for the environment are very similar. About one month after the accident, the radionuclides that survive in the environment are essentially the same: ^{137}Cs, ^{134}Cs, ^{106}Ru and ^{90}Sr, plus, eventually, some traces of activation products and transuranic elements, the latter being significantly present, in most cases, only within 10–30 km from the site of the accident.

By contrast, the fallout and contamination deriving from a severe accident at a nuclear power plant has a very different pattern, strongly dependent on meteorological conditions at the time of the accident. In particular, the fallout is not uniform, being strongly influenced by weather conditions, including the direction and the velocity of the wind and especially the precipitation. Moreover, it is not a global fallout involving

TABLE 12.2   Radioisotope Composition of the Chernobyl Radioactive Releases

Radioisotope	Activity Released (PBq)
Noble Gases	
^{85}Kr	33
^{133}Xe	6500
Volatile Elements	
^{131}I	1760
^{134}Cs	54
^{137}Cs	85
Intermediate Elements	
^{89}Sr	115
^{90}Sr	10
^{103}Ru	170
^{106}Ru	73
^{140}Ba	240
Refractory (Including Fuel Particles)	
^{95}Zr	196
^{99}Mo	170
^{239}Np	196
^{238}Pu	0.035
^{239}Pu	0.030
^{240}Pu	0.042
^{241}Pu	6

Source: Adapted from UNSCEAR, United Nations Scientific Committee on the Effects of Atomic Radiation, *Report to the General Assembly*, with Scientific Annexes, Volume II, Annex J, 2000.

TABLE 12.3 ^{137}Cs Releases to the Atmosphere and the Ocean Due to the Fukushima Accident

	^{137}Cs Released (PBq)
Atmosphere	8.8
Ocean	3–6

Source: Adapted from UNSCEAR, United Nations Scientific Committee on the Effects of Atomic Radiation, Annex A, 2013.

the whole hemisphere; it has a much more limited extent, although in some circumstances, the deposition of the radioactivity can affect a whole region or even an entire continent. For example, the radioactivity deposition following the Fukushima accident (2011) significantly affected an area of about km^2 900, with very high ^{137}Cs deposition (>550 kBq/m^2), but outside a radius of 80 km from the nuclear power plant the ^{137}Cs deposition was much more limited, typically well below 50 kBq/m^2. By contrast, the Chernobyl fallout (1986) affected, albeit in very different ways, the whole of Europe. The highly contaminated zones included large areas of Russia, Ukraine and Belarus, with levels above 1000 kBq/m^2 of ^{137}Cs. Other European countries were also significantly affected by Chernobyl fallout: ^{137}Cs deposition values of the order of 100 kBq/m^2 were measured very far from Chernobyl, in the Scandinavian countries, Poland, Germany, Austria and Northern Italy.

In Table 12.4 the extent of the area affected by ^{137}Cs Chernobyl fallout in the European countries are shown for different contamination levels.

It was estimated that, as a consequence of this fallout, the population of Western Europe also received significant exposure. Committed effective doses of the order of a few mSv were calculated, the main part released in the first year after the accident.

Many other artificial sources whose origins are other than nuclear can enter the environment and give rise to potential radioactive contamination of the biosphere: they are the not-sealed sources, radionuclides in a liquid state, used in particular for medical applications, which were already cited in the previous chapter (Sections 11.4.2 and 11.4.4).

At any rate, all the radioisotopes normally used in nuclear medicine for standard diagnostic procedures (^{99m}Tc, ^{201}Tl, ^{51}Cr, ^{67}Ga, the radioisotopes of iodine ^{123}I, ^{125}I ^{131}I and many others) or the β^+ radiopharmaceuticals used in PET (positron emission tomography: ^{11}C ^{13}N, ^{15}O and ^{18}F) have half-lives so short that their accumulation in the environment in significant quantities is very improbable. Indeed, the liquid radioactive wastes produced by medical use, before being discharged in the environment, are usually stored in large tanks, allowing the complete decay of all the radioisotopes.

Small traces of artificial radionuclides due to medical practices can, however, be found in the water of the rivers, in sediments and in the mud of water treatment plants. In most cases the origin of this slight contamination is metabolic radiotherapy (Section 11.4.4), an anti-cancer therapeutic practice in which radioactive compounds are introduced into the patient's body and deliver high doses to specific organs.

TABLE 12.4 European Areas Affected by Different ^{137}Cs Contamination Levels Due to the Chernobyl Fallout

Country	Extension of the Areas (km^2) with Different Classes of ^{137}Cs Contamination			
	37–185 kBq/m^2	185–555 kBq/m^2	555–1480 kBq/m^2	>1480 kBq/m^2
Russia	49800	5700	2100	300
Belarus	10200	10200	4200	2200
Ukraine	3200	3200	900	600
Sweden	12000			
Finland	11500			
Austria	8600			
Norway	5200			
Bulgaria	4800			
Switzerland	1300			
Greece	1200			
Slovenia	300			
Italy	300			
Moldova	60			

Source: Adapted from UNSCEAR, United Nations Scientific Committee on the Effects of Atomic Radiation, *Report to the General Assembly*, with Scientific Annexes, Volume II, Annex J, 2000.

Therefore, the activities injected in the blood are high; the patients are usually hospitalised for a few days and then sent home with some residual radioactive load in the body. The contaminated excreta of these patients can thus enter the normal sewage system and be dispersed in the environment. Small quantities of ^{131}I (a few Bq/kg) related to this practice have been measured in some river sediment samples, but the radiological impact of this contamination is completely negligible.

The use of high activity sealed sources, both in medicine (brachytherapy) and in industry, is quite common (see Sections 11.4.4 and 11.5). Normally, no significant threat to the environment can arise from these type of sources: The use and disposal of such devices are strictly regulated at national and international levels. Nevertheless, the management of these kinds of sources is a very delicate matter. A few accidents involving this type of source have occurred in the past; the most severe was probably the Goiânia accident, which occurred in Brasil in 1985. A high activity ^{137}Cs sealed medical source (about 50 TBq of caesium chloride salts, a highly soluble and dispersible material) was accidentally disposed of as metal scraps; two people removed the source assembly from the radiation head of the machine. After the source capsule was ruptured, the remnants of the source assembly were sold to a junkyard for scrap, and the owner noticed that the source glowed blue in the dark. Fragments of the source were distributed to several families. Many people were exposed to very high doses and the environment was severely contaminated. Four people died and a number of people who had received large doses suffered acute symptoms of radiation sickness (IAEA Report, 1988).

12.6 Modelling Radioactivity in the Environment: The Birth of Radioecology

The diffusion in the environment of many different radioactive species, both natural and artificial, poses great challenges to scientists and researchers: How to account for the complex behaviour of so many radioactive elements that move in so many different ambient terrestrial, aquatic and atmospheric realms? It became clear early on that the presence of different radioactive species in the environment, each with its own specific chemical characteristics, would interact differently in different environments, and new studies were needed to understand the fate of the radioactive elements in the biosphere, involving biology, chemistry, geology and all the environmental sciences (see for example: Templeton, 1958; Timofeev-Resovskii et al., 1960; Ravera, 1965; Polikarpov 1966; Ravera, 1970). The primary driver for these studies was the need to answer a question: How dangerous is the radioactive contamination of the Earth for mankind? To answer to this question, scientists needed to know about the flow of radionuclides in the biosphere and their paths before they enter the human body.

Obviously, there is no simple answer to such a complex question. However, as usual in science, the theoretical approach followed was to try to simplify the problem to make it manageable. The basic idea was to subdivide the environment to be studied into different parts, called compartments. In each compartment the activity concentration of any radioactive element is assumed to be homogeneous. Each compartment may be connected to any other compartment; the exchange between the compartments is usually modelled by means of constants, named with the letter k, whose physical meaning is the probability per unit of time that a single atom of a given radioactive species moves from one compartment to another. The value of the constant k depends strongly on not only the compartments to be connected but also the different chemical behaviours of the radioactive elements.

In principle, this approach makes it possible to construct a general model describing every type of environment, no matter how complicated, simply by adding new compartments connected to each other by means of the appropriate constants. A general picture of such a description is reported in Figure 12.8, showing, as an example, the matrix description of a forest ecosystem made up of N compartments, connected to each other by means of $(N-1)^2$ constants, k_{ij} (IAEA, 2002).

Following a similar approach, it is possible to derive a mathematical description of any environmental system that is able to give the evolution in time of the activity concentrations of the radionuclides in

1 atmosphere	$K_{1,2}$	$K_{1,3}$	$K_{1,4}$	$K_{1,5}$	$K_{1,6}$	$K_{1,7}$	$K_{1,8}$	$K_{1,9}$	$K_{1,10}$	$K_{1,11}$
$K_{2,1}$	2 trees leaves	$K_{2,3}$	$K_{2,4}$	$K_{2,5}$	$K_{2,6}$	$K_{2,7}$	$K_{2,8}$	$K_{2,9}$	$K_{2,10}$	$K_{2,11}$
$K_{3,1}$	$K_{3,2}$	3 external bark	$K_{3,4}$	$K_{3,5}$	$K_{3,6}$	$K_{3,7}$	$K_{3,8}$	$K_{3,9}$	$K_{3,10}$	$K_{3,11}$
$K_{4,1}$	$K_{4,2}$	$K_{4,3}$	4 living wood	$K_{4,5}$	$K_{4,6}$	$K_{4,7}$	$K_{4,8}$	$K_{4,9}$	$K_{4,10}$	$K_{4,11}$
$K_{5,1}$	$K_{5,2}$	$K_{5,3}$	$K_{5,4}$	5 dead wood	$K_{5,6}$	$K_{5,7}$	$K_{5,8}$	$K_{5,9}$	$K_{5,10}$	$K_{5,11}$
$K_{6,1}$	$K_{6,2}$	$K_{6,3}$	$K_{6,4}$	$K_{6,5}$	6 litter	$K_{6,7}$	$K_{6,8}$	$K_{6,9}$	$K_{6,10}$	$K_{6,11}$
$K_{7,1}$	$K_{7,2}$	$K_{7,3}$	$K_{7,4}$	$K_{7,5}$	$K_{7,6}$	7 organic soil	$K_{7,8}$	$K_{7,9}$	$K_{7,10}$	$K_{7,11}$
$K_{8,1}$	$K_{8,2}$	$K_{8,3}$	$K_{8,4}$	$K_{8,5}$	$K_{8,6}$	$K_{8,7}$	8 mineral soil	$K_{8,9}$	$K_{8,10}$	$K_{8,11}$
$K_{9,1}$	$K_{9,2}$	$K_{9,3}$	$K_{9,4}$	$K_{9,5}$	$K_{9,6}$	$K_{9,7}$	$K_{9,8}$	9 mushrooms	$K_{9,10}$	$K_{9,11}$
$K_{10,1}$	$K_{10,2}$	$K_{10,3}$	$K_{10,4}$	$K_{10,5}$	$K_{10,6}$	$K_{10,7}$	$K_{10,8}$	$K_{10,9}$	10 hypogee plants	$K_{10,11}$
$K_{11,1}$	$K_{11,2}$	$K_{11,3}$	$K_{11,4}$	$K_{11,5}$	$K_{11,6}$	$K_{11,7}$	$K_{11,8}$	$K_{11,9}$	$K_{11,10}$	11 game

FIGURE 12.8 Matrix description of a forest ecosystem: The k_{ij} parameters describe the interaction between the 11 environmental compartments. (Adapted from ANPA SEMINAT, *Long Term Dynamics of Radionuclides in Semi-Natural Environment: Derivations of Parameters and Modelling, Final Report 1996–1999*, ANPA, 2000.)

each compartment. A system of N first order linear differential equations with constant coefficients may be written:

$$\frac{dC_1}{dt} = -(k_{1,2} + k_{1,3} + \cdots + k_{1,N}) \cdot C_1 + k_{2,1} \cdot C_2 + \cdots + k_{N,1} \cdot C_N$$

$$\cdots$$

$$\cdots$$

$$\frac{dC_N}{dt} = k_{1,N} \cdot C_1 + k_{2,N} \cdot C_2 + \cdots - (k_{N,1} + k_{N,2} + \cdots + k_{N,N}) \cdot C_N$$

(12.9)

Because the values of the constants k_{ij} depends on the considered radionuclide, for a complete description of the radioactive contamination of the environment we should write a specific system of equations like (12.9) for each different radionuclide.

The analytical expressions of the solutions of these systems of equations are certainly very complicated, but the mathematics assure us that they can always be calculated, provided that the boundary conditions for each compartment are known – i.e. the numerical values of the activity concentrations at time $t = 0$ are known.

The main difficulties arising during the practical implementation of these kinds of models are the knowledge of the such boundary conditions (namely, the activity concentration values in each compartment) and the proper choice of the values of the k_{ij} constants. In fact, gathering all this information would be a huge and very expensive experimental work. For that reason, systems with a high number of compartments are seldom implemented and models with a smaller number of free parameters to be determined are often preferred. In spite of the simplification, these simpler models are often much more reliable and effective than more general and detailed models, because the free parameters, being fewer, can be determined in a more robust way.

However, there is another more general limitation that prevents wide use of this kind of model: The basic assumption of linearity (first order differential equations) is, in fact, questionable in many real cases. For example, the diffusion processes operating in terrestrial and aquatic environments are usually described with second order differential equations and cannot be adequately modelled with first order kinetics. On the other hand, the introduction of second order differential equations in systems with many compartments interacting with each other would enhance dramatically the mathematical complexity of the problem. In most cases the analytical solutions cannot not be found and the equations must be solved by numerical methods, thus introducing further approximations and uncertainties.

As a consequence of all these difficulties, detailed descriptions of ecosystems with multi-compartment models are rarely implemented. More practical and effective approaches are preferred. Empirical models based on the observation of the time variation of the activity concentrations of the radionuclides in some specific important matrices (soils, sediments, foodstuffs, etc.) can be very useful tools for radiological assessment studies.

For example, the systematic measurements of the ^{137}Cs activity concentration in cow's milk produced in a well-defined alpine pasture, highly contaminated after the Chernobyl fallout, can give an interesting insight into the general behaviour of this particular radionuclide in that ecosystem (Lettner et al., 2009). As a general assumption, we can imagine that all the ^{137}Cs present in the environment is due to the Chernobyl fallout, a very good approximation, especially in high deposition areas.

It is seen from the experimental data that the activity concentrations $C(t)$ can be fitted and modelled in a very simple way: The observed decreasing trend is well described by a single exponential law given by:

$$C(t) = C_0 \cdot e^{-\lambda_{eff} \cdot t}$$

where C_0 is the activity concentration at time zero, in this case the date of the Chernobyl fallout, while λ_{eff} is the effective decay rate, which can be viewed as the sum of two different contributions: $\lambda_{eff} = \lambda_{rad} + \lambda_{eco}$, λ_{rad} being the radioactive decay constant of ^{137}Cs, while λ_{eco} is the 'ecological decay constant' of the radionuclide, a parameter describing the loss of availability of ^{137}Cs due to all the mechanisms other than the physical decay: diffusion, runoff, immobilisation in clay, etc.

We can therefore define the corresponding 'effective half-life' and 'ecological half-life': $t_{1/2eff} = ln2/\lambda_{eff}$ and $t_{1/2eco} = ln2/\lambda_{eco}$. The relationship between these quantities is given by:

$$t_{1/2eff} = \frac{t_{1/2rad} \cdot t_{1/2eco}}{t_{1/2rad} + t_{1/2eco}}$$

The values found for the ecological half-life in a given ecosystem are often very similar in very different matrices, such as sediments, water and fish: In these cases the ecological half-life can be regarded as a general property of the ecosystem, expressing its capability to reduce the pollution, restoring the previous environmental conditions (ARPA Piemonte, 2010).

12.6.1 Concentration Ratios and Transfer Factors

A quantitative description of the translocation of radionuclides from an environmental compartment such as soil, water, etc. to a living organism (plant or animal) can be performed using the concepts of concentration ratios and transfer factors (Carter et al., 1988). The definition and use of these quantities in the scientific literature has not been univocal and some confusion may arise. In our treatment we will follow the definition reported in the IAEA Technical Report Series n° 472 (IAEA, 2002, 2010).

The basic assumption underlying the concept of concentration ratio or transfer factor is the condition of equilibrium between the two considered compartments: plant-soil, water-fish, etc. The activity concentrations of the radionuclides in every compartment are assumed to be constant, except for the radioactive decay. At first sight, this assumption might seem a very crude approximation. However, equilibrium conditions generally apply a few weeks or months after the introduction of a single input of radionuclides into the environment (a typical situation following an accident) or in cases of a continuous release of radioactivity (typical of the controlled discharge of the waste of a nuclear installation). Anyway, the use of the concentration ratios and transfer factors must be considered a practical approach for radioprotection assessments rather than an accurate scientific description of the environmental behaviour of the radionuclides in complex ecosystems.

For that reason, these parameters are usually presented in the available publications as ranges of observed values. The values also depend strongly on the considered ambients: different values are thus provided for temperate, tropical and sub-tropical environments. Specific parameters and values also have to be calculated depending on the different ecosystems: agricultural ecosystems, forests ecosystems, arctic and alpine ecosystems, freshwater ecosystems.

In the following, the definitions of the most important transfer parameters are shown and briefly discussed.

Soil to plant transfer: It is defined as the *concentration ratio* or *transfer factor*, the ratio of the activity concentration in the plant (C_v, Bq/kg, dry weight) to that in the soil (C_s, Bq/kg, dry weight). It is a dimensionless quantity and it is given by:

$$F_v = \frac{C_v}{C_s}$$

The next table shows the mean values and the typical ranges of F_v in temperate environments for different plant groups and for the most significant elements. (Table 12.5)

From these data it can be seen that for cesium and strontium the transfer from soil to plants is much less effective than for transuranic elements (americium and plutonium).

TABLE 12.5 Typical F_v Values (Dimensionless) for Some of the Most Important Elements

Element	Plant Group	Mean Value	Minimum	Maximum
Americium	Cereals	$2.2 \cdot 10^{-5}$	$7.4 \cdot 10^{-7}$	$3.4 \cdot 10^{-2}$
	Leafy vegetables	$2.7 \cdot 10^{-4}$	$4.0 \cdot 10^{-5}$	$1.5 \cdot 10^{-3}$
Cesium	Cereals	$2.9 \cdot 10^{-2}$	$2.0 \cdot 10^{-4}$	$9.0 \cdot 10^{-1}$
	Leafy vegetables	$6.0 \cdot 10^{-2}$	$3.0 \cdot 10^{-4}$	$9.8 \cdot 10^{-1}$
Plutonium	Cereals	$9.5 \cdot 10^{-6}$	$2.0 \cdot 10^{-7}$	$1.1 \cdot 10^{-3}$
	Leafy vegetables	$8.3 \cdot 10^{-5}$	$1.0 \cdot 10^{-5}$	$2.9 \cdot 10^{-4}$
Strontium	Cereals	$1.1 \cdot 10^{-1}$	$3.6 \cdot 10^{-3}$	1
	Leafy vegetables	$7.6 \cdot 10^{-1}$	$3.9 \cdot 10^{-3}$	7.8

Source: Adapted from IAEA, *Handbook of Parameter Values for the Prediction of Radionuclide Transfer in Terrestrial and Freshwater Environments*, Technical Report Series, 472 (2010).

Herbage to animal transfer: It is defined as the *feed transfer coefficient*, as the ratio of the mass (or volume) activity concentration in the animal tissue or product (C_{an}, Bq/kg or Bq/L, fresh weight) divided by the daily intake of the radionuclide (D, Bq per day, Bq/d). Its unit is therefore d/kg or d/L, depending on the unit chosen for C_{an} (B/kg or Bq/L).

$$F_m = \frac{C_{an}}{D}$$

In the following table the F_m data (d/L) for cow's and goat's milk are reported. (Table 12.6)

The values of the feed transfer coefficients F_m are similar for all the elements considered. However, it's worth noting that the values for goat's milk are significantly greater than those for cow's milk.

Transfer in semi-natural ecosystems: It is defined as the *aggregated transfer factor* as the ratio of the mass activity concentration in a specified matrix (C_m, Bq/kg) to the deposition value in soils (I, Bq/m^2):

$$T_{ag} = \frac{C_m}{I}$$

This quantity is very useful, because it allows the calculation of the activity concentration in many products starting from the deposition data (I, Bq/m^2), generally more easily available than the activity concentrations in soils (Bq/kg). In any case, some precautions must be taken using the aggregated transfer factors: high variability of the T_{ag} values were in fact observed, because the relationship between the radionuclide activity concentration in soils and plants is influenced by many factors, including the ecological characteristics of the plants (depth of the roots apparatus, etc.) and the mobility of the radionuclides in soils. The use of the T_{ag} is thus appropriate only for simple screening models. The most reliable use of the T_{ag} values is for a forest ecosystem, where the radionuclides fluxes have stabilised. A large dataset of T_{ag} has been calculated in particular for the ^{137}Cs radioisotopes for many forest organisms: trees, mushrooms, all species of berries and game animals.

Transfer in freshwater ecosystems: Two different *concentration ratios* are usually defined for these ecosystems: the *concentration ratio water-biota*, CR and the *concentration ratio sediment-biota*, CR_{s-b}. The *concentration ratio water-biota*, CR is the ratio of the activity concentration in the receptor biota (C_b, Bq/kg, fresh weight) to the water activity concentration (C_w, Bq/kg or Bq/L):

$$CR = \frac{C_b}{C_w}$$

Its unit can be dimensionless or L/kg, depending on the unit used for the water activity concentration C_w.

TABLE 12.6 F_m Values (d/L) for Cow's and Goat's Milk

Element	Plant Group	Mean Value	Minimum	Maximum
Cesium	Cow's milk	$4.6 \cdot 10^{-3}$	$6.0 \cdot 10^{-4}$	$6.8 \cdot 10^{-2}$
	Goat's milk	$1.1 \cdot 10^{-1}$	$7.0 \cdot 10^{-3}$	$3.3 \cdot 10^{-1}$
Iodine	Cow's milk	$5.4 \cdot 10^{-3}$	$4.0 \cdot 10^{-4}$	$2.5 \cdot 10^{-2}$
	Goat's milk	$2.2 \cdot 10^{-1}$	$2.7 \cdot 10^{-2}$	$7.7 \cdot 10^{-2}$
Strontium	Cow's milk	$1.3 \cdot 10^{-3}$	$3.4 \cdot 10^{-4}$	$4.3 \cdot 10^{-3}$
	Goat's milk	$1.6 \cdot 10^{-2}$	$5.8 \cdot 10^{-3}$	$8.1 \cdot 10^{-2}$

Source: Adapted from IAEA, *Handbook of Parameter Values for the Prediction of Radionuclide Transfer in Terrestrial and Freshwater Environments*, Technical Report Series, 472 (2010).

TABLE 12.7   CR Values (L/kg, Fresh Weight) for Some Freshwater Organisms

Element	Organism	Mean Value	Minimum	Maximum
Cesium	Aquatic plant	$9.7 \cdot 10^1$	$1.9 \cdot 10^0$	$3.3 \cdot 10^3$
	Freshwater invertebrate	$2.3 \cdot 10^1$	$5.4 \cdot 10^{-3}$	$6.1 \cdot 10^4$
	Fish tissue	$3.0 \cdot 10^3$	$7.5 \cdot 10^1$	$2.4 \cdot 10^4$
Iodine	Aquatic plant	$1.3 \cdot 10^2$	$7.9 \cdot 10^1$	$2.7 \cdot 10^2$
	Freshwater invertebrate	$1.7 \cdot 10^1$	$4.0 \cdot 10^{-1}$	$1.3 \cdot 10^3$
	Fish tissue	$6.5 \cdot 10^2$	$1.0 \cdot 10^2$	$4.5 \cdot 10^4$
Strontium	Aquatic plant	$4.1 \cdot 10^2$	$3.9 \cdot 10^1$	$1.9 \cdot 10^3$
	Freshwater invertebrate	$2.7 \cdot 10^2$	$7.7 \cdot 10^1$	$1.3 \cdot 10^3$
	Fish tissue	$1.9 \cdot 10^2$	$2.2 \cdot 10^1$	$7.1 \cdot 10^2$

Source: Adapted from IAEA, *Handbook of Parameter Values for the Prediction of Radionuclide Transfer in Terrestrial and Freshwater Environments*, Technical Report Series, 472 (2010).

The *concentration ratio sediment-biota, CR_{s-b}* is the ratio of the activity concentration of a radionuclide in biota tissues (C_b, Bq/kg, fresh weight) to the activity concentration in sediments (C_{sed}, Bq/kg, fresh weight): It's a dimensionless quantity.

$$CR_{s-b} = \frac{C_b}{C_{sed}}$$

In the next table, some typical concentration ratio values for edible aquatic plants, freshwater invertebrates and fish are shown (Table 12.7).

It can be noticed that the values of *CRs* in freshwater ecosystems are significantly higher than the concentration and transfer factors in other environments: In aquatic environments the concentration processes are definitely more effective.

Many other specific 'concentration factors' have been defined in order to describe the translocation processes of the radionuclides that occur in different ecosystems: They can be used for specific studies and assessments. Those reported here are, however, the most important and frequently used.

There is, however, another important 'concentration ratio' worth considering: the resuspension factor, a parameter that relates the radioactivity deposited on the ground to the activity concentration in the atmosphere. Its importance, in particular for medium- and long-term dose assessment studies, deserves special attention: It will be thus discussed separately in the next sub-section.

12.6.2 Resuspension Factors

The atmospheric resuspension of radionuclides is a phenomenon that consists in the re-injection into the atmosphere of previously deposited radioactivity. The process is driven by the action of wind on surfaces and can act as an additional source of radiation exposure by inhalation after the deposition has finished.

In order to evaluate such exposure, a parameter was introduced: the resuspension factor K, defined as the ratio of the volumetric air concentration C_a (Bq/m^3) to the initial soil deposition I_0 (Bq/m^2) and generally considered a time-dependent function:

$$K(t) = \frac{C_a(t)}{I_0} \quad (m^{-1}) \tag{12.10}$$

In this way, simply knowing the value of K makes it possible to estimate the volumetric air concentrations from the deposition data, usually more easily available than concentration data.

The resuspension factor is thus widely used in all circumstances when volumetric air concentration data are needed and no direct measurements are possible. However, the choice of the proper values of K is usually not a simple task, being quite site-specific and related to the meteorological, geomorphologic and environmental characteristics of the area to be studied. Moreover, several investigations have shown clearly that the values of K are a decreasing function of time. For that reason, K values span several orders of magnitude: typical values in the range 10^{-5}–10^{-10} m^{-1} are reported in the literature for different environmental conditions and the time elapsed since the deposition event (Sehmel, 1980; IAEA, 2010).

Several models for the resuspension factor were proposed and tested in recent years using the Chernobyl accident data; in fact, the widespread contamination of large areas of Europe allowed, especially during the first few years after the accident, the measurement of airborne radionuclide concentration (plutonium and cesium-137, in particular), in very different environmental conditions.

This research led to the formulation and testing of some empirical models commonly used for the prediction of time evolution of the volumetric air concentration $C_a(t)$ in the years following a fallout event. A detailed discussion of these models can be found in the BIOMOVS II final report (Garger, 1999).

The IAEA publication n°472 (2010) suggested the use of three models, corresponding to three different environmental conditions: a) rural conditions, b) urban conditions, c) arid and desert conditions.

For rural conditions the Garland model, which obtained the best scores in the IAEA BIOMOVS II intercomparison, is proposed. It consists of a simple time-decreasing relationship defined as follows:

$$K(t) = \frac{K(0)}{t}$$

where $K(0) = 1.2 \cdot 10^{-6}$ days $\cdot$ m^{-1}, the time t is expressed in days and the formula is valid for $t > 1$ day.

For urban environments the Linsley model, consisting in an exponential decreasing function and a constant asymptotic value, is suggested:

$$K(t) = K(0) \cdot e^{-0.01 \cdot t} + 10^{-9} \text{m}^{-1}$$

where $K(0) = 10^{-6}$ m^{-1} and the time t is expressed in days.

For arid and desert conditions, where resuspension is generally more effective, the Anspaugh model is recommended:

$$K(t) = K(0) \cdot e^{-0.15\sqrt{t}} + 10^{-9} \text{m}^{-1}$$

with $K(0) = 10^{-6}$ m^{-1}.

More complicated models, based on a modification and/or a combination of the above mathematical expressions, were also proposed and tested by other authors in order to find a better agreement with local experimental data. However, no significant improvements in the capability of prediction were achieved. In general, the intercomparison exercises performed in past years showed that the major source of uncertainty for all models is the choice of the initial conditions, i.e. the $K(0)$ values, rather than the differences in the mathematical form of the expressions. The experimentally available data for $K(0)$ generally range between 10^{-5} and 10^{-6} m^{-1} in the first days or months after the fallout event, and are highly site-specific.

All these empirical models provide quite good results for some specific situations, but the lack of knowledge of the underlying physical processes limits the possibility of generalising, in an easy way, the results obtained in a given site to other locations with different environmental characteristics.

From a theoretical point of view, many researchers made the assumption that the relative decrease with time of K is constant, thus explaining the exponential behaviour of most of the current models. However, this would mean that the radioactivity available during the resuspension processes reduces at a constant rate, an assumption that can hardly be considered true, especially in the mid- to long-term period. On the other hand, some experimental data clearly show a deviation from a simple exponential behaviour, being better described by hyperbolic time functions (see, for example, the Garland model) or by the sum of exponential functions.

It is thus clear that, at the moment, none of these current models has broad prediction capabilities (either in time or space) and can be used in different environmental conditions without an *ad hoc* adjustment of the fitting parameters. Moreover, it has widely been recognised that these relatively poor prediction capabilities reflect a poor understanding of the basic physical mechanism underlying resuspension.

Hatano and Hatano (2003) proposed an interesting theoretical model, very different from those described above. This model, based on the observation of the fractal behaviour of the time series of the atmospheric activity concentration measured around Chernobyl some years after the accident, deduces an inverse time dependence of the type $K(t) \sim t^{-4/3}$ that fits quite well with the available experimental data. However, in spite of its mathematical brightness and its sound agreement with some experimental observations, it provides no simple connection with the physical quantities usually involved in the description of resuspension phenomena.

A new, different model has been recently proposed (Magnoni, 2012): This model is based on the assumption of a condition of equilibrium between deposition and re-suspension and takes into account the progressive reduction of availability of radioactivity for resuspension due to the downward migration in soils of the radionuclides (Konshin, 1992; Bossew and Kirchner, 2004). The resulting mathematical expression is given by:

$$K(t) = \frac{\left[1 + \dfrac{v \cdot \Delta z}{2 \cdot D}\right] \cdot \Delta z \cdot K(0)}{2 \cdot \sqrt{\pi \cdot D \cdot t}} \cdot e^{-\frac{v^2 \cdot t}{4D}}$$

where $K(0)$ is the resuspension factor at the time of deposition, v and D are the velocity and diffusion parameters of the diffusion-convection equation describing the downward migration in soils of the radionuclides (the values of which are available in the literature), and Δz is the thickness of the soil available for resuspension, a site-specific parameter that can be estimated experimentally. A value around 55 nm has been measured in a temperate semi-natural environment.

12.7 Effective Dose Evaluation in a Contaminated Environment

A comprehensive evaluation of effective dose delivered to a population living in a contaminated environment needs the knowledge of three different contributions:

1. Irradiation (mainly γ irradiation)
2. Inhalation of radionuclides
3. Ingestion of radionuclides

A very detailed knowledge of the radionuclide contamination of the environment is thus necessary in order to make accurate dose assessments (Polvani, 1990). This is particularly important for the inhalation and ingestion components. In fact, for the irradiation component, very good dose evaluations can be easily performed by measuring the γ dose rate with appropriate instrumentation – pressurised ionisation chambers, proportional counters and compensated Geiger-Mueller counters.

Unfortunately, no direct measurements of the inhalation and ingestion components are possible; the distribution of the radionuclides in the human body leads to a very complex internal irradiation pattern. Models have thus been developed describing the distribution of the different radionuclides in the various organs and calculating the corresponding released doses. The output of these very complex models are the *dose coefficients*, i.e. parameters that allow calculation of the inhalation and ingestion doses in a very simple way. The *dose coefficients* are defined as the dose released per unit of activity intake. They depend, of course, on the specific radionuclide, type of intake (inhalation or ingestion) and the age of the person; specific coefficients are normally calculated for infants, children and adults, as well as for workers. The tables showing these coefficients are periodically revised by the International Commission on Radiological Protection (ICRP) (Section 10.1), and are generally adopted by the official legislative publications of the states (ICRP 119, 2012).

The *effective dose coefficient for inhalation* is thus defined as follows:

$$e_{inh} = \frac{E_{inh}}{Q_{inh}}$$

where Q_{inh} is the quantity of activity inhaled (Bq) and E_{inh} (Sv) is the corresponding effective dose. Similarly, we have:

$$e_{ing} = \frac{E_{ing}}{Q_{ing}}$$

where Q_{ing} is the quantity of activity ingested (Bq) and E_{ing} (Sv) is the corresponding effective dose.

The total effective dose can thus be expressed by the following general formula:

$$E_{tot} = E_\gamma + \sum_j e_{inhj} \cdot Q_{inhj} + \sum_j e_{ingj} \cdot Q_{ingj}$$

in which E_γ is the irradiation component, directly measured by a γ dose-rate instrument, while Q_{inhj} and Q_{ingj} are the inhaled and ingested activities of the generic radioactive element j and e_{inhj} and e_{ingj} are the corresponding dose coefficients. The inhalation and ingestion coefficients are tabulated for most of the radionuclides; the problem of effective dose calculation is thus reduced to the estimation of the inhalation and ingestion activities (ICRP Publication n°119, 2012).

Specific coefficients can also be calculated for other very particular paths of radionuclides introduction that maybe relevant, in particular for workers: wound intake and transcutaneous absorption.

References

ANPA SEMINAT, *Long Term Dynamics of Radionuclides in Semi-Natural Environment: Derivations of Parameters and Modelling*, Final Report 1996–1999, ANPA, 2000.

ARPA Piemonte, *Environmental Radioactivity Report 2006–2009 (In Italian)*, ARPA Piemonte, Torino, 2010.

Bossew P. and Kirchner G., Modelling the vertical distribution of radionuclides in soils. Part 1: The convection-dispersion equation revisited, *Journal of Envirnonmental Radioactivity*, 73(2), 127–150, 2003.

Carter M.W., Harley J.H., Schmidt G.D., and Silini G., Eds, Radionuclides in the Food Chain, Springer-Verlag, Berlin, Heidelberg, New York, London, Paris, Tokyo, 1988.

Danesi P.R. et al., Depleted uranium particles in selected Kosovo samples, *Journal of Environmental Radioactivity*, 64(2–3), 143–154, 2003.

Dutkiewicz V.A. and Husain L., Stratospheric and tropospheric components of [7]Be surface air, *Journal of Geophysical Research*, 90, 5783–5788, 1985.

Facchinelli A., Magnoni M., Gallini L., Bonifacio E., ^{137}Cs contamination from Chernobyl of soils in Piemonte (North-West Italy): spatial distribution and deposition model, *Water, Air and Soil Pollution*, 134, 341–352, 2002.

Garger E.K. et al., Test of existing mathematical models for atmospheric resuspension of radionuclides, *Journal of Envirnonmental Radioactivity*, 42, 157–175, 1999.

Hatano Y. and Hatano N., Formula for the resuspention factor and estimation of the date of surface contamination, *Atmospheric Environment*, 37, 3475–3480, 2003.

https://pugwash.org/1955/07/09/statement-manifesto/

IAEA Report, *The Radiological Accident in Goiânia*, IAEA, Vienna, 1988.

IAEA, Modelling the migration and accumulation of radionuclides in forest ecosystems, IAEA BIO-MASS-1, 2002.

IAEA, Handbook of Parameter Values for the Prediction of Radionuclide Transfer in Terrestrial and Freshwater Environments, *Tecnical Report Series*, 472, 4–10, 2010.

ICRP Publication 119, Compendium of Dose Coefficient Based on ICRP Publication 60, 2012.

Johns H.E. and Cunningham J.R., *The Physics of Radiology*, Published by Charles C Thomas Pub Ltd, Fourth Edition, Springfield, 1983.

Konshin O.V., Applicability of the convection-diffusion mechanism for modelling migration of ^{137}Cs and ^{90}Sr in the soil, *Health Physics*, 63, 291–300, 1992.

Lal D. and Peters B., Cosmic ray produced radioactivity on the Earth, *Handbuch der Physik*, 46/2, 552–616, 1967.

Lettner A. et al., Effective and ecological half life in cow's milk in alpine agriculture, *Radiation Environmental Biophysics*, 48, 47–59, 2009.

Lieser K.H., *Nuclear and Radiochemistry: Fundamentals and Applications*, VCH A wiley Company, Verlagsgesellshaft mbH, Weinheim–New York, 1997.

Magnoni M., Bertino S., Bellotto B., Campi M., Environmental Samples Containing Traces of Depleted Uranium: Theoretical and Experimental Aspects, *Radiation Protection Dosimetry*, 97(4), 337–340, 2001.

Magnoni M., A theoretical approach to the re-suspension factor, *The European Physical Journal EPJ Web of Conferences*, 24, 05008, 2012. DOI: 10.1051/epjconf/20122405008

Matveev L.T., *Fundamentals of General Meteorology Physics of the Atmosphere, Israel Program for scientifica Translation*, Jerusalem, Israel, 1967.

Meitner L. and Frisch O.R., Disintegration of Uranium by Neutrons: A New Type of Nuclear Reaction. *Nature*, 143, 471–472, 1939.

Nazaroff W.W. and Nero A.V., *Radon and its decay Products in Indoor Air*, A Wiley Interscience Publication, John Wiley & Sons, New York, 1988.

NCRP Report 94, *National Council on Radiation Protection and Measurements, Exposure of the Population in the United States and Canada from Natural Background Radiation*, 1987.

OECD NEA Report, 'Chernobyl: Assessment of Radiological and Health Impact, 2002 update; Chapter II – The release, dispersion and deposition of radionuclides', 2002.

Polikarpov G.G., *Radioecology of aquatic organisms*, Reinhold Book Division, New York, 1966.

Polvani, C. *Introduction to Radioprotection (in Italian: Introduzione alla Radioprotezione)*, ENEA, Roma, 1990.

Ravera O., Radioactivity in freshwater organism of some lake of Northern Italy, in *Biological problem in water pollution, Third Seminar, 1962*. U.S. Dept. Health, Education and Welfare, Cincinnati, 195–201, 1965.

Ravera O., *Introduction to radioecology (in Italian: Introduzione allo studio della radioecologia)*, Minerva Medica, Torino, 195–201, 1970.

SCHER – Scientific Committee on Health and Environmental Risks, Opinion on the Environmental and Health Risks Posed by Depleted Uranium, European Commission, Health & Consumer Protection, Directorate General, 2009.

Sehmel G.A., Particle resuspension: a review, *Environment International*, 4, 107–127, 1980.

SIPRI Yearbook 2017, *Armament, Disarmament and International Security*, Oxford University Press, Stockholm, 2017.

Templeton W.L., Fission products and aquatic organisms, in *The effect of pollution on living material.* Institute of Biology, London, September 1958, 125–140.

Timofeev-Resovskii N.V., Timofeeva-Resovskaya E.A., Milyutina G.A., Getsova A.B., Coefficient of accumulation of 16 different elements in freshwater organisms and the effect of EDTA on some of them. *Doklady Akad Nauk SSSR*, 132, 1191–1194, 1960.

UNEP Report, *Depleted Uranium in Kosovo – Post Conflict Environmental Assessment*, 2001.

UNEP Report, *Depleted Uranium in Bosnia-Hergegovina – Post Conflict Environmental Assessment*, 2003.

UNSCEAR, *United Nations Scientific Committee on the Effects of Atomic Radiation, Report to the General Assembly, Annex B*, 1993.

UNSCEAR, *United Nations Scientific Committee on the Effects of Atomic Radiation, Report to the General Assembly, with scientific annexes, Volume II, Annex J*, 2000.

UNSCEAR, *United Nations Scientific Committee on the Effects of Atomic Radiation, Report to the General Assembly, with scientific annexes, Volume I: (Sources), Scientific Annexes JA and JB*, 2008.

UNSCEAR, *United Nations Scientific Committee on the Effects of Atomic Radiation, Annex A*, 2013.

Usoskin I.G. and Kovaltsov G.A., Production of cosmogenic ^{7}Be isotope in the atmosphere: Full 3-D modelling, *Journal of Geophysical Research*, 113, D12107, 2008. doi:10.1029/2007/JD009725.

Vieeze W. and Singh H.B., The distribution of beryllium-7 in the troposphere: implication on stratospheric-tropospheric exchange, *Geophysical Research Letters*, 7, 805–808, 1980.

Yoshimori M., Production and behaviour of beryllium-7 isotope in the upper atmosphere, *Advance in Space Research*, 36, 922–992, 2005.

13

Radon and NORM: Naturally Occurring Radioactive Materials

Cristina Nuccetelli
*Center for Radiation Protection
and Computational Physics*

Rosabianca Trevisi
*2INAIL (National Institute for
Insurance against Accidents at
Work)- Research Sector*

13.1 Introduction: Background and Driving Forces

Every human being is exposed to ionising radiation from natural sources. By natural sources we mean cosmic rays and terrestrial sources that originate from radioactive elements present in the Earth's crust. Naturally occurring radionuclides of terrestrial origin (also called primordial radionuclides or NOR) are present in various amounts in the soil and rocks, in water and in the air: radionuclides with half-lives comparable to the age of the Earth, and their decay products, exist in significant quantities in these media. Many examples of exposure to natural radiation sources are modified by human practices: This is the case with NORM activities, which will be dealt with in Section 14.5.

These natural radionuclides can be responsible for internal or external exposure to ionising radiations. External exposure of the human body is mainly due to the gamma radiation emitted by radionuclides of the ^{238}U and ^{232}Th series and by ^{40}K, while radon, thoron and their decay products are the main sources of internal exposure.

Moreover, exposure to natural sources can occur both outdoors and indoors: natural radionuclides in soils and rocks are responsible for the outdoor background absorbed dose-rate in air at ground level. Outdoors, the population-weighted average absorbed dose-rate in air from terrestrial gamma radiation is 60 nGyh^{-1}, as estimated by UNSCEAR (United Nations Scientific Committee on the Effects of Atomic Radiation) in its 2000 Report (UNSCEAR 2000).

Indoors, the main contribution to human external exposure is gamma radiation emitted by natural radionuclides in building materials, and most internal exposure derives from inhalation of radon and its decay products.

The main aspects have been already described elsewhere (see Chapter 12 and Chapter 13); in this chapter, we summarise the principal aspects of the two most important natural sources.

13.2 Radon and Its Decay Products: Health Effects

Radon is the major contributor to the ionising radiation dose received by the general population from natural sources. Radon is a noble gas that has no colour, smell or taste: It is chemically inert. Radon is a radioactive element: Its isotopes, in particular ^{222}radon and ^{220}radon (also called thoron), are gaseous radioactive products of the decay of the radium isotopes ^{226}Ra (belonging to the chain of ^{238}U) and ^{224}Ra (belonging to the chain of ^{232}Th), respectively, which are present in all terrestrial materials.

The most common isotope, and the one commonly called radon, is ^{222}Rn: It is an alpha emitter with a 3.8-day half-life. Its decay goes through the following sequence:

- ^{222}Rn, 3.82 days, alpha decay
- ^{218}Po, 3.10 minutes, alpha decay
- ^{214}Pb, 26.8 minutes, beta decay
- ^{214}Bi, 19.9 minutes, beta decay
- ^{214}Po, 0.1643 µs, alpha decay
- ^{210}Pb, 22.3 years, beta decay
- ^{210}Bi, 5.013 days, beta decay
- ^{210}Po, 138.376 days, alpha decay
- ^{206}Pb, stable

Uranium and thorium occur naturally in varying levels in all rocks and soils. Some fraction of the radon produced in rocks and soils escapes to the air; therefore, radon is present in the outdoor atmosphere. However, radon gas emanating from rocks and soils tends to concentrate in enclosed spaces like underground mines, workplaces or houses: for this reason, radon is considered an indoor pollutant that is ubiquitously diffused. Indoors there is a mix of radon gas and its decay products, free or attached to aerosols and dust. Thus, simply by breathing, people are exposed to radiation from radon itself and its short-lived radon decay products. From a health point of view, radon gas is not particularly dangerous; its decay products, both attached to aerosols and unattached, represent the risk factor, since they are capable of reaching different portions of the human respiratory tract, where they deposits themselves and interact with biological tissues in the lungs, leading to DNA damage.

Health effects due to radon exposure, most notably lung cancer, have been investigated for several decades, thanks to evidence coming from uranium miners' epidemiological data. The International Agency on Research on Cancer (IARC_WHO 1988) classified radon as a human carcinogen (Group 1) in 1988 and confirmed the classification in 2001 (IARC_WHO 2001).

Radon is the second most important cause of lung cancer after smoking and the leading cause of cancer among non-smokers. The link between exposure to radon and lung cancer is scientifically proven, and even the moderate concentrations found in many homes present an increased risk. Indeed, recent epidemiological studies on indoor radon and lung cancer in Europe, North America and Asia have provided strong evidence of an association between indoor radon exposure and lung cancer, even at the relatively low radon levels commonly found in residential buildings. Current estimates of the proportion of lung cancers attributable to radon range from 3% to 14%, depending on the average radon concentration in the country concerned and the calculation methods (Darby et al. 2005). Epidemiological results indicate that the lung cancer risk increases proportionally with increasing radon exposure, and many people are exposed to low and moderate radon concentrations. There is no known threshold concentration below

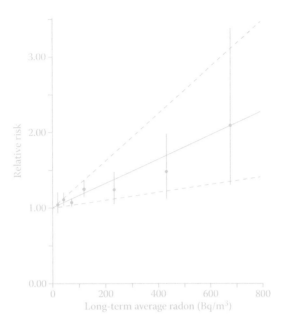

FIGURE 13.1   Relative risk of lung cancer versus long-term average residential radon concentration in the European pooling study. (From Darby, S. et al. 2005. BMJ 330(7485): 223–227.)

which radon exposure presents no risk, thus the majority of lung cancers related to radon are caused by these exposure levels rather than by higher concentrations. The epidemiological findings demonstrated a statistically significant increase of lung cancer risk from prolonged exposure to indoor radon at levels of the order of 100 Bq/m^3; most of the radon-induced lung cancer cases occur among smokers due to the strong combined effect of smoking and radon (Figure 13.1) (WHO 2009).

13.3 Radon: Possible Prevention and Mitigation Systems

High levels of radon in indoor spaces can be reduced by using various corrective actions, well developed and proven to work. The analysis of many experiences helps in the selection of more effective radon prevention and mitigation systems in order to control indoor radon concentrations and reduce the inhabitants' radon exposure. All effective systems consider that soil gas infiltration is the most important source of residential radon.

Other radon sources are building materials, water (especially that extracted from wells), gas and outdoor air. The contribution of different radon sources varies between countries and even regions due to the geological characteristics of the area and the building typology. Finally, attention should also be paid to low ventilation rates, which can decrease the overall quality of indoor air, and to the consequences of energy-saving measures.

Several main exposure mechanisms may be considered:

- Pressure-driven soil gas infiltration (chimney effect and wind effect)
- Emanation of radon from building materials
- Water transport of radon

Since air pressure differences between the soil and the building are the primary driving force for radon entry, radon prevention and mitigation systems usually focus on reversing this pressure difference. This is commonly done through the use of active (fan-powered) or passive (no fan) soil depressurization. There

TABLE 13.1   Common Mitigation Techniques, Performances and Costs[a,b]

Technique	Typical Radon Reduction in [%]	Typical Contractor Installation Costs[€][c]	Typical Annual Operating Costs [€][d]	Notes
ASD[e]: High- to low-porosity subslab	50 to 99	850 to 2700	50 to 275	Subslab suction is placed in a porous stone subslab fill, groundwater control components, and/or perforated sump
ASD[e]: Very low porosity subslab	50 to 99	850 to 2700	50 to 275	Also known as subslab depressurisation
ASD[e]: Submembrane depressurisation	50 to 99	1100 to 2700	50 to 275	In accessible crawlspaces, a membrane is placed over exposed soil and suction is applied under the membrane
Under floor active ventilation	50 to 99	500 to 1600	50 to 275	Uses a fan to pressurise
Under floor passive ventilation	0 to 50	0 to 550 if additional vents added	Variable	Not effective in heating-dominated regions and homes with airtight floors (caution: plumbing may freeze up)
Radon wells	60 to 95	2150 to 4300	Variable	Most effective in very porous soil (such as eskers) May be used to reduce radon entry into multiple houses
Soil pressurisation	50 to 99	500 to 1600	50 to 275	Most effective in very porous soil with moderately elevated soil radon and a very airtight soil-contacted concrete slab
Soil contacted crawlspace pressurisation	50 to 99	500 to 1600	150 to 550	Most effective when the soil-contacted space is relatively airtight and isolated from outdoors and other indoor spaces
Passive ventilation of occupied spaces	Variable/ temporary	None	100 to 750	Significant loss of heated or cooled air; not a permanent mitigation strategy, especially in more severe climates
Active ventilation of occupied spaces	30 to 70	225 to 2700	7 to 550	Ranges from a very small supply fan[f] to a balanced heat recovery ventilator (both operating continuously)

Source: Adapted from World Health Organisation (WHO). 2009. *WHO Handbook on Indoor Radon – A Public Health Perspective.* Edited by Hajo Zeeb and Ferid Shannoun. Geneva: World Health Organisation, ISBN 978-92-4-154767-3.

[a] The data have been reported by USEPA (2003) and have been modified to be similar to those from Finland and the United Kingdom.

[b] The two primary water mitigation techniques are aeration and activated charcoal filtration, which are not listed in this table.

[c] Installation costs may be higher when cosmetic treatments to the house are necessary, when local demand for mitigation is high and/or if there is a shortage of mitigation professionals.

[d] Fan electricity and house heating/cooling loss costs based on assumptions regarding climate (moderate), house size and the local cost of electricity and heating fuel (Bohac et al. 1992).

[e] ASD refers to active soil depressurisation. It is highlighted in this table because ASD is the most common radon mitigation technique.

[f] The small supply fan would be used to slightly pressurise spaces in ground contact.

are several possible systems, and they take into account if there is tight and direct contact between the underlying soil and the building or an unoccupied space. Finally, the addition of radon-proof membranes or barriers can improve the effectiveness of the radon control system.

The introduction of radon control systems is normally cheap and often highly cost-effective compared with other public health interventions. An overview of common mitigation systems, performances and costs drawn up by the World Health Organisation (WHO 2009) is shown in Table 13.1. It is worth noting that mitigation systems can be considered radon prevention systems in newly built houses.

13.4 Radon: International Legislations and Guidelines

A synthetic review of the historical evolution of legislation and radiation protection guidelines about radon is here presented.

13.4.1 Directive 1996/29/Euratom and Radiation Protection 88

The Council Directive 96/29/Euratom (EC 1996) 'laying down basic safety standards for the protection of the health of workers and the general public against the dangers arising from ionising radiation' introduced a Title VII concerning 'work activities … which involve the presence of natural radiation sources and lead to a significant increase in the exposure of workers or members of the public which cannot be disregarded from the radiation protection point of view'. Title VII, in its only two articles, considered all natural radiation sources, such as indoor radon and gamma exposure from NORM, including industrial residues and cosmic radiation. No provisions were given in the Directive, but the guidelines outlined in *Radiation Protection n. 88 'Recommendations for the implementation of Title VII of the European Basic Safety Standards Directive (BSS) concerning significant increase in exposure due to natural radiation sources'* (RP88 in the following, EC 1997) supported Member States in the implementation of Title VII of the Council Directive 96/29/Euratom with the aim of introducing the regulation of exposure to natural sources into the radiation protection system for the first time. Indeed, ICRP Publication 65 (ICRP, 1993), RP88 provided indications about possible workplaces of radiological concern (such as underground sites, tunnels, spas, workplaces in radon-prone areas, etc.) and suggested a radon activity concentration range within which to define national action levels to control radon exposure in workplaces (500–1000 Bq/m^3), etc.

13.4.2 Directive 2013/59/Euratom

Following recommendations given by ICRP in Publications 103 (ICRP, 2010) and 115 (ICRP, 2010), including the Statement on Radon in 2009 (ICRP, 2010), Directive 2013/59/Euratom considered radon exposure at home and at work as an existing exposure situation, since the presence of radon is largely independent of the human activities carried out within the workplace and at home. Consequently, Directive 2013/59/Euratom established reference levels for indoor radon concentration corresponding to $300 \, Bq/m^3$ for both homes and workplaces, strongly stimulating appropriate radon and exposure reduction measures if the national reference level is exceeded, taking into account the principle of optimisation.

13.4.3 IAEA Safety Guide SSG-32

In 2015 the IAEA Safety Guide *'Protection of the Public against Exposure Indoors due to Radon and Other Natural Sources of Radiation'* (IAEA, 2015) provided recommendations for protection of the public against exposure indoors due to natural sources of radiation such as radon and radionuclides of natural origin in materials used for the construction of dwellings, offices, factories and other buildings. Guidance has been provided on the application of the requirements for justification and optimisation of protection

TABLE 13.2 List of Main NORM Industrial Sectors

Industry	Raw Material	Process	By-product
Coal power plant	Coal	Coal burning	Fly ash Bottom ash
Phosphate industry	Phosphorite	Acid attack	Phosphogypsum
Steelworks/metal recycling industry	Iron ore/scrap metal	Melting	Slag
Aluminium processing industry	Bauxite	Chemical	Red mud

by national authorities. Moreover, the Safety Guide provided recommendations and indications for regulatory bodies and national authorities with responsibilities with respect to exposure to radiation from natural sources.

13.5 NORM: Definition

The following internationally accepted definition of NORM appears in the IAEA glossary (IAEA, 2016):
Naturally occurring radioactive material (NORM): Radioactive material containing no significant amounts of radionuclides other than naturally occurring radionuclides*.

- The exact definition of 'significant amounts' would be a regulatory decision.
- Material in which the activity concentrations of the naturally occurring radionuclides have been changed by a process is included in naturally occurring radioactive material (NORM). ....'

The last bullet is particularly interesting from the radiation protection point of view because the main problems with NORM come from industry. Indeed, naturally occurring radionuclides (NOR) are present in the Earth's crust, therefore they are also in the minerals and ores that are used as raw materials in many industrial sectors. Typical examples of minerals widely used by industry are coal, iron ore, phosphorite and bauxite. Industrial processes can concentrate NOR in by-products, and how this occurs depends also on NOR chemical/physical properties. In Table 13.2, by-products which can be considered as NORM of potential radiological concern are listed with their relevant industrial sectors.

Other circumstances occur in which the simple use and storage of materials delivers significant doses to workers and/or members of the public because the materials contain elevated levels of natural radionuclides. Examples might include monazite sands, rare earth ores and the scale which can build up in pipes and valves in some oil, coal or other mineral processing and similar plants. The set of all productive activities that, at least in one phase of the process, can lead to a significant exposure of workers and/or members of the public to NORM will be referred to as 'NORM industries' or 'NORM activities'.

13.6 NORM: Regulation and Radiation Protection

A short review of the historical evolution of regulations about NORM is presented here.

13.6.1 Directive 1996/29/Euratom

In the 1990s the European Commission, after publication of ICRP Publication 60 (ICRP, 1990), started to think about a common approach to natural radioactivity radiological protection. This goal was achieved

---

* 'Radionuclides that occur naturally on Earth in significant quantities. The term is usually used to refer to the primordial radionuclides potassium-40, uranium-235, uranium-238, thorium-232 and their radioactive decay products. Contrasted with radionuclides of artificial origin, anthropogenic radionuclides and human made radionuclides (which all mean the same), and also with artificial radionuclides' (IAEA, 2016).

TABLE 13.3   Rounded General Clearance and Exemption
Levels in kBq/kg (from Table 13.2 of RP 122, Part II)

Natural radionuclides from the U-238 series	0.5
Natural radionuclides from the Th-232 series	0.5
K-40	5

with the Council Directive 96/29/Euratom (EC, 1996) 'laying down basic safety standards for the protection of the health of workers and the general public against the dangers arising from ionising radiation'. The Directive included natural radioactivity in its scope and in Title VII specified 'work activities ... which involve the presence of natural radiation sources and lead to a significant increase in the exposure of workers or members of the public which cannot be disregarded from the radiation protection point of view'. This regulation defined these work activities without providing indications about how to manage this new area of radiation protection.

Transposition of the Directive in national regulations was made possible by the publication of *Radiation Protection n. 88 'Recommendations for the implementation of Title VII of the European Basic Safety Standards Directive (BSS) concerning significant increase in exposure due to natural radiation sources'*, which aimed to ensure a homogeneous approach to Title VII implementation in the European Union Member States (EC, 1997).

Afterwards, two additional guidelines published by the European Commission – RP 122, *'Practical use of the concepts of clearance and exemption – Part II: Application of the concepts of exemption and clearance to natural radiation sources'* (EC 2002) and RP 135, *'Effluent and dose control from European Union. NORM industries. Assessment of current situation and proposal for a harmonised Community approach'* (EC, 2003) – provided tools to manage radiation protection for NORM activities. Indeed, in RP 122, Part II, and RP 135, natural radionuclide activity concentrations were calculated by modelling conservative exposure scenarios and adopting the Article 31 Experts* proposal to set the criteria for exemption-clearance for work activities at an annual effective dose increment of 300 µSv per year for members of the public[†]. In particular, RP 122, Part II proposed general exemption-clearance levels (GCEL) for solid materials; activity concentrations below GCEL (see Table 13.3) should guarantee compliance with the guideline of 300 µSv per year without the need to assess.

13.6.2 Council Directive 2013/59/Euratom

This Directive, under transposition and implementation by EU Member States before February 2018, follows the ICRP recommendations 103 (ICRP, 2007) and adopts a situation-based approach that distinguishes between existing, planned and emergency exposure situations (EC, 2014). As regards NORM activities, a clear evolution in the regulatory approach can be seen: These activities must be considered *a priori* as planned exposure situations (practices). This concept is clearly expressed in Recital 16 of the Directive: 'Protection against natural radiation sources, rather than being addressed separately in a specific title, should be fully integrated within the overall requirements. In particular, industries processing materials containing naturally-occurring radionuclides should be managed within the same regulatory framework as other practices'. This represents a deep change from both the philosophical and operational points of view. After individuation of NORM activities in industrial sectors listed in the Annex ...[‡] of the Directive, Member States shall adopt a graded approach to regulatory control 'commensurate with the magnitude and likelihood of exposures resulting from the practice, and commensurate with the impact that regulatory control may have in reducing such exposures or improving radiological safety': the first level

---

* The Group of Experts established under Article 31 of the Euratom Treaty.

† In this case 'members of the public' includes also workers not classified as exposed.

‡ The list of industrial sectors is indicative and not exhaustive: any Member State can modify the list considering the specific country situation.

TABLE 13.4　Naturally Occurring Radionuclides: Values for Exemption or Clearance for Naturally Occurring Radionuclides in Solid Materials in Secular Equilibrium with Their Progeny

Natural radionuclides from the U-238 series	$1\,\mathrm{kBq\,kg^{-1}}$
Natural radionuclides from the Th-232 series	$1\,\mathrm{kBq\,kg^{-1}}$
K-40	$10\,\mathrm{kBq\,kg^{-1}}$

Source: From table A, Part 2 of European Commission (EC) – Council of the European Union 2014. Council Directive 2013/59/EURATOM of 5 December 2013 laying down basic safety standards for protection against the dangers arising from exposure to ionising radiation, and repealing Directives 89/618/Euratom, 90/641/Euratom, 96/29/Euratom, 97/43/Euratom and 2003/122/Euratom. *Off. J. Eur. Union L* 13: 1–73.)

of regulatory control is the notification, followed by registration or licencing. NORM activities, like other practices, can be exempted. There are two ways to verify the possibility that a NORM industry can be exempted:

1. Solid materials at all stages of the industrial process present an activity concentration less than the values shown in Table 13.4; if this condition is not met,
2. An undertaking, supported by a radiation protection expert, can demonstrate that doses to workers and members of the public are less than the general exemption and clearance criterion in terms of dose of 1 mSv per year.

13.6.3　IAEA BSS

In 2014 IAEA published the last *International Basic Safety Standards* (IAEA BSS; IAEA, 2014). Also in this regulation, for the first time NORM activities are considered planned exposure situations and subjected to a regulatory graded approach.

'3.4 Exposure due to natural sources is, in general, considered an existing exposure situation and is subject to the requirements in Section 13.5. However, the relevant requirements in Section 13.3 for planned exposure situations apply to:

a. Exposure due to material in any practice specified in para. 3.1 (the mining and processing of raw materials that involve exposure due to radioactive material) where the activity concentration in the material of any radionuclide in the uranium decay chain or the thorium decay chain is greater than 1 Bq/g or the activity concentration of ^{40}K is greater than 10 Bq/g;
b. Public exposure due to discharges or due to the management of radioactive waste arising from a practice involving material as specified in (a) above...'

The clear coincidence between IAEA and EU BSS denotes a strict collaboration of the two institutions in elaborating the relevant regulations. Indeed, IAEA representatives participated in many phases of the drafting of EU BSS.

13.7　NORM: Worker and Public Exposure

13.7.1　Workers

The main routes of exposure of workers determined by activities involving NORM are the external exposure to gamma radiation and the internal exposure to contamination after inhalation of dust. Appropriate measures of control aimed at limiting worker exposure may include special placement and management of stored bulk materials, limitation of exposure time and dust control. Radon or thoron presence may be also of radiological concern, and surface contamination may need to be considered. It is worth noting that the highest doses do not necessarily arise when the plant is operating normally. In many circumstances the maximum doses occur during maintenance operations, for example at oil treatment plants or clinker ovens for cement production. Beyond this, however, it is necessary to estimate doses to the workers and eventually apply the regulatory graded approach when NORM activities process non-compliant materials (Table 13.4).

13.7.2 Members of the Public

Exposures of members of the public may originate at different phases of the industrial process: from the final product, from effluents (atmospheric or liquid) from re-use of by-product materials as secondary raw material (e.g. in building materials) or from disposal of solid waste. The principal routes of radiation exposure in members of the public are the external exposure to gamma radiation and the internal exposure due to internal contamination after inhalation and ingestion of dust.

Special sources of external exposure in the public are building materials, which are included in the scope of the Council Directive 2013/59/Euratom in the category of existing exposure. A list of building materials of radiological concern with regard to their emitted gamma radiation is reported in Annex XIII of the Directive. In addition to natural materials, such as tuff, pozzolana and lava, the list includes NORM industry by-products incorporated in building materials, e.g.:

- Fly ash
- Phosphogypsum
- Tin and copper slag
- Red mud
- Residues from steel production

Indeed, in the near future the radiological impact of recycling NORM by-products in the building sector is expected to increase. This is due to the trend in the EU of treating by-products as secondary raw materials in order to reduce landfilling and discourage the use of virgin raw materials (EC, 2010).

References

Bohac, D. et al. 1992. *The energy penalty of sub-slab depressurization radon mitigation systems.* Proceedings of the 1992 International Symposium on Radon and Radon Reduction Technology, US Environmental Protection Agency, (2:7.37-7.55), Research Triangle Park, NC.

Darby, S., Hill, D., Auvinen, A., Barros-Dios J.M., Baysson H., Bochicchio, F., Deo, H. et al. 2005. Radon in homes and risk of lung cancer: collaborative analysis of individual data from 13 European case-control studies. *BMJ*, 330(7485): 223–227.

EC. 2010. European Commission. Communication from the Commission. Europe 2020. A strategy for smart, sustainable and inclusive growth.

European Commission (EC). 1997. *Radiation Protection 88: Recommendations for the implementation of Title VII of the European Basic Safety Standards Directive (BSS) concerning significant increase in exposure due to natural radiation sources.*

European Commission (EC). 2002. *Radiation Protection 122: Practical Use of the Concepts of Clearance and Exemption – Part II Application of the concepts of exemption and clearance to natural radiation sources.* Luxembourg: Office for Official Publications of the European Communities. ISBN 92-894-3315-9.

European Commission (EC). 2003. *Radiation Protection 135: Effluent and dose control from European Union NORM industries: Assessment of current situation and proposal for a harmonised Community approach.* Luxembourg: Office for Official Publications of the European Communities. ISBN 92-894-6361-9.

European Commission (EC) – Council of the European Union 1996 – Council Directive 96/29/Euratom of 13 May 1996 laying down basic safety standards for the protection of the health of workers and the general public against the dangers arising from ionising radiation. Off. J. Eur. Union L 159, 29/06/1996: 1–114.

European Commission (EC) – Council of the European Union 2014. Council Directive 2013/59/ EURATOM of 5 December 2013 laying down basic safety standards for protection against the dangers arising from exposure to ionising radiation, and repealing Directives 89/618/Euratom, 90/641/ Euratom, 96/29/Euratom, 97/43/Euratom and 2003/122/Euratom. Off. J. Eur. Union L 13: 1–73.

International Agency On Research On Cancer – World Health Organisation (IARC-WHO). 1988. *IARC Monographs on the evaluation of carcinogenic risks to humans- Man-made mineral fibres and radon.* Volume 43. Lyon, France: IACR.

International Agency On Research On Cancer – World Health Organisation (IARC-WHO). 2001. *IARC Monographs on the evaluation of carcinogenic risks to humans-Some Internally Deposited Radionuclides.* Volume 78. Lyon, France: IACR.

International Atomic Energy Agency (IAEA). 2014. *Radiation Protection and Safety of Radiation Sources: International Basic Safety Standards IAEA SAFETY STANDARDS SERIES No. GSR Part 3.* Vienna: International Atomic Energy Agency. ISBN 978-92-0-135310-8.

International Atomic Energy Agency (IAEA). 2015. *Protection of the Public against Exposure Indoors due to Radon and Other Natural Sources of Radiation – IAEA SAFETY STANDARDS SERIES No. SSG-32.* Vienna: International Atomic Energy Agency. ISBN 978-92-0-102514-2.

International Atomic Energy Agency (IAEA). 2016. *Iaea Safety Glossary Terminology Used In Nuclear Safety And Radiation Protection 2016 Revision.* Vienna: International Atomic Energy Agency.

International Commission On Radiological Protection (ICRP). 1990. *1990 Recommendations of the International Commission on Radiological Protection.* Publication 60, In: Ann. ICRP 21 (1–3), Oxford: Pergamon Press.

International Commission On Radiological Protection (ICRP). 1993. *Protection Against Radon-222 at Home and at Work.* Publication 65, In: Ann. ICRP 23 (2), Oxford: Pergamon Press.

International Commission On Radiological Protection (ICRP). 2007. *The 2007 recommendations of the international commission on radiological protection.* Publication 103. In: Ann. ICRP 37 (2–4), Oxford: Elsevier.

International Commission On Radiological Protection (ICRP). 2010. *Lung Cancer Risk from Radon and Progeny and Statement on Radon.* Publication 115, In: Ann. ICRP 40(1), Oxford: Elsevier.

United Nations Scientific Committee on the Effects of Atomic Radiation (UNSCEAR). 2000. *UNSCEAR 2000 Report to the General Assembly with Scientific Annexes, Sources and effects of ionising radiation.* New York: UNITED NATIONS. www.unscear.org

United States Environmental Protection Agency. 2003. *Consumer's Guide to Radon Reduction.* USEPA Publication 402-K-03-002, Washington D.C.

World Health Organisation (WHO). 2009. *WHO Handbook On Indoor Radon – A Public Health Perspective.* Edited by Hajo Zeeb and Ferid Shannoun. Geneva: World Health Organisation, ISBN 978-92-4-154767-3.

Index